INTERNATIONAL CENTRE FOR MECHANICAL SCIENCES

COURSES AND LECTURES - No. 330

THE EVALUATION OF MATERIALS AND STRUCTURES BY QUANTITATIVE ULTRASONICS

EDITED BY

J.D. ACHENBACH
NORTHWESTERN UNIVERSITY, EVANSTON

Springer-Verlag   Wien GmbH

Le spese di stampa di questo volume sono in parte coperte da
contributi del Consiglio Nazionale delle Ricerche.

This volume contains 186 illustrations.

In order to make this volume available as economically and as
rapidly as possible the authors' typescripts have been
reproduced in their original forms. This method unfortunately
.has its typographical limitations but it is hoped that they in no
way distract the reader.

ISBN 978-3-211-82441-2   ISBN 978-3-7091-4315-5 (eBook)
DOI 10.1007/978-3-7091-4315-5

PREFACE

Nondestructive evaluation (NDE) procedures are needed for materials processing, as well as for post-process materials testing. They play important roles in product design, analysis of service-life-expectancy, manufacturing and quality control of manufactured products. They are also essential to on-line monitoring of the integrity of structural elements and complex systems. Rational accept and reject criteria should be based on NDE tests. Critical safety, efficiency and operational features of large-scale structures depend on adequate NDE capabilities.

Most methods of nondestructive evaluation provide only limited information. It is, however, frequently not good enough just to detect a flaw or the presence of inferior material properties. Quantitative information is required. This need has given rise to a more rigorous and fundamental approach to nondestructive evaluation called quantitative NDE (QNDE). There are several methods for QNDE. Each method has its advantages and disadvantages for particular applications. The lectures presented in this volume are concerned with quantitative ultrasonic NDE.

The lectures were offered in an Advanced School at the International Centre for Mechanical Sciences, Udine, Italy, September 9-13, 1991. The following topics were discussed: (1) the basic wave propagation theory for ultrasonic NDE; (2) piezoelectric transducers, EMATS and ultrasonic spectroscopy; (3) laser-based ultrasonics; (4) acoustoelasticity; (5) ultrasound in solids with porosity, microcracking and polycrystalline structuring; (6) the determination of mechanical properties of composite materials; and (7) inverse problems and imaging. The lectures covered theory, experiments and applications. The invited lecturers were J.D. Achenbach (Center for Quality Engineering and Failure Prevention, Northwestern University), L. Adler (Ohio State University), P.P. Delsanto (Politecnico di Torino), K.J. Langenberg (University of Kassel), C. Sayers (Schlumberger), C.B. Scruby (AEA Industrial Technology, Harwell Laboratory) and R.B. Thompson (Center for NDE, Iowa State University). J.D. Achenbach served as coordinator.

It is anticipated that the texts of the lectures, written by the invited lecturers and their co-authors, will be of interest to researchers, graduate students and practitioners in the field of quantitative nondestructive evaluation.

J.D. Achenbach

CONTENTS

ULTRASONICS: INTRODUCTION

J.D. Achenbach

Northwestern University, Evanston, IL, USA

ABSTRACT

The propagation of ultrasound in a one-dimensional configuration is discussed. Basic concepts are introduced and relevant governing equations are derived. Reflection and transmission at an interface are discussed. Harmonic waves and Fourier superposition are considered. Energy considerations are presented. The exposition of the theory is supplemented by a number of exercises.

1.1 SOUND IN SOLIDS

Just like in air, sound also propagates in solids. Sound waves in solids are mechanical waves. The range of sound frequencies of human hearing is about 20 Hz to 20 kHz. The term ultrasound is used for frequencies well above 20 kHz. Ultrasonic inspection is generally done at frequencies higher than 0.1 MHz.

Mechanical waves originate in the forced motion of a portion of a deformable medium. As elements of the medium are deformed the disturbance is transmitted from one point to the next and the disturbance, or wave, progresses through the medium. In this process the resistance offered to motion offered by inertia, must be overcome. As the disturbance propagates through the medium it carries along amounts of energy in the forms of kinetic and potential energies. Energy can be transmitted over considerable distances by wave motion. The transmission of energy is effected because motion is passed on from one particle to the next. Mechanical waves are characterized by the transport of energy through motions of particles about an equilibrium position.

Deformability and inertia are essential properties of a medium for the transmission of mechanical wave motions. If the medium were not deformable any part of the medium would immediately experience a disturbance in the form of an internal force or an acceleration upon application of a localized excitation. Similarly, if a hypothetical medium were without inertia there would be no delay in the displacement of particles and the transmission of the disturbance from particle to particle would be effected instantaneously to the most distant particle. Indeed, it will be shown analytically that the velocity of propagation of a mechanical disturbance always assumes the form of the square root of the ratio of a parameter defining the resistance to deformation and a parameter defining the inertia of the medium. All real materials are of course deformable and possess mass and thus all real materials transmit mechanical waves.

Problems of the motion and deformation of substances are rendered amenable to mathematical analysis by introducing the concept of a continuum or continuous medium. In this idealization it is assumed that properties averaged over a very small element, for example, the mean mass density, the mean displacement and the stress, are functions of position and time. Although it might seem that the microscopic structure of real material is not consistent with the concept of a continuum, the idealization produces very useful results, simply because the lengths characterizing the microscopic structure of most materials are generally much smaller than any lengths arising in the deformation of the medium. Even if in certain special cases the microstructure gives rise to significant phenomena, these can be taken into account within the framework of the continuum theory by appropriate generalizations.

Continuum mechanics is a classical subject that has been discussed in great generality in several treatises. The theory of continuous media is built upon the basic concepts of stress, motion and deformation, upon the laws of conservation of mass, linear momentum, moment of momentum, and energy, and on the constitutive relations. The constitutive relations characterize the mechanical and thermal response of a material while the basic conservation laws abstract the common features of all mechanical phenomena irrespective of the constitutive relations.

1.2 ONE-D WAVES IN SOLIDS

Consider the case of longitudinal displacement, i.e., all displacement components vanish except u(x,t). The single strain component is

$$\varepsilon_x = \frac{\partial u}{\partial x}. \tag{1.2.1}$$

Hooke's law gives

$$\sigma_x = \frac{(1-\nu)E}{(1+\nu)(1-2\nu)}\varepsilon_x, \tag{1.2.2}$$

where E is Young's modulus and ν is Poisson's ratio. Now consider an element of length Δx and unit length in the y and z directions. Newton's law gives

$$\frac{\partial \sigma_x}{\partial x}\Delta x = \rho\Delta x\frac{\partial^2 u}{\partial t^2}, \tag{1.2.3}$$

where u is the displacement in the x-direction and ρ is the mass density. Substitution of Eq. (1.2.1) in Eq. (1.2.2) and subsequent substitution in Eq. (1.2.3) yields

$$\frac{\partial^2 u}{\partial x^2} = \frac{1}{c_L^2}\frac{\partial^2 u}{\partial t^2}, \tag{1.2.4}$$

where

$$c_L^2 = \frac{E(1-\nu)}{\rho(1+\nu)(1-2\nu)} \quad . \tag{1.2.5}$$

Equation (1.2.4) is the one-dimensional wave equation.

A general solution to Eq. (1.2.4) can be obtained by introducing the new variables

$$\alpha = t - x/c_L, \qquad \beta = t + x/c_L. \tag{1.2.6a,b}$$

Equation (1.2.4) then becomes

$$\frac{\partial^2 u}{\partial\alpha\partial\beta} = 0. \tag{1.2.7}$$

The general solution of Eq. (1.2.7) is

$$u = f(\alpha) + g(\beta), \tag{1.2.8}$$

or

$$u(x,t) = f(t - x/c_L) + g(t + x/c_L), \qquad \text{(d'Alembert)}. \tag{1.2.9}$$

The terms in Eq. (1.2.9) represent waves propagating in the positive and negative x-directions respectively.

Consider a wave motion defined by

$$u(x,t) = f(t - x/c_L). \tag{1.2.10}$$

The argument of $f( )$ is called the phase of the wave, and c_L is the phase velocity. The shape of the wave does not change as it propagates through the solid, and hence this wave is called nondispersive. Equation (1.2.10) represents a longitudinal wave. Note that

$$\sigma_x = -\rho c_L \dot{u}(x,t), \tag{1.2.11}$$

where $\dot{u}(x,t)$ is the particle velocity. The quantity $Z = \rho c_L$ is called the mechanical impedance or the wave resistance.

———O———

Exercise 1: Consider a half space $x \geq 0$. At time $t = 0$ the surface is subjected to a displacement:

$$x = 0, \quad t = 0: \quad u(0,t) = f(t). \tag{1.2.12}$$

Find an expression for the wave motion generated in the half space.

Exercise 2: Now the surface is subjected to a traction

$$x = 0, \quad t \geq 0: \quad \sigma_x(0,t) = F(t). \tag{1.2.13}$$

Determine $u(x,t)$.

Exercise 3: Two solid half-spaces are perfectly bonded in the plane $x = 0$.

$$x \leq 0: \quad \rho, c_L; \qquad x \geq 0: \quad \rho^A, c_L^A. \tag{1.2.14a,b}$$

A wave is incident from $x < 0$:

$$u^i = f(t - x/c_L). \tag{1.2.15}$$

This wave is reflected and transmitted

$$u^r = Rg(t - x/c_L), \quad u^t = Th(t - x/c_L^A).$$ (1.2.16a,b)

Show that $g(t) = h(t) = f(t)$, and that

$$R = - \frac{\rho^A c_L^A - \rho c_L}{\rho^A c_L^A + \rho c_L} = - \frac{Z^A - Z}{Z^A + Z}, \quad T = 1 - \frac{\rho^A c_L^A - \rho c_L}{\rho^A c_L^A + \rho c_L} = - \frac{2Z}{Z^A + Z}.$$ (1.2.17a,b)

Exercise 4: The same problem can also be solved for an incident stress pulse. Show

$$R_\sigma = \frac{\rho^A c_L^A / \rho c_L - 1}{\rho^A c_L^A / \rho c_L + 1}, \quad T_\sigma = \frac{2 \rho^A c_L^A / \rho c_L}{\rho^A c_L^A / \rho c_L + 1}.$$ (1.2.18a,b)

Investigate the special case $\rho^A = c_L^A \equiv 0$.

Exercise 5: Now consider the case of transverse displacement, $v(x,t)$. All strain components vanish, except

$$\varepsilon_{xy} = \frac{\partial v}{\partial x}.$$ (1.2.19)

Show that $v(x,t)$ satisfies the wave equation

$$\frac{\partial^2 v}{\partial x^2} = \frac{1}{c_T^2} \frac{\partial^2 v}{\partial t^2},$$ (1.2.20)

where

$$c_T^2 = G/\rho,$$ (1.2.21)

here G is the shear modulus. The solution

$$v(x,t) = f(t - x/c_T)$$ (1.2.22)

represents a transverse wave.

1.3 HARMONIC WAVES

A special case of $u(x,t) = f(t - x/c_L)$ is a harmonic wave of the form

$$u(x,t) = U\cos[k(x - x/c_L)].$$ (1.3.1)

For fixed t (a snapshot), $u(x, t)$ is a periodic function of x with wavelength $\lambda = 2\pi/k$.
  Definitions:

amplitude:	U,	(1.3.2)
wavenumber:	$k = 2\pi/\lambda$.	(1.3.3)

For fixed x, $u(x,t)$ is a harmonic function of time, with circular frequency ω, where

$$\omega = kc_L.$$ (1.3.4)

Equation (1.3.1) may also be written as

$$u(x,t) = U\cos(kx - \omega t).$$ (1.3.5)

Definitions:

period: $T = 2\pi/\omega$, (1.3.6)

frequency: $f = 1/T = \omega/2\pi$. (1.3.7)

The frequency f is expressed in Hertz (Hz, 1 Hz = 1 cycle/sec.) The range of human hearing is about 20 Hz to 20 KHz. Sound at higher frequencies is called ultrasound. The particle velocity follows from (1.3.1) as

$$\dot{u}(x,t) = Ukc_L\sin[k(x - c_Lt)].$$ (1.3.8)

Hence

$$(\dot{u}/c_L)_{max} = Uk = 2\pi U/\lambda.$$ (1.3.9)

For linear theory we must have $U/\lambda \ll 1$. Hence

$$\dot{u} \ll c_L.$$ (1.3.10)

It is convenient to use complex notation

$$u(x,t) = U\exp[ik(x - c_Lt)], \quad i = \sqrt{-1}.$$ (1.3.11)

with the understanding that the physical quantity is either the real or imaginary part of Eq. (1.3.11). It easily verified that

$$u(x,t) = \int_{-\infty}^{\infty}U(k)e^{ik(x - c_Lt)}\,dk,$$ (1.3.12)

and

$$u(x,t) = \int_{-\infty}^{\infty}U(\omega)e^{-i\omega(t - x/c_L)}\,d\omega$$ (1.3.13)

are also the solutions of the one-dimensional wave equation (1.2.4). Equations (1.3.12) and (1.3.13) are Fourier integrals. The following relations are useful

$$u(t) = \frac{1}{2\pi}\int_{-\infty}^{\infty}U(\omega)e^{-i\omega t}\,d\omega,$$ (1.3.14)

$$U(\omega) = \int_{-\infty}^{\infty}u(t)e^{i\omega t}\,dt.$$ (1.3.15)

The Fourier integral approach can be used to investigate the propagation of ultrasound in a viscoelastic solid. Let us consider a solid that is viscoelastic in shear according to

$$\sigma_{xy} = G\frac{\partial v}{\partial x} + D\frac{\partial}{\partial t}(\frac{\partial v}{\partial x}).$$ (1.3.16)

This viscoelastic behavior is known as the Kelvin solid. The propagation of a transverse (shear) wave is then governed by

$$G\frac{\partial^2 v}{\partial x^2} + D\frac{\partial}{\partial t}(\frac{\partial^2 v}{\partial x^2}) = \rho\frac{\partial^2 v}{\partial t^2}.$$ (1.3.17)

Now consider a time harmonic wave

$$v(x,t) = V\exp[i(kx - \omega t)].$$ (1.3.18)

Substitution of Eq. (1.3.18) into Eq. (1.3.17) yields the relation

$$k^2 = \frac{\rho\omega^2}{G - i\omega D} = \frac{\omega^2}{(c_T)^2 - i\omega D/\rho} \ . \tag{1.3.19}$$

For real-valued ω the wave number obtained from Eq. (1.3.19) is complex-valued

$$k(\omega) = k_1(\omega) + k_2(\omega), \tag{1.3.20}$$

and thus

$$v(x,t) = V e^{-k_2 x} e^{i(k_1 x - \omega t)}. \tag{1.3.21}$$

Thus, the wave motion decays with distance travelled. For $\omega D/(\rho c_T^2) \ll 1$, we may write

$$k_1(\omega) = \frac{\omega}{c_T}, \quad k_1(\omega) = \frac{D}{2\rho(c_T)^3}\omega^2. \tag{1.3.22a,b}$$

Thus the decay rate increases with frequency.

A more general solution of Eq. (1.3.16) may be written as

$$v(x,t) = \frac{1}{2\pi}\int_{-\infty}^{\infty} V(\omega)e^{i[k(\omega)x - \omega t]} \, d\omega. \tag{1.3.23}$$

Now let us consider the generation of ultrasound in a half-space $x \geq 0$, whose mechanical properties are defined by Eq. (1.3.16), and let

$$x = 0, \quad t \geq 0: \quad v(0,t) = f(t). \tag{1.3.24}$$

Equations (1.3.23) and (1.3.24) yield

$$f(t) = \frac{1}{2\pi}\int_{-\infty}^{\infty} V(\omega)e^{-i\omega t} \, d\omega. \tag{1.3.25}$$

From Eq. (1.3.15), $V(\omega)$ then follows as

$$V(\omega) = \int_{-\infty}^{\infty} f(t)e^{i\omega t} \, dt. \tag{1.3.26}$$

Substitution of Eq. (1.3.26) into Eq. (1.3.23) yields a formal solution to the problem. Integrals of the form (1.3.26) and (1.3.23) can often be evaluated analytically. Results have also been tabulated in Tables of exponential Fourier transforms. There are also approximate methods such as the method of stationary phase that can be used.

———O———

Exercise 6: Consider the superposition of 2 harmonic waves propagating in opposite directions:

$$u(x,t) = U_+ e^{i(kx - \omega t + \gamma_+)} + U_- e^{i(kx + \omega t + \gamma_-)},$$

where U_+, U_- are real-valued amplitudes, γ_+, γ_- are phase angles. Let $U_+ = U_- = U$ and show that the real part becomes

$$u(x,t) = 2U\cos(kx + \tfrac{1}{2}\gamma_+ + \tfrac{1}{2}\gamma_-)\cos(\omega t - \tfrac{1}{2}\gamma_+ + \tfrac{1}{2}\gamma_-).$$

This is a standing wave. Why?

Exercise 7: Now consider 2 waves propagating in the same direction with the same amplitude but slightly different wave numbers (k_1 and k_2) and slightly different frequencies(ω_1 and ω_2). Show that

$$u = C\sin(k_0 x - \omega_0 t),$$

where

$$C = 2U\cos(\Delta k x - \Delta\omega t),$$

and

$$k_0 = \frac{1}{2}(k_1 + k_2), \quad \omega_0 = \frac{1}{2}(\omega_1 + \omega_2),$$

$$\Delta k = \frac{1}{2}(k_1 - k_2), \quad \Delta\omega = \frac{1}{2}(\omega_1 - \omega_2).$$

Here C is the modulation which moves with the group velocity

$$c_g = \Delta\omega/\Delta k.$$

Exercise 8: Consider two half spaces of different materials separated by a layer of springs (spring constant K). Consider normal incidence of harmonic longitudinal waves and determine the reflection and transmission coefficients.

1.4 ENERGY CONSIDERATIONS

Definitions:

Intensity: $\quad I = -\sigma_x \dot{u}$ (1.4.1)

$\qquad\qquad\qquad$ = rate of work on a unit cross section

$\qquad\qquad\qquad$ = power

$\qquad\qquad\qquad$ = rate of energy flow

$\qquad\qquad\qquad$ = energy flux.

Consider a harmonic wave of the type given by Eq. (1.3.11). Then

$$\sigma_x = i\rho c_L \omega u, \qquad\qquad\qquad (1.4.2)$$

and, considering the real parts,

$$I(x,t) = Re[\rho c_L^2 ikUe^{i(kx-\omega t)}]Re[-i\omega Ue^{i(kx-\omega t)}]$$

$$= \rho c_L^2 k\omega U^2 \sin^2(kx - \omega t). \qquad\qquad (1.4.3)$$

It is useful to consider the time averaged intensity, $<I>$, averaged over a period $T = 2\pi/\omega$

$$<I> = \frac{1}{T}\int_{t_1}^{t_1+T} I(x,t)dt = \frac{1}{2}Z\omega^2 U^2, \qquad\qquad (1.4.4)$$

where $Z = \rho c_L$ is the mechanical impedance. Note that $<I>$ is proportional to ω^2 and U^2.

At this point it is a good idea to review the units:

time: second; length: meter; mass: kg

force: Newton; energy: Joule(=N·m); power: Watt(N·m/s)

Note that

$<I> \div watt/m^2$.

In ultrasonics amplitude and intensity ratios are measure in decibels (dB). The following definitions apply

$$\text{ratio in decibels} = 20 \log\frac{U_1}{U_2} dB \qquad (1.4.5)$$

$$= 10 \log\frac{<I_1>}{<I_2>} dB. \qquad (1.4.6)$$

———O———

Exercise 9: Consider the problem of exercise 3 for the special case of an incident harmonic wave. show that the incident, reflected and transmitted waves satisfy conservation of energy.

$$<I^i> = <I^r> + <I^t>. \qquad (1.4.7)$$

Exercise 10: Sometimes energy reflection and transmission coefficients are defined. Show that

$$R_E = \frac{<I^r>}{<I^i>} = R^2, \quad T_E = \frac{<I^t>}{<I^i>} = \frac{Z^A}{Z}T^2 \qquad (1.4.8a,b)$$

Show that

$$R_E + R_T = 1. \qquad (1.4.9)$$

Exercise 11: Consider the transmission from water to Aluminum alloy

$$Z^A/Z = 1.72/0.149 = 11.54.$$

Calculate R_E and T_E. Note that for R_E the same values are obtained for transmission from Al alloy to water.

Calculate the loss in dB for transmission from water into Al alloy.

Answers: .71, .29 and -5.4

According to Eq.(1.3.21) the change in amplitude in a viscoelastic medium is defined by $\exp(-k_2x)$. Usually the symbol α is used rather than k_2. Over a distance d we then have

$$V = V_0 e^{-\alpha d} \quad \text{or} \quad \alpha d = -\ln(V/V_0). \qquad (1.4.10a,b)$$

The quantity α has the dimension nepers/m (Np/m), but α can also be expressed in dB.

Exercise 12: Show that 1 Np = 8.68 dB
Hint: Start with the definition given Eq. (1.4.5), which gives

$$20 \log(e^{\alpha d}) dB/m.$$

Then use the relation $\log\phi = \ln\phi/\ln 10 = 0.434\ln\phi$.

———O———

In general α will always be given in dB/m. Water as well as other materials with low attenuation haves values from 1 to 4 dB/m. The decibel unit can be memorized more easily than the neper because the transition to powers of ten is simpler. Thus, 20 dB is equivalent to a power of ten and an attenuation of 20 dB is a reduction to 1/10th, 40 dB-1/100, 60 dB-1/1000, etc.

1.5 KRAMERS-KRONIG RELATIONS

Under certain quite general conditions the phase velocity and the attenuation are related by the Kramers-Kronig relations. Two forms of these useful relations are

$$\frac{1}{c(\omega)} - \frac{1}{c(0)} = \frac{2\omega^2}{\pi} \int_0^\infty \frac{\alpha(w)}{w^2(w^2 - \alpha^2)} \, dw, \tag{1.5.1}$$

$$\alpha(\omega) = -\frac{2\omega^2}{\pi} \int_0^\infty [\frac{1}{c(w)} - \frac{1}{c(0)}] \frac{dw}{w^2 - \omega^2}. \tag{1.5.2}$$

Thus, if $\alpha(\omega)$ would be known, $c(\omega)$ can be obtained, and vice-versa.

ELASTODYNAMIC THEORY

J.D. Achenbach and Z.L. Li
Northwestern University, Evanston, IL, USA

ABSTRACT

Detailed expositions of elastodynamic theory have been presented in a number of treatises. This lecture is a brief summary of some relevant equations and some general results. The material presented here is based on Refs. [2.1] and [2.2].

2.1 NOTATION

Both indicial notation and vector notation will be used. In a Cartesian coordinate system with coordinates denoted by x_j, the vector $u(x,t)$ where t indicates the time variable is represented by

$$u = u_1 i_1 + u_2 i_2 + u_3 i_3. \qquad (2.1.1)$$

Here i_j $(j=1,2,3)$ are the base-vectors. Since summations of this type occur frequently, it is convenient to introduce the *summation convention*, whereby a repeated Roman subscript implies a summation. Equation (2.1.1) may then be rewritten as

$$u = u_j i_j. \qquad (2.1.2)$$

The following notation is used for the field variables :

position vector	x	(coordinates x_i)	(2.1.3)
displacement vector	u	(components u_i)	(2.1.4)
small-strain tensor	ε	(components ε_{ij})	(2.1.5)
stress tensor	σ	(components σ_{ij})	(2.1.6)

Partial derivatives are indicated by

$$\frac{\partial u_i}{\partial x_j} = u_{i,j}. \qquad (2.1.7)$$

For example, the divergence of the displacement vector u may be written as

$$\text{div } \mathbf{u} = \mathbf{\nabla \cdot u} = u_{i,i} \,. \tag{2.1.8}$$

The Kronecker delta is defined as

$$\delta_{ij} = \begin{cases} 1, & i = j, \\ 0, & i \neq j. \end{cases} \tag{2.1.9}$$

2.2 GOVERNING EQUATIONS

Let the field defining the displacements in a solid be denoted by $\mathbf{u}(x,t)$. The deformation of the solid is described by the small-strain tensor $\mathbf{\epsilon}$ with components

$$\epsilon_{ij} = \frac{1}{2}(u_{i,j} + u_{j,i}). \tag{2.2.1}$$

It is evident that $\epsilon_{ij} = \epsilon_{ji}$.

Related to a surface element with a unit outward normal $\mathbf{n}$, as shown in Fig. 2.1, the traction $\mathbf{t}$ is defined as the force per unit area. The component of $\mathbf{t}$ can be expressed as

$$t_j = \sigma_{ij} n_i, \tag{2.2.2}$$

where σ_{ij} is the stress tensor which is defined as the component in the x_j-direction of the traction on the surface with the unit normal i_i.

The principle of balance of linear momentum leads to the equation of motion

$$\sigma_{ji,j} + f_i = \rho \ddot{u}_i, \tag{2.2.3}$$

where $\mathbf{f}$ is the body force per unit volume, and $(\dot{\,})$ indicates the derivative with respect to time. The principle of balance of angular momentum leads to the conclusion that σ_{ij} is a symmetric tensor

$$\sigma_{ij} = \sigma_{ji}. \tag{2.2.4}$$

The relation between stresses and displacement gradients is expressed by Hooke's law:

$$\sigma_{ij} = C_{ijkl} u_{k,l}, \tag{2.2.5}$$

where C_{ijkl} denotes the tensor of elastic moduli. For a homogeneous isotropic linearly elastic solid we have

$$C_{ijkl} = \lambda \delta_{ij}\delta_{kl} + \mu(\delta_{ik}\delta_{jl} + \delta_{il}\delta_{jk}), \tag{2.2.6}$$

where λ and μ are Lame's elastic constants. By using (2.2.1), Eq.(2.2.5) reduces to

$$\sigma_{ij} = \lambda \epsilon_{kk}\delta_{ij} + 2\mu\epsilon_{ij} \tag{2.2.7}$$

If Eq.(2.2.7) is substituted into Eq. (2.2.3), we obtain the displacement equations of motion:

$$(\lambda + \mu)u_{j,ji} + \mu u_{i,jj} + f_i = \rho \ddot{u}_i. \tag{2.2.8}$$

In vector notation Eq. (2.2.8) may be written as

$$(\lambda + \mu)\mathbf{\nabla \nabla \cdot u} + \mu \nabla^2 \mathbf{u} + \mathbf{f} = \rho \, \ddot{\mathbf{u}}. \tag{2.2.9}$$

On the surface S of a body, boundary conditions must be prescribed. The following boundary conditions are most common:

(i) Displacement boundary conditions: the three components u_i are prescribed on the

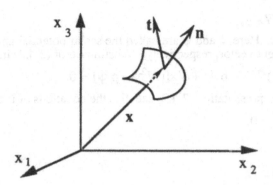

Fig. 2.1. A surface element with traction vector **t** and normal **n**.

boundary S.

(ii) Traction boundary conditions: the three traction components t_i are prescribed on the boundary S.

(iii) Displacement boundary conditions on part S_1 of the boundary and traction boundary conditions on the remaining part $S-S_1$.

To complete the problem statement we define initial conditions at $t = 0$:

$$u_i(\mathbf{x},0) = \overset{o}{u}_i(\mathbf{x}),$$

(2.2.10)

$$\dot{u}_i(\mathbf{x},0^+) = \overset{o}{v}_i(\mathbf{x}).$$

(2.2.11)

2.3 HARMONIC WAVES

A displacement wave $\mathbf{u}(\mathbf{x},t)$ is called harmonic wave if at any position $\mathbf{x}$, $\mathbf{u}$ is time-harmonic with time period $T = 2\pi/\omega$, where ω is the circular frequency,

$$\mathbf{u}(\mathbf{x},t + T) = \mathbf{u}(\mathbf{x},t).$$

(2.3.1)

For a linearly elastic system, all field variables are time-harmonic if the displacement **u** is time-harmonic. In complex notation all field variables of a harmonic wave have the common time-harmonic factor $\exp(-i\omega t)$. This common time-harmonic factor is usually omitted and the equation of motion (2.2.3) then reduces to

$$\sigma_{ij,j} + f_i = -\rho\omega^2 u_i.$$

(2.3.2)

2.4 DISPLACEMENT POTENTIALS

In the absence of body forces the displacement equations of motion (2.2.9) become

$$(\lambda + \mu)\nabla\nabla\cdot\mathbf{u} + \mu\nabla^2\mathbf{u} - \rho\ddot{\mathbf{u}} = 0.$$

(2.4.1)

The displacement equations of motion couple the three displacement components. A more convenient approach is to express the components of the displacement vector in terms of derivatives of potentials. These potentials satisfy uncoupled wave equations. Let us consider a decomposition of the displacement vector of the form

$$\mathbf{u} = \nabla\phi + \nabla\times\boldsymbol{\psi}, \tag{2.4.2}$$

where we also set $\nabla\cdot\boldsymbol{\psi} = 0$. Here, ϕ and $\boldsymbol{\psi}$ are called the scalar potential and the vector potential of the displacement vector, respectively. Substitution of (2.4.2) into (2.4.1) yields

$$\nabla[(\lambda + 2\mu)\nabla^2\phi - \rho\,\ddot{\phi}] + \nabla\times[\mu\nabla^2\boldsymbol{\psi} - \rho\,\ddot{\boldsymbol{\psi}}] = 0. \tag{2.4.3}$$

Clearly, the displacement representation (2.4.2) satisfies the equations of motion if

$$\nabla^2\phi - \frac{1}{c_L^2}\ddot{\phi} = 0, \tag{2.4.4}$$

and

$$\nabla^2\boldsymbol{\psi} - \frac{1}{c_T^2}\ddot{\boldsymbol{\psi}} = 0, \tag{2.4.5}$$

where

$$c_L^2 = (\lambda+2\mu)/\rho \quad \text{and} \quad c_T^2 = \mu/\rho. \tag{2.4.6a,b}$$

2.5 PLANE WAVES

A plane displacement wave propagating with phase velocity c in a direction defined by the unit propagation vector $\mathbf{p}$ can be represented by

$$\mathbf{u} = f(\mathbf{x}\cdot\mathbf{p} - ct)\mathbf{d}, \tag{2.5.1}$$

where $\mathbf{d}$ is a unit vector defining the direction of the displacement. By substituting the components of (2.5.1) into Hooke's law (2.2.7), the components of the stress tensor are obtained as

$$\sigma_{lm} = [\lambda\delta_{lm}(d_j p_j) + \mu(d_l p_m + d_m p_l)]f'(x_n p_n - ct), \tag{2.5.2}$$

where a prime denotes a derivative of the function $f(\ )$ with respect to the argument.

Substituting the expression for the plane wave, Eq. (2.5.1), into the displacement equations of motion (2.4.1) and by employing the relations

$$\nabla\cdot\mathbf{u} = (\mathbf{p}\cdot\mathbf{d})f'(\mathbf{x}\cdot\mathbf{p} - ct)$$

$$\nabla\nabla\cdot\mathbf{u} = (\mathbf{p}\cdot\mathbf{d})f''(\mathbf{x}\cdot\mathbf{p} - ct)\mathbf{p}$$

$$\nabla^2\mathbf{u} = f''(\mathbf{x}\cdot\mathbf{p} - ct)\mathbf{d}$$

$$\ddot{\mathbf{u}} = c^2 f''(\mathbf{x}\cdot\mathbf{p} - ct)\mathbf{d},$$

we obtain

$$[(\lambda+\mu)(\mathbf{p}\cdot\mathbf{d})\mathbf{p} + \mu\mathbf{d} - \rho c^2\mathbf{d}]f''(\mathbf{x}\cdot\mathbf{p} - ct) = 0,$$

hence

$$(\mu - \rho c^2)\mathbf{d} + (\lambda + \mu)(\mathbf{p}\cdot\mathbf{d})\mathbf{p} = 0. \tag{2.5.3}$$

Since $\mathbf{p}$ and $\mathbf{d}$ are two different unit vectors, Eq. (2.5.3) can be satisfied in two ways only:

$$\text{either} \quad \mathbf{d} = \pm\mathbf{p}, \quad \text{or} \quad \mathbf{p}\cdot\mathbf{d} = 0.$$

If $\mathbf{d} = \pm\mathbf{p}$, Eq. (2.5.3) yields

$$c^2 = c_L^2 = (\lambda+2\mu)/\rho. \tag{2.5.4}$$

In this case the motion is parallel to the direction of propagation and the wave is therefore called a *longitudinal* wave or an L wave.

If $\mathbf{d} \neq \pm\mathbf{p}$, both terms in (2.5.3) have to vanish independently, yielding

$$\mathbf{p} \cdot \mathbf{d} = 0, \quad \text{and} \quad c^2 = c_T^2 = \mu/\rho. \tag{2.5.5a,b}$$

Now the motion is normal to the direction of propagation, and the wave is called a *transverse* wave or a T wave. The displacement can have any direction in a plane normal to the direction of propagation, but usually we choose the (x_1, x_2) plane to contain the vector $\mathbf{p}$ and we consider motions which are in the (x_1, x_2) plane or normal to the (x_1, x_2) plane. The transverse motion in the (x_1, x_2) plane is called a "vertically" polarized transverse wave or a TV wave, while the transverse motion normal to the (x_1, x_2) plane is called a "horizontally" polarized transverse wave or a TH wave.

A plane harmonic displacement wave propagating with phase velocity c and propagation vector $\mathbf{p}$ is represented by

$$\mathbf{u} = \mathbf{A}\exp[ik(\mathbf{x} \cdot \mathbf{p} - ct)], \tag{2.5.6}$$

where the amplitude $\mathbf{A}$ is independent of x and t, and $k = \omega/c$ is the wave number. Here ω is the circular frequency. Eq. (2.5.6) clearly is a special case of (2.5.1).

2.6 REFLECTION AND TRANSMISSION OF PLANE HARMONIC WAVES AT A PLANE FLUID-SOLID INTERFACE

An ideal fluid is a medium that does not offer resistance to shear deformation. In such a medium we define the pressure as

$$p = -\lambda_F \nabla \cdot \mathbf{u}, \tag{2.6.1}$$

where λ_F is the coefficient of compression of the fluid. In an ideal fluid only the scalar potential exists, thus

$$\mathbf{u} = \nabla \phi, \tag{2.6.2}$$

where ϕ satifies the equation of motion:

$$\nabla^2\phi - \frac{1}{c_F^2}\ddot{\phi} = 0, \quad c_F^2 = \frac{\lambda_F}{\rho_F}, \tag{2.6.3a,b}$$

here ρ_F and c_F are the mass density and wave velocity of the fluid, respectively.

A plane harmonic wave is of the form

$$\mathbf{u} = \mathbf{U}\exp[i(k\mathbf{x} \cdot \mathbf{p} - \omega t)] = \mathbf{U}\exp[ik(\mathbf{x} \cdot \mathbf{p} - ct)]. \tag{2.6.4}$$

Substituting (2.6.2) and (2.6.4) into (2.6.1) and by using (2.6.3) we have

$$p = -\lambda_F\nabla^2\phi = -\rho_F\ddot{\phi} = \rho_F\omega^2\phi. \tag{2.6.5}$$

An important problem is the reflection and transmission of ultrasound across a fluid-solid interface. The geometry is shown in Fig. 2.2. The displacement incident wave which emanates from infinite depth in the fluid propagating toward the fluid-solid interface is represented by

$$\mathbf{u}^i = \mathbf{U}\mathbf{p}^i\exp[i(k_F(\mathbf{x} \cdot \mathbf{p}^i - c_F t)], \tag{2.6.6}$$

where U and k_F are the amplitude and wavenumber of the incident wave, respectively, and

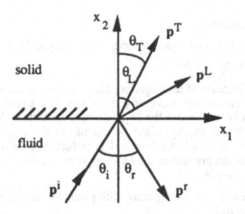

Fig. 2.2. Incident reflected and transmitted waves at a plane fluid-solid interface.

$$\mathbf{p}^i = \sin\theta_i \, \mathbf{i}_1 + \cos\theta_i \, \mathbf{i}_2. \tag{2.6.7}$$

The presence of the fluid-solid interface produces reflected and transmitted waves propagating away from the interface. The reflected wave is represented by

$$\mathbf{u}^r = RU\mathbf{p}^r\exp[i(k_F(\mathbf{x}\cdot\mathbf{p}^r - c_Ft)], \tag{2.6.8}$$

where R is the reflection coefficient, and

$$\mathbf{p}^r = \sin\theta_r \, \mathbf{i}_1 - \cos\theta_r \, \mathbf{i}_2. \tag{2.6.9}$$

Let λ, μ and ρ be the Lamé elastic constants and mass density of the solid, respectively. The displacement transmitted waves in the solid may consist of both longitudinal (L) and transverse (TV) waves, represented by

$$\mathbf{u}^L = T^LU\mathbf{p}^L\exp[i(k_L(\mathbf{x}\cdot\mathbf{p}^L - c_Lt)], \tag{2.6.10}$$

and

$$\mathbf{u}^T = T^TU\mathbf{d}^T\exp[i(k_T(\mathbf{x}\cdot\mathbf{p}^L - c_Tt)], \tag{2.6.11}$$

respectively. Here T^L and T^T are amplitude ratios of transmitted L and TV waves to the incident wave, c_L and c_T are longitudinal and transverse wave velocities in the solid, $k_L = \omega/c_L$, $k_T = \omega/c_T$, and

$$\mathbf{p}^L = \sin\theta_L \, \mathbf{i}_1 - \cos\theta_L \, \mathbf{i}_2 \tag{2.6.12}$$

$$\mathbf{p}^T = \sin\theta_T \, \mathbf{i}_1 - \cos\theta_T \, \mathbf{i}_2 \tag{2.6.13}$$

$$\mathbf{d}^T = - \cos\theta_T \, \mathbf{i}_1 + \sin\theta_T \, \mathbf{i}_2 \tag{2.6.14}$$

By employing Hooke's law (2.2.7) the stresses of the incident, reflected, and transmitted waves are obtained as

$$p^i = - i \, k_F\rho_Fc_F^2U\exp[i(k_F(\mathbf{x}\cdot\mathbf{p}^i - c_Ft)], \tag{2.6.15}$$

$$p^r = - i \, k_F\rho_Fc_F^2RU\exp[i(k_F(\mathbf{x}\cdot\mathbf{p}^r - c_Ft)], \tag{2.6.16}$$

$$\sigma_{22}^L = ik_L[(\lambda + 2\mu)d_2^Lp_2^L + \lambda d_1^Lp_1^L] \, T^LU\exp[i(k_L(\mathbf{x}\cdot\mathbf{p}^L - c_Lt)], \tag{2.6.17}$$

$$\sigma_{21}^L = ik_L\mu[d_2^L p_1^L + d_1^L p_2^L] \, T^L U exp[i(k_L(\mathbf{x} \cdot \mathbf{p}^L - c_L t)], \qquad (2.6.18)$$

$$\sigma_{22}^T = ik_L[(\lambda + 2\mu)d_2^T p_2^T + \lambda d_1^T p_1^T] \, T^T U exp[i(k_T(\mathbf{x} \cdot \mathbf{p}^T - c_T t)], \qquad (2.6.19)$$

$$\sigma_{21}^T = ik_T\mu[d_2^T p_1^T + d_1^T p_2^T] \, T^T U exp[i(k_T(\mathbf{x} \cdot \mathbf{p}^T - c_T t)], \qquad (2.6.20)$$

respectively.

The relations governing the reflection and transmission depend on the boundary conditions at $x_2 = 0$. If there is no cavitation at the fluid-solid interface, the normal displacement will be continuous. Since an ideal fluid cannot support shear we have $\sigma_{21} = 0$ and the normal tractions are continuous.

Note that the x- and t- dependence is in the exponential terms of the expressions of the wave fields only. This dependence on x and t must be the same for all terms. Hence

$$k_F c_F = k_F c_F = k_T c_T = \omega,$$

$$k_F \sin\theta_i = k_F \sin\theta_r = k_L \sin\theta_L = k_T \sin\theta_T.$$

We conclude that

$$\theta_r = \theta_i, \quad \sin\theta_L = \frac{c_L}{c_F} \sin\theta_i, \quad \sin\theta_T = \frac{c_T}{c_F} \sin\theta_i, \qquad (2.6.21a,b,c)$$

also

$$\sin\theta_L = \frac{c_L}{c_T} \sin\theta_T. \qquad (2.6.22)$$

Equations (2.6.21) and (2.6.22) are known as Snell's law. Substituting the results of (2.6.21) and (2.6.22) into the boundary conditions we have three algebraic equations for three unknowns : R, T^L and T^T. These equations can be solved.

Note that for angles larger than certain critical angles, there are no longer real valued solutions for θ_L and θ_T.

A complete solution for the problem discussed in this section can be found in Refs. [2.3] and [2.4].

Example : interface of water and aluminum ($c_T/c_L = 0.488$, $c_L/c_F = 4.26$),

critical angles:　$(c_L/c_F) \sin\theta_i = 1 \rightarrow \sin\theta_i^{cL} = c_F/c_L \rightarrow \theta_i^{cL} = 13.56°$,

$$(c_T/c_F) \sin\theta_i = 1 \rightarrow \theta_i^{cT} = 29.2°.$$

For $\theta_i > \theta_i^{cL}$, $\cos\theta_L = [1-\sin^2\theta_L]^{1/2} = i [(c_L \sin\theta_i/c_F)^2 - 1]^{1/2}$. Thus

$$\mathbf{u}^L = T^L U \, \mathbf{p}^L \, exp[ik_L(x_1\sin\theta_L - c_L t)] \, exp[-k_L x_2\{(c_L\sin\theta_i/c_F)^2 - 1\}^{1/2}]$$

Note that $\mathbf{p}^L$ is complex valued and that the transmitted wave decays with x_2.

2.7 RAYLEIGH SURFACE WAVES

The free surface of an elastic half-space can support waves that decay with depth (surface waves). These wave were discovered by Rayleigh.

Consider the case $u_3 \equiv 0$, and $\partial/\partial x_3 \equiv 0$, and consider a half-space $x_2 \geq 0$. At $x_2 = 0$ we have

$$\sigma_{22} = 0 \quad \text{and} \quad \sigma_{21} = 0. \tag{2.7.1}$$

Now consider displacement potentials of the forms

$$\phi(x_1, x_2, t) = \Phi \exp[-\gamma_L x_2] \exp[ik(x_1 - c_R t)], \tag{2.7.2}$$

$$\psi_3(x_1, x_2, t) = \Psi \exp[-\gamma_T x_2] \exp[ik(x_1 - c_R t)], \quad (\psi_1 = \psi_2 = 0). \tag{2.7.3}$$

The real parts of γ_L and γ_T are supposed to be positive so that the displacements decrease with increasing x_2 and tend to zero as x_2 increases beyond bounds. Substituting (2.7.2) into the wave equation for ϕ, and (2.7.3) into the wave equation for ψ we obtain

$$\gamma_L = (k_R^2 - k_L^2)^{1/2} \quad \text{and} \quad \gamma_T = (k_R^2 - k_T^2)^{1/2}, \tag{2.7.4}$$

where $k_R = \omega/c_R$.

By substituting (2.7.2)-(2.7.4) into (2.4.2) we obtain the coresponding expressions for the displacements. Subsequent substitution of the results into Hooke's law (2.2.7) yields expressions for σ_{22} and σ_{21}. Substitution of the expressions for σ_{22} and σ_{21} into the boundary conditions (2.7.1) yields two homogeneous equations for the constants Φ and Ψ. A nontrivial solution of the system of equations exists only if the determinant of the coefficients vanishes, which leads to the following equation for c_R

$$(2 - c_R^2/c_T^2)^2 - 4(1 - c_R^2/c_L^2)^{1/2}(1 - c_R^2/c_T^2)^{1/2} = F(c_R) = 0. \tag{2.7.5}$$

It can be shown, see Ref. [2.1], that Eq. (2.7.5) has only one root for c_R^2, and that this root is real valued.

The following approximation has been found useful

$$c_R \approx \frac{0.862 + 1.14\nu}{1 + \nu} c_T. \tag{2.7.6}$$

Note that c_R does _not_ depend on the frequency.

2.8 BASIC SINGULAR SOLUTION

Basic singular solutions are of importance in the formulation of integral equations. For time-harmonic motion, the basic singular displacement solution $u_{ik}^G(x - y)$ is defined as the ith component of the displacement field at point x in an infinite domain due to an unit time-harmonic concentrated force applied in the k-direction at point y. This solution satisfies the following equation of motion

$$(\lambda + \mu)u_{jk,ji}^G + \mu u_{ik,jj}^G + \rho\omega^2 u_{ik}^G = -\delta_{ik}\delta(x - y), \tag{2.8.1}$$

where $\delta(\ )$ is the Dirac delta function. The solution to Eq.(2.8.1) can be found in Ref. [2.2]. It may be sumarized as follows:

$$u_{ik}^G(x - y) = A[U_1(r)\delta_{ik} - U_2(r)\frac{\partial r}{\partial x_i}\frac{\partial r}{\partial x_k}], \tag{2.8.2}$$

where $r = |x - y|$ and

(i) for the three-dimensional case:

$$A = 1/(4\pi\mu), \tag{2.8.3}$$

$$U_1 = [1 + \frac{i}{k_T r} - \frac{1}{(k_T r)^2}]\frac{e^{ik_T r}}{r} - (\frac{k_L}{k_T})^2 [\frac{i}{k_L r} - \frac{1}{(k_L r)^2}]\frac{e^{ik_L r}}{r}, \qquad (2.8.4)$$

$$U_2 = [1 + \frac{3i}{k_T r} - \frac{3}{(k_T r)^2}]\frac{e^{ik_T r}}{r} - (\frac{k_L}{k_T})^2 [\frac{3i}{k_L r} - \frac{3}{(k_L r)^2}]\frac{e^{ik_L r}}{r}, \qquad (2.8.5)$$

(ii) for the two-dimensional case:

$$A = i/(4\pi), \qquad (2.8.6)$$

$$U_1 = H_0^{(1)}(k_T r) - \frac{1}{k_T r} [H_1^{(1)}(k_T r) - \frac{k_L}{k_T}H_1^{(1)}(k_L r)], \qquad (2.8.7)$$

$$U_2 = H_0^{(1)}(k_T r) - (\frac{k_L}{k_T})^2 H_0^{(1)}(k_L r) - \frac{2}{k_T r} [H_1^{(1)}(k_T r) - \frac{k_L}{k_T}H_1^{(1)}(k_L r)]. \qquad (2.8.8)$$

Here $H_n^{(1)}( )$ is the Hankel function of the first kind of the nth order.

The corresponding basic singular stress solution $\sigma_{ijk}^G(\mathbf{x} - \mathbf{y})$ follow from Hooke's law .

2.9 THE DYNAMIC RECIPROCAL THEOREM (BETTI-RAYLEIGH)

The dynamic reciprocal theorem presents a relation between two elastodynamic states of the same body. For time-harmonic problems it relates two sets of displacements and stresses both satisfying the equation of motion (2.3.2) and Hooke's law (2.2.7), but with possibly different distribution of body forces and different boundary conditions.

Let $\{f_i^A , u_i^A\}$ and $\{f_i^B , u_i^B\}$ denote two elastodynamic states. By using the equation of motion (2.3.2) we find

$$f_i^B u_i^A - f_i^A u_i^B = - u_i^A(\sigma_{ij,j}^B + \rho\omega^2 u_i^B) + u_i^B(\sigma_{ij,j}^A + \rho\omega^2 u_i^A)$$

$$= \sigma_{ij,j}^A u_i^B - \sigma_{ij,j}^B u_i^A$$

$$= (\sigma_{ij}^A u_i^B - \sigma_{ij}^B u_i^A)_{,j} - (\sigma_{ij}^A u_{i,j}^B - \sigma_{ij}^B u_{i,j}^A).$$

Note that by using Hooke's law (2.2.7) we may write

$$\sigma_{ij}^A u_{i,j}^B - \sigma_{ij}^B u_{i,j}^A$$

$$= [\lambda u_{k,k}^A \delta_{ij} + \mu(u_{i,j}^A + u_{j,i}^A)] u_{i,j}^B - [\lambda u_{k,k}^B \delta_{ij} + \mu(u_{i,j}^B + u_{j,i}^B)] u_{i,j}^A = 0.$$

Thus

$$f_i^B u_i^A - f_i^A u_i^B = (\sigma_{ij}^A u_i^B - \sigma_{ij}^B u_i^A)_{,j} . \qquad (2.9.1)$$

Integration of the relation (2.9.1) over a region V with boundary S yields after application of Gauss' theorem

$$\int_V (f_i^B u_i^A - f_i^A u_i^B) \, dV = \int_S (u_i^B \sigma_{ij}^A - u_i^A \sigma_{ij}^B) \, n_j \, dS. \qquad (2.9.2)$$

Here $\mathbf{n}$ denotes the unit vector along the outward normal to S. The equation (2.9.2) is the reciprocal identity for time harmonic waves. The reciprocal identity is very useful to derive integral representations for the displacement field.

2.10 INTEGRAL REPRESENTATION

The reciprocal theorem derived in Section 2.9 can be used to derive an integral representation for the field generated when a steady-state time-harmonic wave is scattered by an obstacle of arbitrary shape in an elastic solid. The obstacle may be a void or it may be an inclusion. Let V be the bounded domain of the obstacle, and let S be its boundary. Let S_R be a sphere with radius R centered at the origin, and let V_R denote the domain interior to S_R. the radius R is chosen so large that S_R completely surrounds S. A bounded sub-domain D of V_R-V contains sources of elastodynamic radiation which generate a wave incident on the obstacle. The obstacle then generates scattered waves, and thus it is considered as a source of secondary radiation. The geometry is shown in Fig. 2.3.

For scattering problems the total field is usually written as the sum of the incident field and the scattered field:

$$u_i^t = u_i^{in} + u_i^{sc} \quad \text{and} \quad \sigma_{ij}^t = \sigma_{ij}^{in} + \sigma_{ij}^{sc} . \tag{2.10.1a,b}$$

Here u_i^{in} and σ_{ij}^{in} represent the displacement and stress fields in the solid in the absence of the obstacle V. Clearly, the scattered field satisfies homogeneous governing equations, and appropriate conditions on S.

We now consider the reciprocal theorem given by (2.9.2) for the domain V_R-V, and for the following states:

$$\{f_i^A , u_i^A \} = \{0, u_i^{sc}\}$$

and

$$\{f_i^B , u_i^B \} = \{\delta_{ik}\delta(x - y), u_{ik}^G(x - y)\}.$$

Thus state A corresponds to the scattered field, while state B is the basic singular solution for a point force applied at the point of observation y in the k-direction.

For $y \in V_R$-V, $y \notin S$ and $y \notin S_R$, application of (2.9.2) now yields

$$\varepsilon(y)u_k^{sc}(y) = -\int_S [u_{ik}^G (x-y)\sigma_{ij}^{sc}(x) - u_i^{sc}(x)\sigma_{ijk}^G(x-y)]n_j(x)dA(x) + I_k, \tag{2.10.2}$$

where

$$\varepsilon(y) = \begin{cases} 1, & y \in V_R-V, \\ 0, & y \in V, \end{cases}$$

and

$$I_k = \int_{S_R} [u_{ik}^G(x -y)\sigma_{ij}^{sc}(x) - u_i^{sc}(x)\sigma_{ijk}^G(x -y)]n_j(x)dA(x) . \tag{2.10.3}$$

Here the normals n are as indicated in Fig. 2.3. In particular it should be noted that the normal on the surface S of the obstacle points into the surrounding solid.

From the Sommerfeld radiation condition for elastodynamic problems, it can be shown that $I_k \to 0$ as $R \to \infty$, see e.g., Ref.[2.2]. Thus, for a point of observation not on the boundary of the obstacle, (2.10.2) reduces to

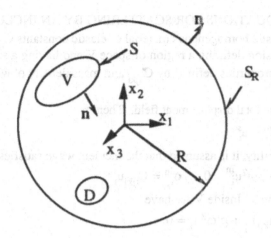

Fig. 2.3 Obstacle of volume V and boundary S in a homogeneous, isotropic, linearly elastic
       solid.

$$\varepsilon(\mathbf{y})u_k^{sc}(\mathbf{y}) = \int_S [u_i^{sc}(\mathbf{x})\; \sigma_{ijk}^G(\mathbf{x}-\mathbf{y}) - u_{ik}^G(\mathbf{x}-\mathbf{y})\; \sigma_{ij}^{sc}(\mathbf{x})]n_j(\mathbf{x})dA(\mathbf{x}),$$

$$\text{for } \mathbf{y} \in V_R\text{-}V, \;\; \mathbf{y} \notin S. \qquad (2.10.4a)$$

Sometimes it may be more convenient to change the integration variable from x to y:

$$\varepsilon(\mathbf{x})u_k^{sc}(\mathbf{x}) = \int_S [u_i^{sc}(\mathbf{y})\; \sigma_{ijk}^G(\mathbf{y}-\mathbf{x}) - u_{ik}^G(\mathbf{y}-\mathbf{x})\; \sigma_{ij}^{sc}(\mathbf{y})]n_j(\mathbf{y})dA(\mathbf{y}),$$

$$\text{for } \mathbf{x} \in V_R\text{-}V, \;\; \mathbf{x} \notin S, \qquad (2.10.4b)$$

Equation (2.10.4) is not immediately useful in solving scattering problems, since the
scattered fields are unknown on S.

2.11 INTEGRAL EQUATIONS FOR SCATTERING BY A VOID

The integral representations for the displacement fields of the scattered wave can be
used to derive a set of integral equations for the unknown displacements u_i^{sc} on the boundary
of the obstacle.   For a traction-free void, the limit of $\mathbf{x} \notin V$, $\mathbf{x} \notin S \to \mathbf{x} \in S$ in Eq.(2.10.4b)
yields the integral equation

$$u_k^{sc}(\mathbf{x}^S) = \lim_{\mathbf{x} \to \mathbf{x}S} \int_S u_i^{sc}(\mathbf{y})\; \sigma_{ijk}^G(\mathbf{y}-\mathbf{x})\; n_j(\mathbf{y})dA(\mathbf{y})$$

$$= \int_S u_i^{sc}(\mathbf{y})\; \sigma_{ijk}^G(\mathbf{y}-\mathbf{x}^S)\; n_j(\mathbf{y})dA(\mathbf{y}), \qquad (2.11.1)$$

and the limit of $\mathbf{x} \in V$, $\mathbf{x} \notin S \to \mathbf{x} \in S$ in Eq.(2.10.4b) yields the integral equation

$$\lim_{\mathbf{x} \to \mathbf{x}S} \int_S u_i^{sc}(\mathbf{y})\; \sigma_{ijk}^G(\mathbf{y}-\mathbf{x})\; n_j(\mathbf{y})dA(\mathbf{y}) = \int_S u_i^{sc}(\mathbf{y})\; \sigma_{ijk}^G(\mathbf{y}-\mathbf{x}^S)\; n_j(\mathbf{y})dA(\mathbf{y}) = 0. \qquad (2.11.2)$$

In (2.11.1) and (2.11.2) $\mathbf{x}^S$ denotes a point on the surface of the void.

2.12 INTEGRAL EQUATIONS FOR SCATTERING BY AN INCLUSION

In a linearly elastic homogeneous material of elastic constants C_{ijkl} and mass density ρ, there exsists an inclusion defining a region of space V and having a surface S. The material inside V has elastic properties defined by C'_{ijkl} and mass density ρ' which differ from those of the host material.

Let u_i denote the total displacement field. Then

$$u_i = u_i^{in} + u_i^{sc}. \tag{2.12.1}$$

Without loss of generality, it is assumed that the incident wave satisfies (everywhere)

$$\sigma_{ij,j}^{in} + \rho\omega^2 u_i^{in} = 0, \quad \sigma_{ij}^{in} = C_{ijkl}u_{k,j}^{in} \tag{2.12.2a,b}$$

where C_{ijkl} are constants. Inside V we have

$$(C'_{ijkl}u_{k,l})_{,j} + \rho'\omega^2 u_i = 0. \tag{2.12.3}$$

Let

$$C'_{ijkl} = C_{ijkl} + \delta C_{ijkl}, \quad \rho' = \rho + \delta\rho, \tag{2.12.4a,b}$$

where δC_{ijkl} and $\delta\rho$ are assumed to depend on the spatial coordinates. Then inside V

$$C_{ijkl}u_{k,lj} + \rho\omega^2 u_i = -(\delta C_{ijkl}u_{k,l})_{,j} - \delta\rho\omega^2 u_i. \tag{2.12.5}$$

By using (2.12.1) and the equation of motion (2.12.2a) for the incident wave, we can subsequently write

$$C_{ijkl}u_{k,lj}^{sc} + \rho\omega^2 u_i^{sc} = \begin{cases} -(\delta C_{ijkl}u_{k,l})_{,j} - \delta\rho\omega^2 u_i & \text{in V} \\ 0 & \text{outside V} \end{cases} \tag{2.12.6}$$

Note that the right-hand side can be considered as a body force. Consequently the solution may formally be written as

$$u_m^{sc}(x) = \int_V u_{im}^G(x-y)(\delta C_{ijkl}u_{k,l})_{,j} \, dV(y) + \omega^2 \int_V \delta\rho u_{im}^G(x-y)u_i(y)dV(y). \tag{2.12.7}$$

Rewrite

$$u_{im}^G \frac{\partial}{\partial y_j}(\delta C_{ijkl}u_{k,l}) = \frac{\partial}{\partial y_j}(u_{im}^G \delta C_{ijkl}u_{k,l}) - (\frac{\partial}{\partial y_j}u_{im}^G)\delta C_{ijkl}u_{k,l} \tag{2.12.8}$$

The volume integral of the first term in the right-hand side can be converted into an integral over the surface S by the use of Gauss' theorem. That integral will, however, vanish if δC_{ijkl} vanishes on S. We then have

$$u_m^{sc}(x) = -\int_V \delta C_{ijkl}(\frac{\partial}{\partial y_j}u_{im}^G)u_{k,l}dV + \omega^2 \int_V \delta\rho u_{im}^G u_i dV. \tag{2.12.9}$$

The last equation is also valid when δC_{ijkl} is uniform inside V (i.e., δC_{ijkl} does not vanish on S). Then

$$u_m^{sc}(x) = -\delta C_{ijkl}\int_V (\frac{\partial}{\partial y_j}u_{im}^G)u_{k,l}dV + \omega^2\delta\rho \int_V u_{im}^G u_i dV. \tag{2.12.10}$$

This equation is difficult to solve rigorously. An approximate solution can, however, easily be obtained by the Born approximation. In this approximation the field inside the scatterer is taken as the incident field. The approximation is valid for weak scatterers.

For further details we refer to Refs. [2.5]. To obtain a numerically exact solution it is more convenient to work with boundary integral equations, i.e., systems of singular integral equations over S. These equations and a boundary element method for their solution have been discussed in Ref. [2.6].

References

2.1. Achenbach, J. D.: Wave Propagation in Elastic Solids, North Holland / Elsevier, Amsterdam / New York 1973.
2.2. Achenbach, J. D., Gautesen,, A. K. and McMaken, H.: Ray Methods for Waves in Elastic Solids, Pitman Advanced Publishing Program, Boston 1982.
2.3. Krautkrämer, J. and Krautkrämer, H.: Ultrasonic Testing of Materials, Springer Verlag, Berlin 1983.
2.4. Ultrasonic Inspections, in Metals Handbook: Vol.17, Nondestructive Evaluation and Quality Control, ASM International, Metals Park, OH, USA 1989, 231-277.
2.5. Gubernatis, J. E., Domany, E., Krumhansl, J. A. and Huberman, M.: The Born approximation in the theory of the scattering of elastic waves by flaws, J. Appl. Phys, 48(1977), 2804-2811 and 2812-2819.
2.6. Kitahara, M., Nakagawa, K. and Achenbach, J. D.: Boundary-integral equation method for elastodynamic scattering by a compact inhomogeneity, Computational Mechanics, 5(1989), 129-144.

ULTRASONIC SPECTROSCOPY

L. Adler
The Ohio State University, Columbus, OH, USA

ABSTRACT

Ultrasonic Spectroscopy is a relatively new method to ultrasonic nondestructive evaluation. The method is based on the Fourier frequency components of an ultrasonic signal. Since most wave-materials interaction is frequency dependent, this technique proved to be more useful for quantitative studies. In this paper a general ultrasonic spectroscopy system is described and analyzed as a linear time-invariant system.

1 INTRODUCTION

Ultrasonic spectroscopy is the study of ultrasonic waves resolved into their Fourier frequency components. Since most material-wave interaction is frequency dependent the purpose of ultrasonic spectroscopy is to determine frequency dependent material properties. These might be of geometrical origin e.g. flaw size, shape, orientation etc., or material properties e.g. attenuation and velocity dispersion. Ultrasonic nondestructive evaluation before 1960's were carried out using mostly narrowband ultrasonic instrumentation. The value to use broadband ultrasonic pulse was recognized by Gericke [1]. He developed a system to utilize multifrequency (broadband) ultrasonic pulses using contact transducers in a pulse-echo mode. Gericke also appears to be the first to use the term Ultrasonic Spectroscopy. In analogy with light spectroscopy he postulated that additional interpretation could be obtained corresponding to the "color" of the waves after interaction with the materials. Gericke obtained "white ultrasound" by exciting a highly damped ceramic transducer. His results indicated qualitatively that ultrasonic spectrum of an echo contains information related to configuration of a defect in a solid and to microstructure of materials. Other investigators refined ultrasonic spectroscopic instrumentation by introducing immersion systems, and using multiple transducers through transmission and pitch-catch ultrasonic spectroscopy [2], and broadened the list of applications [3]. In addition to flaw characterization, and applying it for assessment of solid-state bonds [4], it has proven to be useful for determining surface properties and [5] subsurface gradient [6], porosity in cast aluminum [7], and in composites [8], monitoring polymer curing [9], as well as, measurement of frequency dependent velocity and attenuation. In the following section the components used to build a basic ultrasonic spectroscopy system will be described. In subsequent chapters examples of applications of ultrasonic spectroscopy in nondestructive evaluation will be given.

2 SYSTEM MODEL

Although there are many possible designs for an ultrasonic spectrum analysis system, each contains provisions for: 1) generating ultrasound, 2) receiving a portion of the ultrasound that has interacted with the test material, and 3) analyzing the received wave to determine its magnitude (and sometimes phase) at a number of frequencies. Fig. 1 illustrates the major components of a typical spectroscopic system.

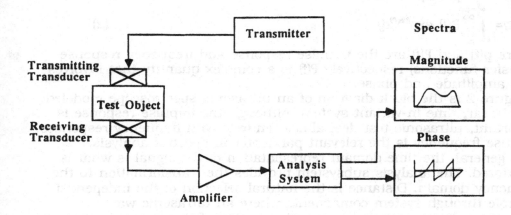

Figure 1. Generalized ultrasonic spectroscopy system.[see ref. 3]

In this system configuration, an electrical waveform generated by the transmitter is applied to the transmitting transducer. Conversion of the electrical energy into mechanical energy occurs within the transducer, producing an ultrasonic wave. As the wave propagates through the test material, interactions of the ultrasonic energy with the material alter the amplitude, phase, and direction of the wave. A receiving transducer intercepts a portion of the ultrasonic energy and conversion occurs from mechanical to electrical energy. Because the electrical signal is usually small, an amplifier is used to increase its amplitude. The purpose of the analysis system, which follows the amplifier, is to sort out the signals characteristic to the ultrasonic interactions within the material and to present their amplitude and phase spectra. Since experimental artifacts must be removed from displayed spectra it has been proposed that the spectroscopic instrument be modelled as a linear time-invariant system.

The behavior of a linear, time-invariant system is completely described by its impulse response(in the time domain) or its frequency response(in the frequency domain). The two descriptions of system response are equivalent and are linked by the continuous time Fourier transform pair.(see eqs. 1 and 2)

$$P(f)= \int_{-\infty}^{\infty} p(t)exp^{(-j2\pi ft)}dt \qquad (1)$$

$$p(t)= \int_{-\infty}^{\infty} P(f) exp^{(j2\pi ft)}df \qquad (2)$$

Where p(t) and P(f) are the impulse response and frequency response (transfer functions) respectively P(f) is a complex quantity that involves both amplitude and phase.

Figure 2 is the block diagram of an ultrasonic spectroscope modeled as a linear, time in-variant system. Although the impulse response is important, ultrasonic test deal almost entirely with frequency responses because frequency is the relevant parameter in spectral analysis.

In general, the time domain representation of the signal is what is monitored. The analysis subsystem provides the transformation to the frequency domain. Distance is the natural selection of the independent variable through system components where an ultrasonic wave propagates. Distance may be converted to time if the wave velocity is known. System components in which ultrasonic waves propagate are shown with impulse response in terms of time-varying quantities and distance.

An ideal spectroscopic system emits ultrasonic pulses whose spectra are uniform over a large range of frequencies (such as 20 kHz to 500 MHz). For the model in fig. 2 this implies that:

$$V_1(f)B_1(f)X_1(f)C_1(f)=T(f)= constant \qquad (3)$$

For the frequency range of interest.
A receiving section whose response is uniform over the same frequency range is also desirable.
That is:

$$C_2(f)X_2(X)B_2(f)A(f)......=R(f)= const \qquad (4)$$

If the ideal system just described is used to test a material with impulse response p(t), the output of the spectrum analysis system $V_2(f)$ is P(t) multiplied by a constant. However, such an ideal spectroscopic system is not strictly realizable over the wide-frequency range mentioned earlier.

The linear, time-invariant model provides a method of correcting for the nonuniform frequency response in nonideal systems if the transfer functions of the systems components are known. Coupling the

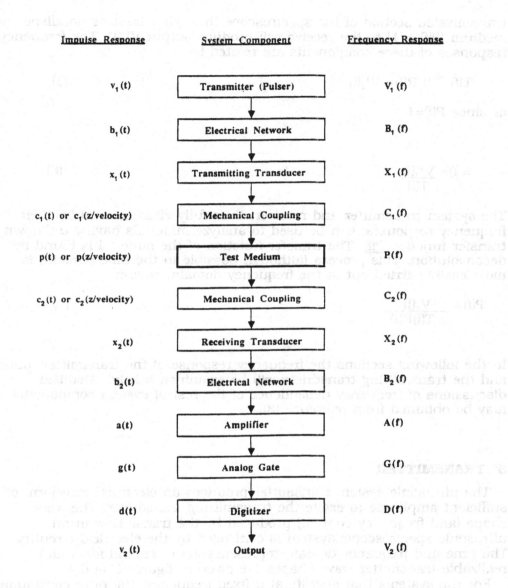

Impulse Response	System Component	Frequency Response
$v_1(t)$	Transmitter (Pulser)	$V_1(f)$
$b_1(t)$	Electrical Network	$B_1(f)$
$x_1(t)$	Transmitting Transducer	$X_1(f)$
$c_1(t)$ or c_1 (z/velocity)	Mechanical Coupling	$C_1(f)$
$p(t)$ or p (z/velocity)	Test Medium	$P(f)$
$c_2(t)$ or c_2 (z/velocity)	Mechanical Coupling	$C_2(f)$
$x_2(t)$	Receiving Transducer	$X_2(f)$
$b_2(t)$	Electrical Network	$B_2(f)$
$a(t)$	Amplifier	$A(f)$
$g(t)$	Analog Gate	$G(f)$
$d(t)$	Digitizer	$D(f)$
$v_2(t)$	Output	$V_2(f)$

Figure 2. Elements of an ultrasonic spectroscopic system modeled as a linear, time-invariant system. [see ref. 3]

transmission section of the spectroscope through a lossless nondispersive medium $[P(f) = 1]$ to the receive will produce output $V'_2(t)$. The frequency responses of these components are related by

$$T(f)\ P(f)\ R(f) = V'_2(f) \tag{5}$$

or since $P(f)=1$

$$R(f) = \frac{V'_2(f)}{T(f)} \tag{6}$$

The system transmitter and receiver, thus fully characterized by their frequency responses, can be used to analyze materials having unknown transfer function $P(f)$. The transfer function of the material is found by deconvolution. This process (although possible in the time domain) is most easily carried out in the frequency domain, where:

$$P(f) = \frac{V_2(f)}{T(f)R(f)} \tag{7}$$

In the following sections the frequency response of the transmitter (pulse) and the transmitting transducer will be examined briefly. Detailed discussions of frequency dependence of the rest of system components may be obtained from reference (3).

3 TRANSMITTER

The ultrasonic system transmitter produces an electrical waveform of sufficient amplitude to excite the transmitting transducer. The wave shape (and frequency content) produced by the transmitter in an ultrasonic spectroscopic system is controlled by the electrical circuitry. The time and frequency domain representation of several ideal and realizable transmitter wave shapes are given in figures 3 and 4.

For the systems that operate at a fixed frequency, the pure continuous wave sinusoid is used. Note that this waveform contains only one frequency if it is infinitely extended in time. For cases in which the sinusoid exits for a finite length of time, the frequency domain representation includes more than a single frequency. This type of waveform is represented as a sine wave modulated by a rectangular envelope. Since some spectroscopic systems emit tone bursts, the effect of varying the width of the rectangular envelope should be examined. Figure 5 illustrates the effect of width variation on the frequency content of a sinusoidal burst. The width of the main frequency lobe increases as the length of the rectangular modulating waveform decreases and is independent of the sine wave frequency. The advantages of a narrow frequency output have to be considered against the poor range resolution of a pulse with long time duration.

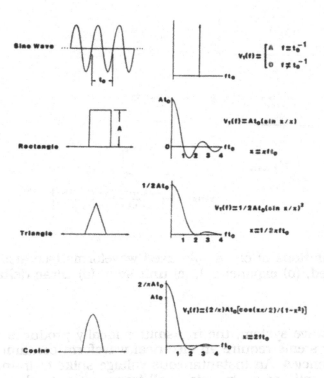

Figure 3. Representations of commonly used waveforms (see also Figure 4); (a)sine wave, (b)rectangle waveform, (c) triangle waveform and (d) cosine

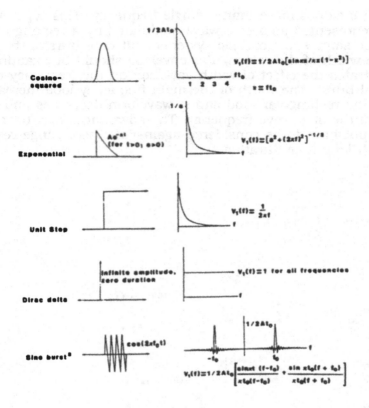

Figure 4. Representations of commonly used waveforms (see also Figure 3): (a) cosine squared, (b) exponential, (c) unit step, (d) Dirac delta and (e) sine burst.

In a continuous wave system, the transmitter ideally produces a single frequency. Pulsed systems require an electrical waveform containing a wide range of frequencies. An instantaneous voltage spike of infinite amplitude contains uniform magnitude of all frequencies. Such a waveform (called the Dirac delta function) cannot be realized, but pulses having very short duration and large amplitudes can be produced.

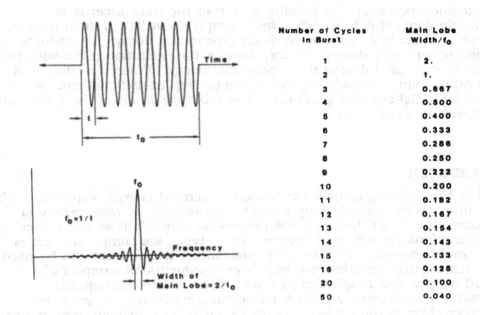

Number of Cycles in Burst	Main Lobe Width/f_0
1	2.
2	1.
3	0.667
4	0.500
5	0.400
6	0.333
7	0.286
8	0.250
9	0.222
10	0.200
11	0.182
12	0.167
13	0.154
14	0.143
15	0.133
16	0.125
20	0.100
50	0.040

Figure 5. Effect of pulse duration of a sinusoid upon the spectrum [see Ref 3]

A cosine squared pulse has a bandwidth greater than that of a rectangular pulse with the same duration t_0. However, the circuitry required to produce this signal is more complex. A step pulse with a single transition is more attractive because of superior range resolution and the large bandwidth. The bandwidth and transition time of the step are related by:

$$B = 1/2t_0 \tag{8}$$

where B = the signal bandwidth (megahertz), and
    t_0 = the transition time of the step (microseconds).

For instance, for a transition time of 5 ns, the pulse has a bandwidth of 100 MHz.

A common procedure for pulsing an ultrasonic transducer is to momentarily connect it to a high voltage supply them let it return to zero. This type of pulse is a unit step with an exponential decay. Transducer excitation occurs only during the transition of the excitation pulse with no appreciable excitation during the exponential decay. Pulser waveforms of shapes other than sinusoids (square waves or step transitions) can be obtained by a digital pulse generator. The arbitrary shape of the pulse can be designed by a computer.

4 TRANSDUCER

The transmitting transducer converts electrical energy, supplied by the pulser, into mechanical vibratory energy - ultrasound. A wide variety of transducer types exist; however, the discussion which follows will center on piezoelectric (and ferroelectric) devices. other types are competitive under special conditions only. For instance, noncontact transducers can be used in high temperature or otherwise hostile environments on samples of awkward shapes and rough surfaces, as well as on moving objects. Electrostatic transducers, although non-contacting by nature, must be placed very close to the surface to achieve sufficient sensitivity required in most NDE applications. Electromagnetic acoustic transducers (EMATs) are more practical but still require that the sensors be placed within a few millimeters of the object. Unfortunately, their limited sensitivity and usually narrow bandwidth render them less useful in spectroscopic applications. Thanks to the recent development of inexpensive, rugged, portable YAG lasers, laser generation of ultrasound by pulsed infrared irradiation has become a very effective NDE tool. The acoustical pulse is produced by thermal expansion and surface vaporization, and the bandwidth of the generated pulse can be in excess of 100 MHz. Laser detection is also feasible by heterodyne interferometry, but the sensitivity of the technique still leaves much to be desired.

4.1 Frequency Dependence of Piezoelectric Transducers

One possible approach to the problem of determining a transducer's transfer characteristics is to begin from first principles and consider the mathematics relating the electrical and elastic properties of the piezoelectric materials.

Consider a piezoelectric disk of thickness l and with surface area of the circular face equal to S. The disk, much larger in diameter than thickness, is assumed to be made of a material having non electrical conductivity and which produces no ultrasonic attenuation. The transducer is operated in a thickness expander mode to produce longitudinal ultrasonic waves propagating in the -x and +x directions.

The acoustic impedance of the transducer is

$$Z = \rho_1 c_1 \qquad (9)$$

where ρ_1 is the density and c_1 is the ultrasonic velocity. Subscripts 2 and 3 are assigned to the material properties, respectively, behind (-x) and in front (+x) of the transducer. Additionally, the two bounding media are considered to be infinitely extended. Because only longitudinal waves are considered, the problem becomes one dimensional. The amplitude reflection coefficients (from transducer to bounding media) are given by the formulas in Figure 6.

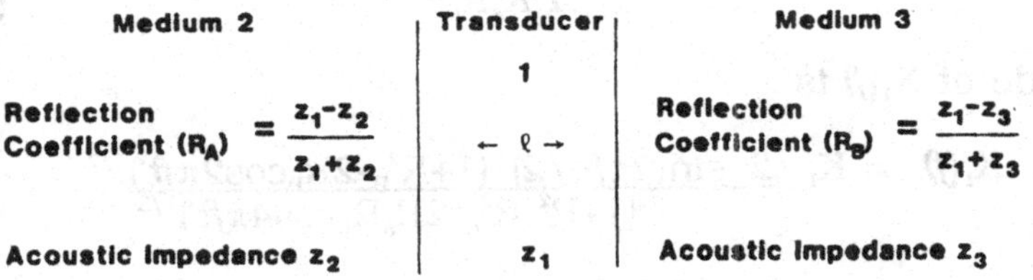

Medium 2	Transducer	Medium 3
	1	
Reflection Coefficient (R_A) $= \dfrac{z_1 - z_2}{z_1 + z_2}$	$\leftarrow \ell \rightarrow$	Reflection Coefficient (R_B) $= \dfrac{z_1 - z_3}{z_1 + z_3}$
Acoustic Impedance z_2	z_1	Acoustic Impedance z_3

Figure 6. Transducer bounded by two media. The reflection coefficients at each boundary are shown.

The derivation proceeds by considering the equation describing the direct piezoelectric effect:

$$T = C(\delta u / \delta x) - hD \qquad (10)$$

where T is the stress on a face perpendicular to the x axis, u is the particle displacement in the x direction, D is the displacement, C is the elastic stiffness at constant electric displacement in the x direction, and h is the piezoelectric coefficient at constant strain. Once an initial (Dirac delta) excitation, $D\delta(t)$, occurs, the disk's behavior is determined solely by the propagation characteristics of the ultrasound in the transducer and the loading at its faces. The complete analysis of this situation leads to the following transfer function for ultrasonic waves radiating into medium 2:

$$X_1(s) = \frac{-h(1+R_A)(1-e^{-t's})(1-R_B e^{-t's})}{2Z_1(1-R_A R_B e^{-2t's})} = \frac{P(s)}{V_1(s)} \tag{11}$$

where t' is the transit time of an ultrasonic pulse through the transducer (i.e., $1/c_1$), $P(s)$ and $V_1(s)$ are the Laplace transforms of the pressure (force/area) and driving voltage, respectively, and s is the Laplace operator.

Let us examine the frequency-dependent part of $X^1(f)$. The frequency response may be written:

$$X_1(f) = K_1 \frac{(1-e^{-j2\pi ft'})(1-R_B e^{-j2\pi ft'})}{1-R_A R_B e^{-j4\pi ft'}} \tag{12}$$

The magnitude of $X_1(f)$ is

$$X_1(f) = K_1 \frac{2 \sin(2\pi ft'/2)(1+R_B^2-2R_B\cos2\pi ft')^{1/2}}{(1+R_A^2 R_B^2-2R_A R_B\cos4\pi ft')^{1/2}} \tag{13}$$

Notice that $X_1(f)$ is periodic (period=$2\pi/t'$). Additionally, the transfer function goes to zero for

$$2\pi f = n(2\pi/t'), \quad n=0,1,2,... \tag{14}$$

that is, at frequencies

$$f = n/t' = nc_1/l \tag{15}$$

In addition to these minima, $X_1(f)$ has maxima, where

$$2\pi f = (2n+1)\pi/t' \tag{16}$$

$$f=(2n+1)/2t'$$ (17)

Thus the transducer demonstrates a frequency response with variations. The amplitude of the modulation is determined by the reflection coefficients at the transducer boundaries.

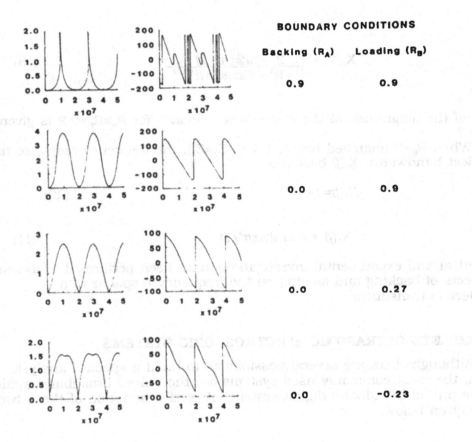

Figure 7. Frequency response of a transducer disk for a variety of backing and loading materials.[see ref 3]

Figure 7 contains plots of transducer response (as calculated using Eq. 12) for several boundary conditions. Notice that the response broadens for appropriate backing.

For the case when the reflection coefficients R_A and R_B are the same (=R), the expression $X_1(f)$ simplifies considerably:

$$X_1(f) = \frac{1 - e^{-j2\pi ft'}}{1 + Re^{-j2\pi ft'}} \tag{18}$$

and

$$X_1(f) = \frac{2 \sin(2\pi ft'/2)}{(1 + R^2 + 2R\cos 2\pi ft')^{1/2}} \tag{19}$$

A plot of the magnitude of the transducer response for $R_A=R_B=0.9$ is given in Figure 7.

When $R_B=0$ (matched backing, i.e., $Z_1=Z_2$), the frequency response has its widest bandwidth. $X_1(f)$ becomes

$$X_1(f) = 1 - e^{-j2\pi ft'} \tag{20}$$

and

$$X_1(f) = 2 \sin(2\pi ft'/2) \tag{21}$$

Theoretical and experimental investigations have been performed to assess the effects of backing and loading on the frequency response of a piezoelectric transducer.

5 COMPLETE ULTRASONIC SPECTROSCOPIC SYSTEMS

Although there are several possibilities to build a spectral analysis system, the most commonly used systems are the pulsed broadband analog and the pulsed broadband digital system. A brief description of these two will be given below.

5.1 Broadband Analog Pulsed System

In a broadband pulsed system, a short ultrasonic signal containing a

very wide range of frequencies interrogates the test object (see Fig. 8).
Elements of the transmitting subsystem are matched to produce the widest
bandwidth and most uniform response. The signal from the broadband
receiving subsystem is connected to an analog spectrum analyzer.

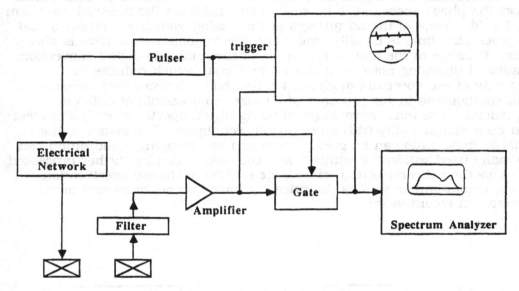

Figure 8 Broadband pulsed ultrasonic spectroscopic system, utilizing digital
signal analysis. [see ref 3]

    Ordinarily, the analysis is performed by a superheterodyne instrument
which is used as a continuously tuned selective receiver. At each repetition
of the signal to be analyzed, only one spectral component is filtered out and
measured by the superheterodyne receiver. The center frequency is slowly
changed between repetitions. Adjacent spectral components are measured
from subsequent signals, and the overall spectrum is displayed by
interleaving a large number of such measurements to obtain a more-or-less
continuous spectrum. Since the repetition frequency is usually on the order
of 100 to 1000Hz in NDE applications, a fairly detailed spectrum of 100
spectrum lines takes approximately 0.1 - 1 s. This analysis speed makes it
possible to monitor dynamic processes in "real time" unless the changes are
exceptionally fast. One disadvantage of this technique with respect to digital
analyzers is that phase information is not available unless phase sensitive
synchronous detection is applied instead of the commonly used amplitude

measurement.

5.2 Broadband Digital Pulsed System

Incorporation of digital data analysis into a broadband pulsed system allows the phase spectrum to be easily extracted from the received waveform (see fig. 9). Some of the advantages of the analog system are retained, but the processing time is usually longer and the frequency resolution is also lower. Because of the limited resolution of the analog-to-digital conversion, broadband digitizing noise is added to the signal, which reduces the selectivity of the spectrum analyzer, i.e. its ability to accurately measure weak components in the presence of stronger components of different frequencies. The main advantages of using digital spectrum analyzers stem from their compatibility with other digital techniques. For instance, the signal-to-noise ratio can be greatly increased by averaging, anodization can be readily used within the window, and the results can be further processed (e.g. smoothened, normalized, stored, etc.). Also, technical specification (speed, resolution) as well as the price tag have become more and more attractive in recent years.

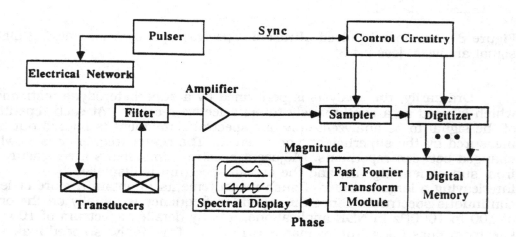

Figure 9. Broadband pulsed ultrasonic spectroscopic system utilizing analog signal analysis. [see ref 3]

REFERENCES

1. Gericke, O.R.: Determination of the Geometry of Hidden Defects by Ultrasonic Pulse Analysis Testing, Journal of the Acoustical Society of America,35 (1963), 364.

2. Whaley, H.L. and l. Adler: Flaw Characterization by Ultrasonic Frequency Analysis, Materials Evaluation, 29 (1971), 182.

3. Fitting, D.E. and l. Adler: Ultrasonic Spectral Analysis for Nondestructive Evaluation, Plenum Press, New York, 1981.

4. Nagy, P.B. and L. Adler: Ultrasonic NDE of Solid-State Bonds: Inertia and Friction Welds. Journal of NonDestructive Evaluation, 7 (1988), 199.

5. de Billy, M., J. Doucet and G. Quentin: Angualar Dependence of the Backscattered Intensity of Acoustic Waves from Rough Surfaces, Ultrasonics International, 1975, 218.

6. Doyle, P.A and C.M. Scala: Crack Depth measurement by Ultrasonics, Ultrasonics, 16 (1978), 164.

7. Adler, L., J.H. Rose and C. Mobley: Ultrasonic Method to Determine Gas Porosity in Aluminum Alloy Castings: Theory and Experiment, Journal of Applied Physics, 59 (1986), 336.

8. Hsu, D.K. and S.M. Nair: Evaluation of Porosity in Graphite-Epoxy composite by Frequency Dependence of Ultrasonic Attenuation, Review of Progress in Quantitative Nondestructive Evaluation, 6B (1987), 1185.

9. Rohklin, S.I., D.K. Lewis, R.F. Graff and L. Adler: Real Time Study of Frequency Dependence Attenuation and Velocity of ultrasonic Waves During The Curing Reaction of Epoxy Resin, J. Acoust. Soc. Am., 79 (1986), 1786.

SCATTERING OF ULTRASONIC WAVES BY CRACKS

J.D. Achenbach and Z.L. Li
Northwestern University, Evanston, IL, USA

ABSTRACT

Mathematical techniques to analyze scattering of ultrasound are discussed. Particular attention is devoted to scattering by cracks. Results are presented for time-domain finite difference calculations and for the ray method. The derivation of a boundary integral equation for the crack-opening-displacement according to rigorous elastodynamic theory and the numerical solution of this equation by the boundary element method are discussed in some detail. When the crack-opening-displacement is known, the scattered field can be obtained by the use of an integral representation.

4.1 MATHEMATICAL MODELING

Mathematical modeling of ultrasonic wave scattering provides valuable quantitative information for methods to detect and characterize cracks. Even though both the geometrical configuration and the process of ultrasonic wave propagation must be simplified to accommodate a mathematical approach, the characteristic features of the scattering phenomenon can be maintained. Results obtained by the modeling approach are essential to the development of measurement models, as discussed in more detail in lecture 7.

In recent years numerous results have become available for fields generated by scattering of ultrasonic waves by flaws. Solutions are available for scattering by voids, inclusions, internal cracks, macrocrack-microcrack configurations, arrays of cracks and surface breaking cracks. Most results are in the frequency domain, but they can be converted to the time domain by application of the Fast Fourier Transform. Recently numerical results have been developed directly in the time domain.

In this lecture we will discuss scattering of ultrasound by cracks. Analytical and numerical results are generally obtained for a perfect mathematical crack. The faces of a perfect mathematical crack are smooth and infinitesimally close, but they are specified not to interact with each other. From the analytical point of view a perfect mathematical crack is a surface in space which does not transmit tractions. The model is acceptable for a real crack, provided that the latter's faces are slightly separated and that the length characterizing crack-face roughness is much smaller than the dominant wavelengths of an incident pulse of ultrasonic wave motion.

There are few exact closed-form analytical solutions to elastodynamic scattering problems. In a rigorous approach, often the best that can be done is to reduce the mathematical formulation to a form which is suitable for numerical work. Approximate closed-form solutions can be obtained at high frequencies (Kirchhoff approximation or geometrical diffraction theory) or at low frequencies (Rayleigh approximation). However these approximations are of limited value for scattering in the mid-frequency range in which characteristic wave-lengths of an incident displacement pulse are of the same order of magnitude as a characteristic length of the flaw. In that range it is necessary to use numerical methods; for example, numerical schemes to solve systems of governing integral equations, finite difference techniques, finite element techniques, T-matrix methods, etc. The validity but also the efficiency of various techniques depends on the dimensionless wave numbers ka. Figure 4.1 indicates the useful ranges of these techniques.

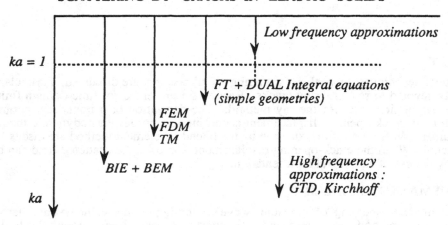

FREQUENCY-DOMAIN TECHNIQUES TO ANALYZE
SCATTERING BY CRACKS IN ELASTIC SOLIDS

Low frequency approximations

ka = 1

*FT + DUAL Integral equations
(simple geometries)*

*FEM
FDM
TM*

BIE + BEM

*High frequency
approximations :
GTD, Kirchhoff*

ka

k: wavenumber = $2\pi/\lambda$ = ω/c,   a: characteristic length of crack.
example: titanium $c_L \approx 6.10^3 m/s$, $k_L a = (\omega/c_L)a = (2\pi f/c_L)a$,
* $f = 10Hz$, $a = 1mm$, $k_L a \approx 1$.*

Fig. 4.1. Analytical and numerical techniques for scattering by cracks, and their ranges of applicability.

As shown in Ref. [4.1], the effect of a crack on an incident field of ultrasonic wave motion can be nicely displayed in full-field snapshots, with the aid of time-domain finite difference calculations. Some results are shown in section 4.2. For crack characterization, only the fields at a few well selected points of observation are required. These are generally more efficiently calculated by the use of the boundary element method, which can also be used when the crack is not perfect.

In the frequency domain, the exact formulation of elastodynamic scattering problems results in the statement of a set of singular integral equations. For scattering by cracks, the unknown quantity in the integral equations is the crack-opening displacement. Generally, the integral equations must be solved numerically, preferably by the boundary element method. Once the crack-opening displacement has been determined the scattered field at an arbitrary point can be calculated by the use of an integral representation for the scattered field.

Real cracks, particularly fatigue and stress corrosion cracks have rough faces, which may contact each other. Sometimes there is not a single crack, but rather a configuration of a principal crack and an adjoining satellite crack, e.g., a macrocrack and a neighboring microcrack. It also frequently happens that other smaller scatterers such as voids and inclusions are located near a crack. The secondary scattering from these inhomogeneities will affect the overall scattered field. Still another complicating factor may be introduced by the presence of small zones of different material properties at the crack tips. Ref. [4.2] shows some analytical results which account for these complicating features of real cracks, with a view towards predicting their effects in actual testing situations.

The calculation of the fields of stress and deformation in a cracked body under static loading conditions, is mathematically almost completely analogous to scattered field calculations for ultrasonic wave incidence. From the point of view of fracture mechanics, the near-field is of interest while for ultrasonic scattering the far-field is most relevant. In fracture mechanics the emphasis is on stress-intensity factors which define the fields near the edge of a crack. It is , however, not surprising that the stress intensity factors generated by static service loads can be related to the scattered field generated by the incidence of ultrasonic pulses. This relation, and its relevance to strength assessment of a body containing a crack has been discussed in Ref. [4.3].

4.2 TIME-DOMAIN FINITE DIFFERENCE CALCULATIONS

We will first present solutions that have been obtained by the finite difference method. The finite difference method used is explicit in time. The details have been presented in Ref. [4.1]. In Ref. [4.1], use is made of new difference formulations for some of the special nodes contained in the numerical domain. A radiation condition which allows free transmission of energy from the numerical domain for body waves as well as surface waves was successfully employed.

The configuration is shown Fig. 4.2. A transducer on the upper face of the plate generates a transverse wave which propagates toward the surface-breaking crack on the lower face of the plate. All fields are taken as independent of the direction normal to the figure, which reduces the problem to one of two-dimensional plane-strain scattering.

The transducer has been simulated by choosing the surface tractions in such a way that they produce a beam of transverse wave motion, of finite width and finite duration, which traverses the slab at a 45° angle, and is directed towards the mouth of the surface-breaking crack. A time delay is incorperated into the incident pulse so that at the onset of the numerical calculations, displacements and particle velocities may be set equal to zero.

To complete the formulation of the problem, conditions at the ends of the plate must be specified. As the plate is infinite in extent, the displacement waves are outgoing. Hence, a radiation condition is employed in the truncated numerical domain.

The numerical results are displaced in two ways, (1) spatial displacement distributions (snapshots) at a specified time, and (2) time histories of the normal component of the particle velocity at the midpoint of the transducer/plate interface. A

spatial displacement distribution is obtained by depicting displacements by vectors
emanating from the corresponding nodal points, but only at nodes where the displacement
magnitude is greater than a specified value. This yields a snapshot of the displacement
distribution generated by scattering of the incident pulse, both in magnitude and direction.
The normal component of the particle velocity at the midpoint of the transducer/plate
interface has been chosen for the time-history display.

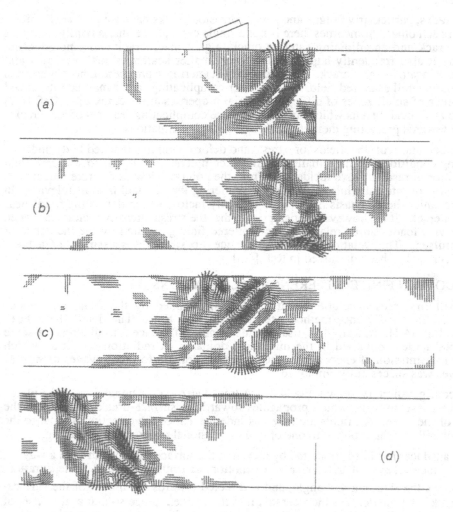

Fig. 4.2. Snapshots for a homogeneous plate with a crack: (a) t = 13.5µs;
(b) t = 20.25 µs; (c) t = 27 µs; (d) t = 40.5 µs.

Figure 4.2 shows snapshots for the case of a homogeneous plate with a crack. The four pictures represent the displacement field at times 13.5 µs, 20.25 µs, 27 µs, and 40.5 µs. Only displacements with a magnitude greater than 0.054 (scaled with respect to the maximum amplitude of the incident wave) are shown, while displacements having a magnitude greater than 0.18 are represented by vectors with an arrow in the direction of the displacement.

The values of time increment Δt and length increment h are $\Delta t = 0.09$ (scaled with respect to τ) and h = 0.05 (scaled with respect to the crack length). The calculations were carried out for $\tau = 1\mu s$, crack depth a = 1.27 cm, plate thickness H = 2.54 cm, $c_L = 6\times10^5$ cm/s, and $c_T = 3\times10^5$ cm/s. The midpoint of the transducer/plate interface is defined by x_m = (-2,2). The right and left artificial boundaries are positioned at $x_{BR} = 2.5$ and $x_{BL} = -6.5$, respectively.

The transducer/plate interface is represented by the line parallel to the upper surface of the plate. The incident displacement field is illustrated in Fig. 4.2a, in which the transverse pulse, the Rayleigh surface wave, and the cylindrically spreading displacements caused by the edges of the transducer can be seen. The surface breaking crack at the bottom surface causes a reflection of the 45° shear pulse while leaving the surface wave on the upper plate surface relatively unchanged. the final snapshot shows that the radiation conditions applied at the artificial boundaries are effective. It is noted that the Rayleigh surface wave is also traveling out of the domain on the right side. Figure 4.2d also shows the re-reflection of the transverse pulse from the upper plate surface and the surface wave on the lower surface.

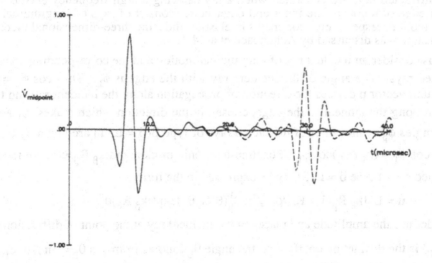

Fig. 4.3. Time histories of the normal velocity at the midpoint of the transducer.
Solid line: no crack; dashed line: crack present.

Figure 4.3 compares the time histories of the normal velocity at the midpoint of the transducer/plate interface, for the cases with and without a crack. The most obvious differences are the sharp peaks centered around 30 μs for the plate with a crack. This pulse represents the reflection of the transverse pulse by the crack faces. The incident pulse is represented by the peak centered around 7.5μs. A small peak near 25 μs can be identified as the diffraction from the crack tip. The difference in arrival times of the two signals can be used to compute the depth of the crack by a simple formula, see e.g. Refs. [4.4]-[4.6].

4.3 RAY METHOD

In ray theory it is assumed that disturbances propagate along straight or curved rays, and that the interaction of rays with inhomogeneities follows simple geometrical rules which can be established on the basis of solutions to canonical problems. If the rules are known, then rays can be traced and (in principle) the signals that propagate along all rays passing through a point of observation can be superimposed to yield the complete field. The geometrical aspects of ray theory have intuitive appeal, and they are relatively simple. From the mathematical point of view, ray theory gives an expansion which has asymptotic validity with respect to "high" frequency or "small" time after arrival of a disturbance.

The scattering of a bundle of rays by a crack-like flaw follows relatively simple rules. At sufficiently high frequencies, diffraction at certain points on the crack edge, which have been called the "flash points", produces the dominant part of the scattered field. the flash points emit bundles of diffracted waves which propagate towards a point of observation. The basic theory has been presented in Ref. [4.7].

The theory of elastodynamic crack-edge diffraction is based on the result that two cones of diffracted rays are generated when a ray carrying a high-frequency elastic wave strikes the edge of a crack. the inner and outer cones consist of rays of longitudinal and transverse motion, respectively. for cracks in elastic solids, the three-dimensional theory of edge diffraction was discussed by Achenbach et al [4.7].

Let us consider an incident ray of longitudinal motion and the corresponding cones of diffracted rays. The angle of the incident ray with the edge is ϕ_L. Thus $\cos \phi_L = \mathbf{p} \cdot \mathbf{t}$, where the unit vector $\mathbf{p}$ defines the direction of propagation along the incident ray and $\mathbf{t}$ is a unit vector along the tangent to the edge, chosen in the direction which makes ϕ_L acute. The half-angles ϕ_β of the cones of diffracted rays of type $\beta(\beta = L, T)$ are given by ϕ_L and ϕ_T, where $\cos \phi_T = (c_T / c_L)\cos \phi_L$. For time-harmonic motion and $k_\beta R_\beta >> 1$, the field on the diffracted ray of type $\beta = L,T$ may be expressed in the form

$$\mathbf{u} = U_0[k_\beta R_\beta(1 + R_\beta /\rho_\beta)]^{-1/2}D_\beta^L(\theta;\phi_L,\theta_L)\exp(ik_\beta R_\beta)\mathbf{d}^\beta, \tag{4.3.1}$$

where U_0 defines the amplitude and phase on the incident ray at the point of diffraction, $D_\beta^L(\theta;\phi_L,\theta_L)$ is the diffraction coefficient, the angle θ_L follows from $\cos \theta_L = \mathbf{p} \cdot \mathbf{n} /\sin \phi_L$, $\mathbf{d}^\beta$ is the unit vector which defines the displacement direction, R_β is the distance along a diffracted ray measured from the point of diffraction and ρ_β is the distance from the point of diffraction to the other caustic. An explicit expression for ρ_β is

$$\rho_\beta = -a \sin \phi_\beta / [a \, d\phi_\beta/ds + \cos \theta], \tag{4.3.2}$$

where s is arc length measured along the edge and a is the signed radius of curvature of the edge.

The diffraction coefficient $D_\beta^L(\theta;\phi_L,\theta_L)$ can be obtained by solving the canonical problem of diffraction of plane waves by a semi-infinite crack. For crack faces which are free of surface tractions (an empty crack) the solutions to canonical problems, including detailed listings of the diffraction coefficients have been given in Ref. [4.7].

In the direct problem, the positions of the source and the receiver, which are defined by x_S and x_Q, respectively, are known. Also known is the location of the edge of the crack. The location of the flash point D, with position vector x_D, must be determined. For L-L diffraction, it follows that x_D satisfies the relation

$$t \cdot \{|x_D - x_S|(x_Q - x_D) - |x_Q - x_D|(x_D - x_S)\} = 0, \tag{4.3.3}$$

where the unit vector t is tangent to the edge at $x = x_D$. When the position of D has been determined, the total ray length may be computed as

$$R = R_i + R_L = |x_D - x_S| + |x_Q - x_D|. \tag{4.3.4}$$

Solutions to crack-scattering problems obtained by ray methods can be found in Ref. [4.7].

Comparison of theoretical ray theory results with experimental results has been given in Ref. [4.8]. Interesting practical applications can be found in Ref. [4.9].

In the direct problem the incident wave and the geometrical configuration are known. for a given point of observation the positions of the flash points on the crack edge can then be determined by geometrical considerations, and the scattered field can subsequently be determined by direct ray tracing. If the geometrical configuration is unknown, but information is available on the diffracted field, an inverse ray tracing procedure can be used to determine the flash points on the crack edge from which diffracted signals have emanated.

In recent papers, see e.g. [4.10], two analytical methods have been developed to map the edge of a crack by the use of data for diffraction of elastic waves by the crack-edge. The methods of Ref.[4.10] are based on elastodynamic ray theory and the geometrical theory of diffraction, and they require as input data the arrival times of diffracted ultrasonic signals. The first method maps flash points on the crack edge by a process of triangulation with the source and receiver as given vertices of the triangle. By the use of arrival times at neighboring positions of the source and/or the receiver, the directions of signal propagation, which determine the triangle, can be computed. This inverse mapping is global in the sense that no a-priori knowledge of the location of the crack edge is necessary. The second method is a local edge mapping which determines planes   relative to a known point close to the crack edge. Each plane contains a flash point. The envelope of the planes maps an approximation to the crack edge.

In Ref. [4.10] the material containing the crack was taken as a homogeneous, isotropic and linearly elastic solid. More recently, extensions to include anisotropy of the material have been given in Ref. [4.11].

Mathematical details and a fairly detailed error analysis can be found in Ref. [4.10]. The reference also includes applications of the methods to synthetic data. It is of particular interest that the local mapping technique allows for an iteration procedure whereby the result of a computation suggests an improved choice of the base point which in the subsequent iteration yields a better approximation to the crack edge. A comparison with experimental data has been given in Ref.[4.12].

4.4. FREQUENCY-DOMAIN ANALYSIS

4.4.1. Integral Equations for Crack Problems

Consider the scattering problems of time-harmonic waves by cracks. When external loads are applied to a solid, a crack becomes a surface of discontinuity. Geometrically, a crack is a limiting case of a void. Let A denote the surface of a void in the solid. For a crack of arbitrary shape, we write $A = A^+ + A^-$, where A^+ is the insonified side of the crack and A^- is the shadow side. Also n_j is the unit normal vector of A with direction shown in Fig. 4.4.

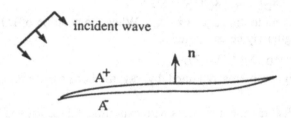

Fig. 4.4. A plane wave incident on a crack of arbitrary shape.

For a crack the integral representation for the scattered field, Eq.(2.10.4b), can be rewritten as

$$u_k^{sc}(x) = \int_{A^+} \sigma_{ijk}^G(y - x)n_j^+(y)u_i^{sc}(y)dA(y) + \int_{A^-} \sigma_{ijk}^G(y - x)n_j^-(y)u_i^{sc}(y)dA(y)$$

$$= \int_{A^+} \sigma_{ijk}^G(y - x)n_j(y)\Delta u_i^{sc}(y)dA(y), \quad \text{for x inside the solid, } x \notin A \quad (4.4.1)$$

where $n_j^+(y)= n_j(y)$, $n_j^-(y)= -n_j(y)$, and

$$\Delta u_i^{sc}(y) = u_i^{sc}(y^+) - u_i^{sc}(y^-), \tag{4.4.2}$$

here $y^+ \in A^+$, $y^- \in A^-$. It should be noted that $\Delta u_i^{sc} = \Delta u_i$, where u_i is the total displacement field, since the incident field is continuous across the crack faces. A system of integral equations can not be obtained directly from Eq. (4.4.1) because the limiting process of $x \to x^A$, where x^A is a point on A, will produce a degenerate system of boundary integral equations.

Substituting equation (4.4.1) into Hooke's law (2.2.7) and using

$$t_p^{sc}(x) = - \sigma_{pq}^{sc}(x)n_q(x),$$

we obtain

$$t_p^{sc}(x) = -C_{pqkl}n_q(x) \int_{A^+} \sigma_{ijk,l}^G(y - x)n_j(y)\Delta u_i^{sc}(y)dA(y),$$

$$\text{for x inside the solid, } x \notin A^+. \tag{4.4.3}$$

A system of boundary integral equations without degeneracy can be derived from equation (4.4.3) by taking the limit $x \to A^+$, and by applying the traction-free boundary conditions. We obtain

$$\lim_{x \to x^A} t_p^{sc}(x) = t_p^{sc}(x^A) = -t_p^{in}(x^A)$$

$$= \lim_{x \to x^A} \{ -C_{pqkl}n_q(x) \int_{A^+} \sigma_{ijk,l}^G(y - x)n_j(y)\Delta u_i^{sc}(y)dA(y) \}$$

$$= -C_{pqkl}n_q(x^A) \lim_{x \to x^A} \{ \int_{A^+} \sigma_{ijk,l}^G(y - x)n_j(y)\Delta u_i^{sc}(y)dA(y) \}$$

$$= -C_{pqkl}n_q(x^A) \int_{A^+} \sigma_{ijk,l}^G(y - x^A)n_j(y)\Delta u_i^{sc}(y)dA(y) \}. \qquad (4.4.4)$$

In the last step of Eq. (4.4.4), the limit $x \to x^A$ has been brought inside the integration. In a strict sense this interchange is not valid. In fact the system of boundary integral equations obtained in this manner is hypersingular when the observation point y and the source point x coincide, since the terms $\sigma_{ijk,l}^G(y - x^A)$ behaves as

$$\sigma_{ijk,l}^G(y - x^A) \approx \begin{cases} 1/r^2, & \text{two-dimensional,} \\ 1/r^3, & \text{three-dimensional,} \end{cases} \quad \text{as } r \to 0, \qquad (4.4.5)$$

where $r = |y - x^A|$. Thus the expression in the right-hand side of (4.4.4) is not integrable.

Several methods have been proposed to overcome this difficulty, but all of them have the same basic idea: before taking the limit, the integral with the hypersingular integrand is transformed to several integrals with singular integrands of lower order which are integrable after the limiting process $x \to x^A$. Two strategies have been employed. For the first strategy, before taking the limit $x \to x^A$ integration-by-parts is used to reduce the higher-order singularities and then the limit $x \to x^A$ is taken. In this manner a system of boundary integral equations is obtained, but for both the crack-opening displacements $\Delta u_i^{sc}(y)$ and the dislocation densities (derivatives of the crack-opening displacement) $\Delta u_{i,j}^{sc}(y)$ as unknowns. The process is referred to as a regularization of the (higher-order singular) integrals. This method was proposed by Budiansky and Rice [4.13] for solving a 3-D flat crack scattering problem. The integration-by-parts regularization procedure has also been used by Sládek and Sládek [4.14] and Nishimura and Kobayashi [4.15]. For a 2-D crack configuration, analogous formulations have been proposed by Tan [4.16], Schmerr [4.17] and Zhang and Achenbach [4.18] and [4.19].

The integration-by-parts regularization procedure is easily implemented for a flat (3-D problem) or straight (2-D problem) crack, but it becomes quite cumbersome for curved cracks. Furthermore, different forms of the regularized BIE's are obtained through the nonunique integration-by-parts procedure, though they are equivalent (see Cruse, [4.20].

Zhang and Achenbach [4.21] proposed a new approach to deal with this problem. By employing an elastodynamic conservation integral, instead of using the Betti-Rayleigh reciprocity relation, they derived the boundary integral equations for crack configurations naturally without any integration-by-parts procedure. The result is

$$t_p^{in}(x) = C_{pqkl}n_q(x) \int_{A^+} \{[\Delta u_{m,n}^{sc}(y)\sigma_{mnl}^G(y - x) - \rho\omega^2\Delta u_i^{sc}(y)\, u_{il}^G(y - x)]\delta_{jk}$$

$$- \Delta u_{i,k}^{sc}(y)\sigma_{ijl}^G(y - x)n_j(y)\}dA(y), \qquad\qquad x \in A^+. \qquad (4.4.6)$$

Equation (4.4.6) is valid for a crack of arbitrary shape and it does not involve a hypersingularity. All integral equations obtained by various integration-by-parts procedures are special cases of Eq. (4.4.6). The disadvantage of Eq. (4.4.6) is that it requires not only the computation of the crack-opening displacements, but also of the generally irrelevant displacement derivatives.

The main idea of the alternate strategy can be explained as follows. First, A^+ is divided into two parts: $A^+ = A_c^+ + A_s^+$ where A_s^+ is a small area on A^+ with x^A inside it, for example with x^A located at its geometrical centroid. In actual numerical calculations, A_s^+ can be chosen as one element. We can write Eq.(4.4.4) as

$$t_p^{sc}(x^A) = -C_{pqkl}n_q(x^A)\int_{A_c^+}\sigma_{ijk,l}^G(y - x^A)n_j(y)\Delta u_i^{sc}(y)dA(y) \; -C_{pqkl}n_q(x^A)\lim_{x \to x^A} I,$$

$$(4.4.7)$$

where

$$I = \int_{A_s^+}\sigma_{ijk,l}^G(y - x)n_j(y)\Delta u_i^{sc}(y)dA(y), \quad x \notin A_s^+. \qquad (4.4.8)$$

It is well known that the elastodynamic Green's function can be expressed as

$$\sigma_{ijk}^G = \sigma_{ijk}^{GD} + \sigma_{ijk}^{GS}, \qquad\qquad (4.4.9)$$

where σ_{ijk}^{GS} is the corresponding elastostatic Green's function which is hypersingular but the dynamic part σ_{ijk}^{GD} is regular. Now assuming that $\Delta u_i^{sc}(y)$ is differentiable and its derivatives $\Delta u_{i,m}^{sc}(y)$, , satisfy a Hölder condition at $y = x^A$, $\Delta u_i^{sc}(y)$ can be represented by the following expansions

$$\Delta u_i^{sc}(y) = \Delta u_i^{sc}(x^A) + \Delta u_{i,m}^{sc}(x^A)(y_m - x_m^A) + O(r^{1+\alpha}), \qquad (4.4.10)$$

where $r = |y - x^A|$ and $\alpha > 0$. Substitution of (4.4.9) and (4.4.10) into (4.4.8) yields

$$I = \int_{A_s^+}\sigma_{ijk,l}^{GD}(y - x)n_j(y)\Delta u_i^{sc}(y)dA(y)$$

$$+ \Delta u_{i,m}^{sc}(x^A)\int_{A_s^+}\sigma_{ijk,l}^{GS}(y - x)n_j(y)(y_m - x_m^A)dA(y)$$

$$+ \Delta u_i^{sc}(x^A)\int_{A_s^+}\sigma_{ijk,l}^{GS}(y - x)n_j(y)dA(y), \quad x \notin A_s^+. \qquad (4.4.11)$$

In the limit $x \to x^A$, the first integral of (4.4.11) is regular and the second integral is found to be integrable in the sense of a Cauchy Principal Value (CPV) integral. By using the equation of static equilibrium, $\sigma_{ijk,j}^{GS} = 0$, the third integral of (4.4.11) can be rewritten as

$$\int_{A_s^+} \sigma_{ijk,l}^{GS}(y-x)n_j(y)dA(y) = \int_{A_s^+} [\sigma_{ijk,l}^{GS}(y-x)n_j(y) - \sigma_{ijk,j}^{GS}(y-x)n_l(y)]dA(y)$$

$$= \int_{\partial A_s^+} e_{mjl}\, \sigma_{ijk,l}^{GS}(y-x)v_m(y)dS(y), \qquad (4.4.12)$$

where e_{mjl} is the alternating symbol, ∂A_s^+ denotes the edge of A_s^+ and v_m is the unit normal vector of ∂A_s^+. In the last step of deriving (4.4.12), Stokes' integral theorem has been used. Since x^A is not on the edge of A_s^+, the line integral on the right-hand side of (4.4.12) will remain regular in the limit $x \to x^A$. By using a Taylor series expansion of the Green's function, the first two integrals of (4.4.11) and the line integral (4.4.12) can be evaluated analytically for simple shapes of A_s^+. This method was used by Budreck and Achenbach, [4.22], to solve the problem of intermediate and short wave length scattering from a penny-shaped and elliptic crack.

The key point of this method, the derivation of (4.4.12), is to convert the surface integral on A_s^+ which is hypersingular when $x \to x^A$ where $x \in V$ and $x^A \notin \partial A_s^+$ to a line integral around A^+ which is regular as $x \to x^A$ by the use of Stokes' theorem. Thus, in the approach of Budreck and Achenbach [4.22], Stokes' theorem is used not only to reduce the order of singularity by transforming a derivative on the fundamental solution to the density function, i.e., the usual regularization as in the integration-by-parts method, but also to convert surface integrals to line integrals.

The multi-region method avoids the occurrence of hypersingularity of the integral equations for crack problems, because each crack face is modeled as part of the boundary of a separate region. The disadvantage of the multi-region method is that additional boundaries are introduced. For an interface crack problem, i. e., a crack between dissimilar media, this is, however, the only practical method available.

4.4.2. Shape Functions for Crack Opening Displacements

Special care must be taken in the numerical implementation to account for the square root behavior of the crack-face displacements near the edges of cracks. Consider a typical element which shares a boundary with the crack edge. In a local coordinate system $(\xi_1 \xi_2 \xi_3)$ the element is shown in Fig.4.5. The element lies in the $\xi_1 \xi_3$-plane, where the ξ_1 axis is normal to the crack edge. The shape functions must have the following form:

$$\Delta u_i^{sc} \to C_i \rho^{1/2}, \quad \text{as} \quad \rho = \overline{OT} \to 0, \qquad (4.4.20)$$

where C_i is a constant proportional to the local stress intensity factors.

For a flat elliptical crack lying in the plane $y_3 = 0$, with semi-major and minor axes a_1 and a_2, the following crack opening displacement was used by Budreck and Achenbach [4.22]

$$\Delta u_m^{sc}(y) = f(y)\Delta u_m^{sc}(y_g). \qquad (4.4.21)$$

Here y_g denotes the geometrical centroid of the q-th element. The shape function $f(y)\equiv 1$ was used for all elements except those sharing a boundary with the crack edge. For edge elements, the shape function was taken as

$$f(y) = \left\{ [1 - (\tfrac{y_1}{a_1})^2 - (\tfrac{y_2}{a_2})^2] / [1 - (\tfrac{y_{1q}}{a_1})^2 - (\tfrac{y_{2q}}{a_2})^2] \right\}^{1/2}.$$
(4.4.22)

For BIE's involving both the crack-opening displacement $\Delta u_i^{sc}(y)$ and its derivatives, the dislocation density $\Delta u_{i,j}^{sc}(y)$, it is particularly useful to employ an a-priori assumed functional form of $\Delta u_i^{sc}(y)$, so that its differentiation can be carried out explicitly. This methodology was essentially used by Nishimura and Kobayashi [4.15], who employed cubic spline functions over a circular crack.

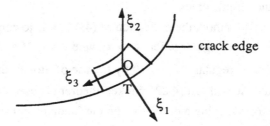

Fig. 4.5. Local coodinates for an element on the crack edge.

4.5 OTHER APPROXIMATIONS

According to the representation integral given by Eq. (4.4.1), the scattered field can be calculated when the crack-opening-displacement (COD) is known. However, the COD itself is part of the problem and must be determined from the integral equation given by Eq. (4.4.4). Equation (4.4.1) does, however, suggest that an approximation to the scattered field can be obtained by an educated guess of the COD. For example the COD might be selected as the displacement of the insonified crack face as if the wave was specularly reflected from an unbounded free surface. This approximation is known as the Kirchhoff approximation. It applies at high frequencies and is best in the zone of specular reflection. Details can be found in Ref. [4.7]. Another approximation would be to substitute the quasi-static COD, i.e. the COD due to the stresses of the incident field at zero frequency. This approximation is valid at very low frequencies. A better low frequency approximation is obtained by expanding the fields in a polynomial series of the frequency. Substitution in Eq. (4.4.4) yields a set of integral equations for the coefficients that can be solved analytically. Details are given in Ref. [4.23]

References

4.1   Scandrett, C. L. and Achenbach, J. D.: Time-domain finite difference calculations for interaction of an ultrasonic wave with a surface-breaking crack, Wave Motion, 9(1987), 171-190.

4.2   Achenbach, J. D., Sotiropoulos, D. A. and Zhang, C.,: Effect of near-tip

inhomogeneities on scattering of ultrasonic waves by cracks, Metallurgical
Transactions A, 20A(1989), 619-625.

4.3   Achenbach, J. D., Sotiropoulos, D. A. and Zhu H.: Characterization of cracks from
       ultrasonic scattering data, J. Appl. Mech., 54(1987), 754-760.

4.4   Doyle, P. A. and Scala, C. M.: Crack depth measurement by ultrasonics: a review,
       Ultrasonics, 16(1978), 164-170.

4.5   Ogura, Y.: Height determination studies for planar defects by means of ultrasonic
       testing, The Non- Destructive Testing J. of Japan, 1(1983), 22-29.

4.6   Harumi, K., Ogura, Y. and Uchida, M.: Ultrasonic Defect Sizing: Japanese Tip Echo
       Handbook, Printed Micro Printing Co, Ltd, Japan, 1989.

4.7   Achenbach, J. D., Gautesen,, A. K. and McMaken, H.: Ray Methods for Waves in
       Elastic Solids, Pitman Advanced Publishing Program, Boston 1982.

4.8   Achenbach, J. D., Adler, L., Lewis, D. K. and McMaken, H.: Diffraction of ultrasonic
       waves by penny-shaped cracks in metals: theory and experiment, J. Acoust. Soc.
       Am., 66(1979, 1848-1856.

4.9   Coffey, J. M. and Chapman, R. K.: Application of elastic scattering theory for smooth
       flat cracks to the quantitative prediction of ultrasonic defect detection and sizing,
       Nuclear Energy, 22(1983), 319-333.

4.10   Norris, A. N. and Achenbach, J. D.: Mapping of a crack edge by ultrasonic methods,
       J. Acoust. Soc. Am., 72(1982), 264-272.

4.11   Norris, A. N.: Inverse ray tracing in elastic solids with unknown anisotropy, J.
       Acoust. Soc. Am., 73(1983), 421-426.

4.12   Achenbach, J. D., Norris, A. N., Ahlberg, L. and Tittmann, B. R.: Crack mapping by
       ray methods, in: Review of Progress in Quantitative Nondestructive Evaluation, 2B
       (Eds. D. O. Thompson and D. E. Chimenti), Plenum press, New York, 1982, 1097-
       1116.

4.13   Budiansky, B. and Rice, J. R.: An integral equation for dynamic elastic response of
       an isolated 3-D crack, Wave Motion, 1(1979), 187-192.

4.14   Sladek, V. and Sladek, J.: Transient elastodynamic three-dimensional problems in
       cracked bodies, Appl. Math. Modeling, 8(1984), 2-10.

4.15   Nishimura, N. and Kobayashi, S.: An improved boundary integral equation method
       for crack problems, in: Proc. IUTAM Symposium on Advanced Boundary Element
       Methods (Ed. T. A. Cruse), Springer, 1987.

4.16   Tan, T. H.: Diffraction Theory for Time-Harmonic Elastic Waves, Doctoral
       dissertation, Delft University of Technology, 1975.

4.17   Schmerr, L. W.: The scattering of elastic waves by isolated cracks using a new
       integral equation method, in: Review of Progress in Quantitative Nondestructive
       Evaluation, 1 (Eds. D. O. Thompson and D. E. Chimenti), Plenum Press, New York,
       1982, 511-515.

4.18   Zhang, C. and Achenbach, J. D.: Scattering by multiple crack configurations, J. Appl.
       Mech., 55(1988), 104-110.

4.19   Zhang, C. and Achenbach, J. D.: Numerical analysis of surface-wave scattering by
       the boundary element method, Wave Motion, 10(1988), 365-374.

4.20   Cruse, T. A.: Fracture mechanics, in: Boundary Element Method in Mechanics (Ed.
       D. E. Beskos), North-Holland, 333-365, 1987.

4.21   Zhang, C. and Achenbach, J. D.: A new boundary integral equation formulation for
       elastodynamic and elastostatic crack analysis, J. Appl. Mech., 56(1989), 284-290.

4.22   Budreck, D. E. and Achenbach, J. D.: Scattering from three-dimensional planar
       cracks by the boundary integral equation method, J. Appl. Mech., 55(1988), 405-412.

4.23   Roy, A.: Diffraction of elastic waves by an elliptic crack-II, Int. J. Engng. Sci.,
       25(1987), 155-169.

FLAW CHARACTERIZATION
BY ULTRASONIC SPECTRUM ANALYSIS

L. Adler
The Ohio State University, Columbus, OH, USA

ABSTRACT

In this paper the problem of flaw characterization by ultrasonic spectroscopy method is addressed. Recent analytical works on scattering of elastic waves by obstacles embedded in solid body is summarized. Both two dimensional (crack like) defects and three dimensional (voids of inclusions) are considered. Experimental ultrasonic scattering date from artificial defects in metals are compared favorably to theoretical predictions.

1 INTRODUCTION

Recent interest in quantitative flaw characterization in metals has stimulated several developments in the study of ultrasonic scattering from discontinuities in solids. The characterization of a discontinuity (size, orientation, shape, etc.) from the information obtained from scattered ultrasonic waves is a problem of great interest.

A new approach started by Gericke[1] was the development of a contact ultrasonic spectrum analysis system to characterize the geometry of hidden flaws in solids. He used a broadband rather than a single frequency ultrasonic pulse and he found that the spectral components of the echoes were strongly affected by the geometry of the discontinuity. Whaley and Cook[2] developed and immersion spectrum analysis system. Adler and Whaley[3] carried out a systematic study of the spectral response of scattered waves from circular reflectors in water by using this system. They developed a model to measure the size and orientation of circular reflectors. In this model the edges of disks are assumed to be Huygens sources of the scattered field in analogy to the Sommerfeld approach to diffraction of light waves by an aperture in a screen. In the course of these investigations it became evident that quantitative flaw characterization will be possible only if elastic wave interaction with known discontinuities can be described in a simple but accurate manner. The so called inverse problem can be attacked only if the direct problem is understood, i.e., if it is known how the parameters of the scattered ultrasonic field are affected by known discontinuities. The need to analyze ultrasonic scattering data on the basis of simple theoretical predictions became obvious.

Mathematically, ultrasonic wave scattering in solids is treated by using the equations of elastodynamic theory, with boundary conditions appropriate to the problem. Exact solutions exist only for a few idealized configurations. Several approximate solutions have been developed recently. The problem is classified as to whether the discontinuities are bounded by smooth closed surfaces such as for spheroids, or by planar surfaces having sharp edges, as for crack-like discontinuities. Another classification of the theory is based on the applicability in the region of ka where k is the wave number and a is some dimension of the discontinuity. For ka <1 and for spheroidal discontinuities a theory based on the born approximation was developed by Gubernatis, Domany, and Krumhansl[4]. Other approximate theories in the region were also developed.

For crack-like planar defects in the region ka>1, Adler and Lewis[5] used Keller's geometrical diffraction theory to analyze their experimental data. Achenbach and Gautesen[6] extended geometrical diffraction theory to the three-dimensional diffraction by cracks in solids, where wave motions are governed by a scalar and a vector wave equation, whose solutions are coupled by the boundary conditions on the diffracting obstacles. Subsequently Achenbach, Adler, Lewis and McMaken[7] applied geometrical theory of diffraction for penny shaped cracks in metals and used a simple inversion algorithm to invert the spectral data to crack size. This method was extended later to elliptical cracks by Adler and Achenbach[8]. In this paper theoretical and experimental results will be presented in both regions of ka $\geq$ 1 and ka $\leq$ 1.

2 THEORETICAL RESULTS

The scattering of ultrasonic waves by flaws in a solid is schematically illustrated on Fig. 1. An incident wave is scattered by a flaw and the scattered wave is received at a point which is described by polar angle Θ and azimuthal angle ϕ. It is convenient to present theoretical results in three-dimensional display. The choices of parameters are the scattered amplitudes vs. frequency and scattering polar angle for fixed azimuthal angle of amplitude vs. polar angles and azimuthal angles for a fixed frequency.

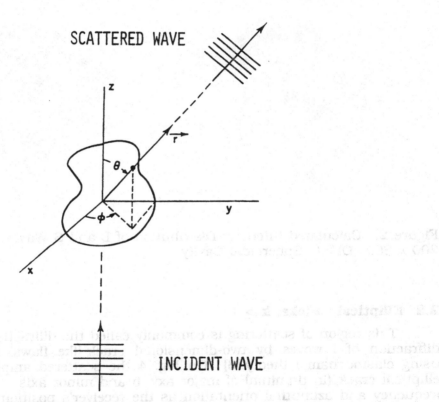

Figure 1: Scattering of Ultrasonic Waves from a Discontinuity.

2.1 Ellipsoidal Cavities. ka $\leq$ 1

For the region of ka $\leq$ 1 and for ellipsoidal cavities the Born approximation calculation was used.[4] For an incident L wave the scattered L and S wave amplitudes are plotted on Fig. 2. The flaw is an oblate spheroidal cavity in titanium with dimensions of 200µ x 400µ. For the frequency range from 3 to 7 MHz the corresponding ka is from 0.6 to 1.4. On Fig. 3 the calculated amplitude field is plotted which is scattered from the same oblate spheroidal cavity at the frequency of 5.1 MHz. The polar angle changes from 120° to 30° and the azimuthal angle changes form 0° to 360°.

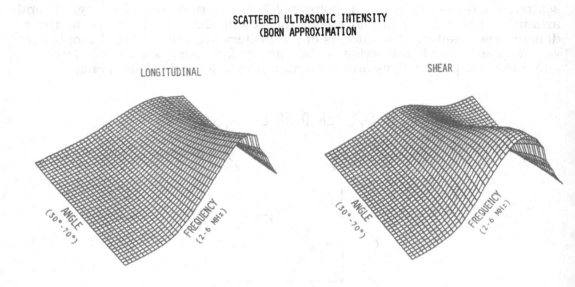

SCATTERED ULTRASONIC INTENSITY
(BORN APPROXIMATION

LONGITUDINAL   SHEAR

Figure 2: Calculated Intensity Distribution of L and S Waves Scattered from 200 x 800   Oblate Spheroidal Cavity.

2.2 Elliptical Cracks. ka≥ 1

This region of scattering is commonly called the diffraction region. The diffraction of l waves by two-dimensional crack-like flaws was calculated using elastodynamic theory.[6] On Fig. 4 the scattered amplitude from an elliptical crack (in titanium) of major axis b and minor axis a are plotted vs. frequency and azimuthal orientation as the receiver's position changed from along the major to the minor axis. The polar angle is 60°. The four different plots correspond to aspect ratios (b/a) of 1,2,4, and 8. The major axis b in each case is 2500 . The spectra along the minor axis shows periodicity which is inversely proportional to the aspect ratio. The other theoretical prediction is that the energy scattered along the major axis is always less than the energy along the minor axis. The difference is increasing with increasing aspect ratios.

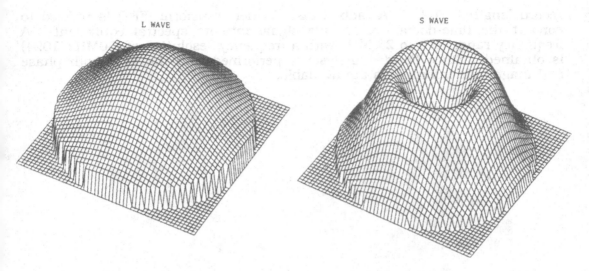

Figure 3: Calculated Scattering Field Profile for Incident L wave of 5.1 MHz from 200 x 800 Oblate Spheroidal Cavity. (a) Scattered L Wave (b) Scattered S Wave

3 EXPERIMENTAL

3.1 Spectrum Analysis System

A configuration of the ultrasonic spectroscopic system used in these experiments is a broad-band pulsed system utilizing digital spectrum analysis (Fig. 5.). A fast-rise-time, high-voltage spike (from an SCR pulser) excites a 15 MHz Panametrics Videoscan immersion transducer. The pulser, connecting cable and transducer are matched to achieve a wide-band ultrasonic pulse. The ultrasound travels along a 3-inch water path and enters a titanium disk having an embedded void.

A second broadband transducer, 3 inches from the disk, intercepts a portion of the ultrasonic energy scattered from the simulated defect. Because the amplitude of the resultant electrical signal is small, a wideband, low-noise amplifier is employed to boost the signal. The amplified pulse is sampled at MHz and digitized to 8-bit resolution by a Biomation 8100 transient recorder. The digital record is then transferred to a minicomputer. Signal averaging may be used to enhance the signal-to-noise ratio.

Analysis of the ultrasonic signal is performed digitally with the minicomputer. First, every other sample point is deleted, giving and effective sample rate of 50 MHz. Since the transducers, cabling and amplifier band-limit the signal to approximately 20 MHz, a sample rate of 50 MHz is sufficient to prevent aliasing. The section of the data record containing the signal of interest is extracted. Next, zeroes are appended to bring the total record length to 1024. A radix-2 fast Fourier transform (FFT) is utilized to convert the time-domain ultrasonic signal into its spectral equivalent. A frequency range of 0 to 25 MHz with a frequency resolution of (50MHz/1024) is obtained. Since Fourier analysis is performed with the FFT, both phase and magnitude information are available.

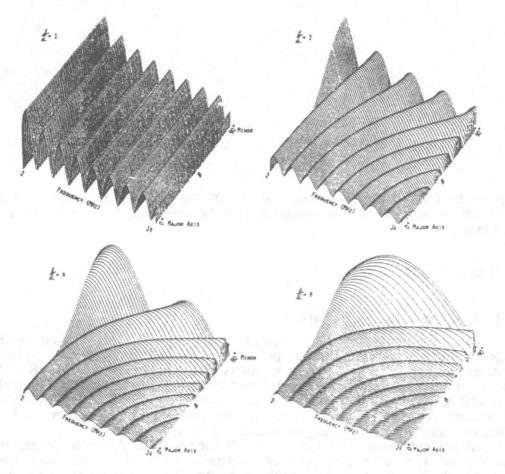

Figure 4: Theoretical Amplitude Spectra of Normal Incidence L Waves Scattered from Elliptical Cracks. Polar Angle is 60°.

The output of the ultrasonic spectroscopic system is that of the wave diffraction convolved with the spectroscopic system response. Deconvolution may be performed, if desired, to remove modifications caused by the equipment. Although the effects of hardware (pulser, transducer, etc.) on system response are predominant, the modifications of the ultrasonic pulse by attenuation in the water and titanium, partitioning at the water-titanium boundary as well as radiation coupling effects (diffraction corrections), cannot be neglected.

3.2 Experimental Technique and Procedure

The technique used is shown of Fig. 6. The sample, which is a titanium alloy disk (2.5 x 10 cm) with a flat surface, is immersed in water. The transmitter launches a longitudinal wave toward the liquid-solid interface at an angle. For nonnormal incidence both L and T waves are produced in the metal. The cavity can be insonified either by the L wave or by the T wave with incident angle α. At the cavity, the waves are scattered and mode conversions occurs. The scattered waves are received and can be analyzed separately due to their separation in time. The defects studied were cracks (circular and elliptical) and cavities (oblate and prolate spheroidal) in diffusion-bonded titanium. Defect dimensions ranged from 200µ to 2500µ.

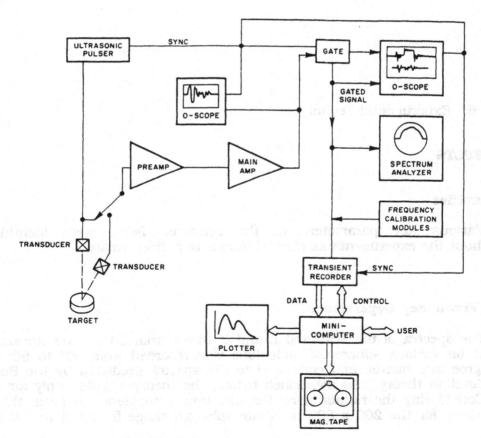

Figure 5: Experimental System.

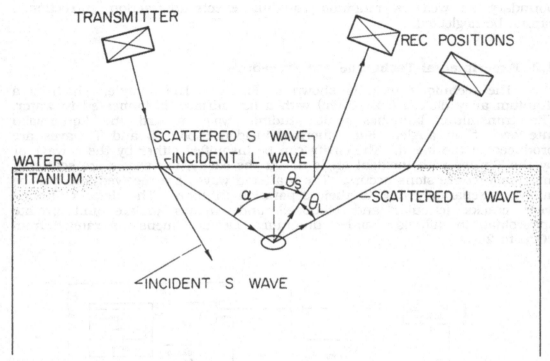

Figure 6: Experimental Technique

4 RESULTS

4.1 Cavities

Various key parameters of the scattered fields were identified throughout the experiments as characteristics of a given cavity.

4.1.1 Frequency Dependence

The spectra of the scattered L and S waves from an L wave normally incident on various spheroidal inclusions was recorded from 30° to 60° in two degree increments, and compared to the spectra predicted by the Born approximation theory. As mentioned before, the theory should apply for ka < 1. Considering the radius along the direction of incidence to be a, the L wave values for the 200 x 800 m oblate spheroid range from 0.2 to 0.8 for frequencies of 2 to 8 MHz for normal incidence. Figure 7 compares theory with experiment for this oblate spheroid for two angles of both L and S scattered waves. The theory has been corrected by the transfer function of the ultrasonic spectroscopic system. The agreement for both L and S wave scattering is reasonably good.

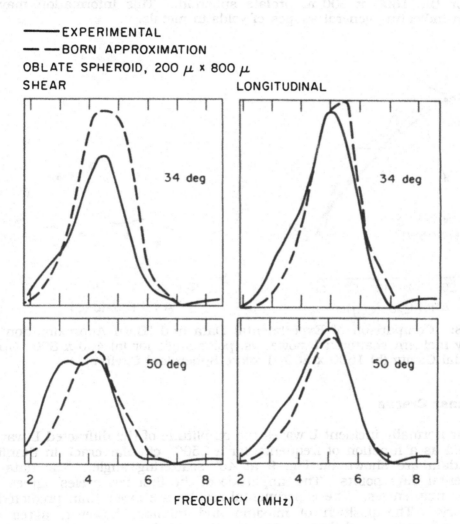

Figure 7: Power Spectra for L-S and L-L Scattering.

4.1.2 Scattered Power Variation as an Indicator of Geometrical Shape

The Born theory predicts that the power decreases with increased polar angle for the oblate spheroids and increases with increasing polar angle for the prolate spheroids, and this behavior was observed experimentally. Figure 8 shows the variation of total power with angle for three azimuthal angles for the 400 x 800 m oblate spheroid for normal incidence. Figure 8 is the average of these azimuthal values as a function of angle for the 1600 x 800 m prolate spheroid. This information may be useful in indicating general shapes of voids in metals.

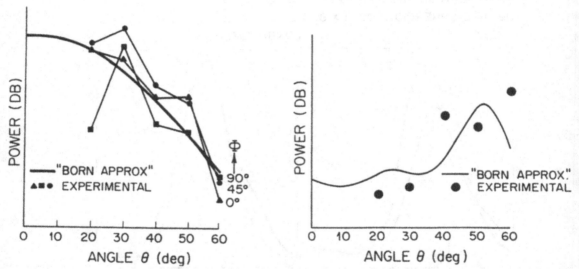

Figure 8: Comparison of Experimental Data and "Born Approximation" for normally incident scattered L power vs. polar angle for (a) 400 x 800 Oblate Spheroidal Cavity (b) 1600 x 800 Prolate Spheroidal Cavity.

4.2 Planar Cracks

For normally incident L waves the amplitude of the diffracted L wave is calculated as a function of frequency for a 2500 circular crack in titanium. Typical data are shown on Fig. 9 at 45° scattering angle. The data are experimental data points. The amplitude of the first few cycles agrees well at higher frequencies. The experimental results are lower than predicted by the theory. The position of maxima and minima, however, agree well between experiment and theory. The locations of the maxima are significant for the inversion process as will be shown in the next section. On Fig. 10, both theoretical and experimental amplitude spectra are shown for different azimuthal angles. The agreement between theory and experiment is good for the larger angles and poor for the smaller angles, i.e., near the major axis. A possible explanation of this phenomenon is the available scattered energy is higher (as predicted by Fig. 4) as one approaches the minor axis and therefore the analysis is more meaningful.

4.2.1 Determination of Crack Size from the Diffraction Spectra

It has been shown earlier by Achenbach, Adler, Lewis, and McMaken[7] that one can carry out a simple inversion procedure on characteristics of the amplitude spectra. The dimension of the crack

$$a = \frac{C_l}{2\sin\Theta(\Delta f_{max})_{av}}$$

where Θ is the scattering angle, C_L is the longitudinal velocity in the solid, and $(\Delta f_{max})_{av}$ is the average frequency spacing between two consecutive frequency maxima. table 1 illustrates the results collected from several measurements for four different size elliptical cracks. The measured and actual dimensions are in very good agreement.

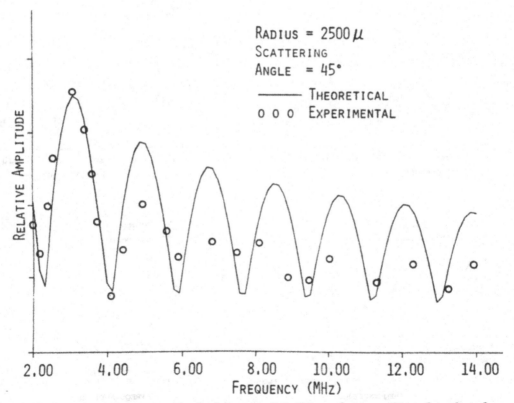

Figure 9: Amplitude Spectra of Scattered L Wave from a Circular Crack.

TABLE 1

Determination of Major (B) and Minor Axis (A) of Elliptical Cracks in Titanium from Scattered Amplitude Spectra

(B) in Microns		(A) in Microns	
Actual	Measured	Actual	Measured
2500	2498	2500	2498
2500	2587	1250	1291
2500	2358	625	830
2500	----	312	285

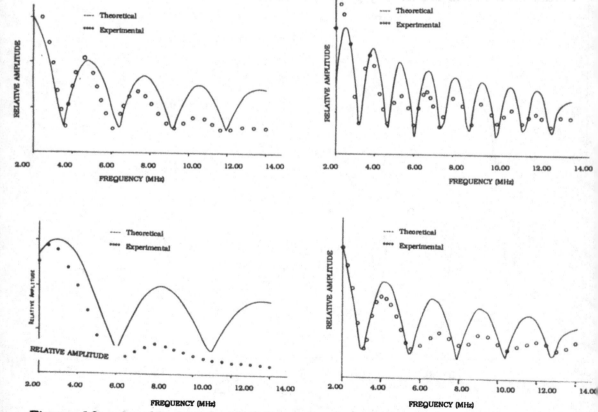

Figure 10: Amplitude Spectra of Scattered 1 Waves from a 2500 x 1250 Elliptical Crack in Titanium Along different Azimuthal Directions. Polar Angle 60°.

REFERENCES

1.   O.R. Gericke: Determination of the Geometry of Hidden Defects by Ultrasonic Pulse Analysis Testing, J. Acoust. Soc. Am., 35, (1963) ,364.

2.   H.L. Whaley and K.V. Cook: Ultrasonic Frequency Analysis, Mater. Eval., 28 (1970),61-66.

3.   L. Adler and H.L. Whaley: Interference Effect in a Multifrequency Ultrasonic Pulse Echo and its Application to Flaw Characterization, J. Acoust, Soc. Am., 51 (1972), 881.

4. E. Domany, J.A. Krumhansl and S. Teitel: Quasi-static Approximation to the Scattering of Elastic Waves by a Circular Crack, J. Appl. Phys., 49 (1978), 2599.

5.   L. Adler and D.K. Lewis: Diffraction Model for Ultrasonic Frequency Analysis and Flaw Characterization, Mat. Eval., 35 (1977), 51.

6.   J.D. Achenbach and A.K. Gautesen: Geometrical Theory of Diffraction for Three-D Elastodynamics, J. Acoust. Soc. Am.,61 (1977), 413.

7.   J.D. Achenbach, L. Adler, D.K. Lewis, and H. McMaken: Diffraction of Ultrasonic Waves by Penny-Shaped Cracks in Metals: Theory and Experiment, J. Acoust. Soc. Am., 66 (1979), 1848.

8.   L. Adler and J.D. Achenbach: Elastic WAve Diffraction by Elliptical Cracks: Theory and Experiment, J. of Nondestructive Evaluation, 1 (1980), 87.

ELECTROMAGNETIC - ACOUSTIC TRANSDUCERS
(EMATs)

R.B. Thompson
Iowa State University, Ames, IA, USA

ABSTRACT

Electromagnetic-acoustic transducers (EMATs) are devices which can excite and detect ultrasonic waves with no contact to the surface of metal parts. The physical principles governing the operation of these devices are reviewed. Included are discussions of mechanisms of transduction, practical probe geometries, systems design, radiation characteristics in nonmagnetic and magnetic materials, and guided mode generation. The paper concludes with discussion of typical applications.

1   INTRODUCTION

By far the most common type of ultrasonic transducer utilizes piezoelectric elements to excite and detect the ultrasonic energy [1,2]. These devices have a number of attractive features, including broad bandwidth, high efficiency and low cost. However, in a number of specialized applications, such as measurements at elevated temperature and when the part is moving with respect to the transducer, the piezoelectric transducers suffer serious limitations because of their need to be mechanically coupled to the part. Typically, this coupling is achieved by either immersing the transducer and part in water or pressing the transducer against the part, coupled by a thin fluid layer. Variabilities in this coupling can influence a measurement in an uncontrolled manner, or hostile conditions such as part motion may make it difficult to establish. Hence there is considerable motivation to develop other types of transducers which require no couplant. Here we consider one example, the electromagnetic acoustic transducer

(EMAT).[3-6] These are relatively well developed devices, which require
no couplant but generally must be in close proximity to the part. A
second important class of noncontact probes, based on laser interactions
with the material [7,8], is discussed in other lecture notes in this
book.[9]

In anticipation of the material to be presented, it can be noted
that the couplant free operation of EMATs affords a number of advantages.
Included are the possibility for operation at elevated temperatures, in
vacuum, on a moving part, or through loose paint or scale. In addition,
EMATs have the capability of producing tailored radiation patterns and
exciting unusual wave types such as SH waves, and they can be used to
make precise velocity measurements without disturbing the surface. These
advantages must be traded against relatively low efficiencies which
sometimes lead to large probes with poor resolution, the need for high
power, low noise electronics and associated feed-through problems.
Moreover, these devices can only operate on electrically conducting or
magnetic materials. To minimize these limitations, it is important to
pay particular attention to EMAT and circuit design.

2.   EMAT PRINCIPLES

A transducer is a device for converting electrical signals to
ultrasonic signals (transmitter) and ultrasonic signals to electrical
signals (receiver). In an EMAT, this coupling is accomplished via direct
electromagnetic interactions. The fundamental elements of an EMAT, a
magnet producing a static field and a wire carrying a dynamic current,
are shown in Fig. 1. Also shown are the fields that exist in the
transmit mode.

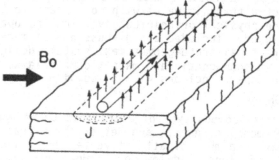

Fig. 1.   Schematic of single element of EMAT transducer showing applied
current I, induced eddy currents $\vec{J}$, magnetic induction $\vec{B}_0$, and
body forces $\vec{f}$. (after Ref. [5])

First consider the mechanism of coupling in the transmit mode. When
a dynamic current is passed through the coil and the EMAT is placed near
a metal surface, eddy currents $\vec{J}$ are induced in a thin, near-surface
layer whose penetration is given by the electromagnetic skin depth,

$$\delta = \sqrt{\frac{1}{\omega\mu\sigma}} \tag{1}$$

where ω is an assumed time harmonic during current, μ is the magnetic permeability and σ is the electrical conductivity. In the presence of a magnetic flux density, $\vec{B}_0$, these eddy currents will experience Lorentz forces,

$$\vec{f}_L = \vec{J} \times \vec{B}_0. \tag{2}$$

Via a variety of interactions whose details lie outside the scope of this lecture, these forces are transmitted to the lattice and serve as a source of ultrasonic waves.

If the material is also magnetic, these forces will be augmented by other coupling mechanisms. One of the most important of these, magnetostriction, is illustrated in Fig. 2. As a consquence of spin-orbit coupling, application of an applied field will cause the sample to change length. The static magnetic field of an EMAT creates such a response. In addition, the dynamic currents passing through the coil create dynamic magnetic fields which may be thought of as modulating this bias. The result is a tendency for a unit element of the material to alternately lengthen and shorten about the length established by the bias field. In practice, the response is complicated by the inertial effects of the surrounding material. These will be discussed more fully later in these notes. When all of these considerations are introduced the net result is an additional set of driving forces which launch ultrasonic waves.

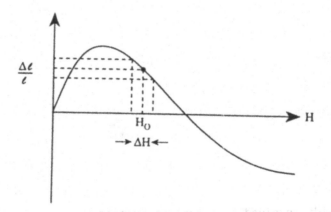

Fig. 2. Mechanisms of EMAT operation on magnetic materials. The dynamic magnetic fields, superimposed on the static bias, cause the material to alternately lengthen and shorten, and ultrasonic waves are launched as a reaction. The particular response shown schematically is characteristic of polycrystalline iron.

The same structure can detect as well as excite ultrasonic waves, as shown in Fig. 3. Part (a) illustrates the non-magnetic case. Incident wave motion $\vec{u}$ causes material to cut lines of magnetic flux, thereby inducing eddy currents, $\vec{J}$, in the material surface. These eddy currents inductively couple to the pick-up coil through dynamic magnetic fields $\vec{H}$, thereby creating a received voltage.

In a magnetic material the corresponding processes are shown in part (b). The dynamic strains of an ultrasonic wave cause modulation of magnetization which again inductively couples to coil. In both the magnetic and non-magnetic cases, the transduction is reciprocal. For the non-magnetic case, a good analogy is the relationship of an electric motor and generator. In the former, currents flowing in the presence of magnetic fields create a mechanical motion. In the latter, motion through a magnetic field induces a voltage.

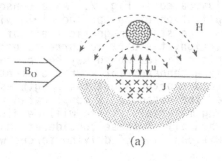

(a)

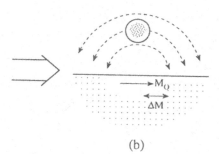

(b)

Fig. 3. Fields involved in reception process
        (a) Non-magnetic metal   (b) Magnetic material

3. PRACTICAL PROBES

A practical probe consists of a coil of wire and a structure for producing a bias magnetic field, as shown conceptually in Fig. 4.

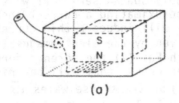

(a)

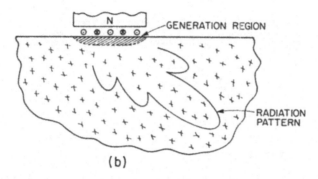

(b)

Fig. 4.   Conceptual view of an EMAT transducer. (after Ref. [5]).
          (a) Internal structure of particular probe; (b) radiation
          pattern.

The waves are generated in reaction to forces which are exerted on the
material in a region concentrated near the surface, and decay
exponentially at a rate determined by the electromagnetic skin depth, δ.
The details of the radiation patterns are controlled by exactly the same
parameters that govern the radiation of piezoelectric or other
transducers. However, a desirable feature of EMATs is that, by
controlling the shape of the coil and magnet, it is possible to tailor
the driving force distribution and hence the radiation pattern. To aid
in the discussion of these possibilities, we will adopt the coordinate
system shown in Fig. 5.

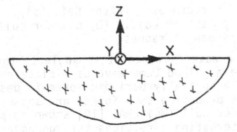

Fig. 5.   Coordinate system used in analysis of EMAT transducers.

As an example, Fig. 6 shows elementary EMAT structures which are
used to produce a variety of wavetypes. Part (a) shows the simplest, in
which the coil consists of a closely spaced series of wires carrying
current in the same direction, approximating a current sheet. When
driven by a dynamic current, this will induce a thin current layer in the
sample, penetrating to a depth given by the electromagnetic skin depth δ,
[See Eq. (1)]. In accordance with Eq. (2) the Lorentz forces will be in
the x-z plane. In the most common cases, they will be oriented to launch
either longitudinal waves ($\vec{B}_0 \| \hat{a}_x$) or transverse waves ($\overline{B}_0 \| \hat{a}_z$). These
translationally invarient forces will launch waves propagating normal to
the material surface.

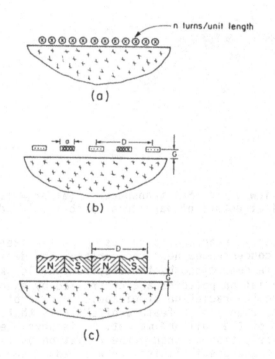

(a)

(b)

(c)

Fig. 6.   Elementary EMAT structures: (after Ref. [5]).
          (a) uniformly polarized coil; (b) meander coil;
          (c) periodic permanent magnet.

In order to launch a wave inclined at an angle to the material
surface, or to couple to a guided Lamb wave of a plate or a Rayleigh wave
on an isolated surface, a spatially periodic force is desired. Two
common examples of EMATs producing such forces are shown in the remainder
of Fig. 6. In the meander coil (MC) EMAT [10] shown in part (b), the
current is passed in alternating directions through adjacent wire

elements with a period D, and is generally placed in a uniform magnetic field. Such a structure can couple to plate waves propagating at the angle θ with respect to the normal, given by

$$\theta_{L,T} = \sin^{-1}\left(\frac{D}{f}C_{L,T}\right) \tag{3}$$

where f is the excitation frequency and $C_{L,T}$ is the wave speed of the excited wave which is indicated by the subscript L (longitudinal) or T (transverse). As for the uniform coil, the driving force will be in the x-y plane. In general, both wave types will be radiated with relative amplitudes determined by the force direction and the elastodynamic response of the material as discussed in the next section. An interesting practical consequence of Eq. (3) is that the angle of the wave may be scanned by varying the frequency.[11]

A problem with the MC EMAT is that is is impossible to excite horizontally polarized shear (SH) waves polarized in the $\hat{a}_y$ direction since the directions of $\vec{f}$ are restricted to the x-z plane. To overcome this, the periodic permanent magnet (PPM) EMAT [12] was designed, as shown in Fig. 6c. By alternating the polarity of slab magnets and passing a uniform current sheet in the $\hat{a}_x$ direction, spatially periodic forces polarized in the $\hat{a}_y$ direction can be produced, coupling to inclined SH waves (whose angles again satisfy Eq. (3)) or guided SH waves in plates.

Practical EMATs are composed of finite sets of wires and magnets based on general structures such as those shown in Fig. 6. Figure 7 shows cross-sections of a number of specific examples. The motivation for those are generally obvious from the previous discussion. An exception is the spiral EMAT, illustrated in part (a). Because of its axial symmetry (spiral coil and cylindrical magnet), this structure will produce a radially polarized transverse wave. Such beams are not good for the detection of small flaws because of the presence of an on-axis null in the radiation pattern. However, they have found important application in thickness gaging because of the low inductance of the flat coil, which simplifies certain electronic problems.

4. SYSTEM CONSIDERATIONS

An EMAT operates like any other transducer in a measurement system. However, because of the relatively low efficiency of the transduction process, care must be taken in optimizing the electronic circuitry to maximize the transfer impedance. This low efficiency is partially offset by the fact that the signal levels are very reproducible and predictable. Hence, transducer design calculations are an important part of building an EMAT system. Some useful relations are developed below.

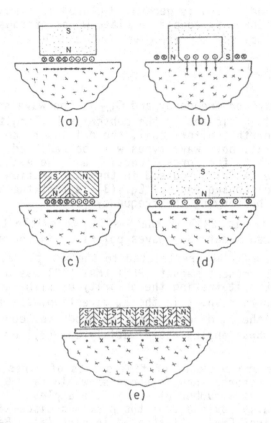

Fig. 7. Cross-sectional views of practical EMAT configurations. (a)
Spiral coil EMAT exciting radially polarized shear waves
propagating normal to surface. (b) Tangential field EMAT for
exciting plane polarized longitudinal waves propagating normal
to surface. (c) Normal field EMAT for exciting plane polarized
shear waves propagating normal to surface. (d) Meander coil
(MC) EMAT for exciting obliquely propagating L or SV waves,
Rayleigh waves, or guided modes of plates. (e) Periodic
permanent magnet (PPM) EMAT for exciting grazing or obliquely
propagating SH waves or guided SH modes of plates. (after Ref.
[5]).

A useful engineering parameter in the design and evaluation of EMATs
is the transfer impedance, Z_{ba} , which is defined as

$$Z_{ba} = \frac{V_o}{I_T}$$

(4)

where I_T is the current delivered to the transmitting EMAT and V_o is the open current received voltage, as defined in the equivalent circuit in Fig. 8a. However, the derivation of expressions for this parameter is more convenient in a transmission line representation of the same system, as shown in Fig. 8b.

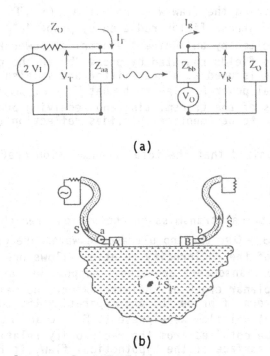

(a)

(b)

Fig. 8. Conceptual view of an EMAT measurement
   (a) Discrete component equivalent circuit as used to describe laboratory measurements.
   (b) Transmission line model as needed to apply reciprocity theorem.

The basis of the analysis is an electromechanical reciprocity relation developed by Auld.[13] Consider a system in which each EMAT is connected by a coaxial cable of impedance Z_o to a signal generator or receiver. Consider the electrical signals in the transmission lines to be represented by travelling waves propagating in the positive and negative s-directions. Define Γ_{aa} to be the voltage reflection coefficient, equal to the ratio of the voltage of the reflected travelling wave to that of the incident travelling wave at terminal a, and define Γ_{ba} to be the voltage transmission coefficient, equal to the ratio of the voltage of the transmitted travelling wave at terminal b to that of the incident travelling wave. From the reciprocity theory, Auld shows that the change in the transmission coefficient, $\delta\Gamma_{ba}$, induced by the presence of a flaw, is equal to

$$\delta\Gamma_{ba} = \Gamma'_{ba} - \Gamma^\circ_{ba} = \frac{-1}{4P}\int_{S_F}(\vec{v}_1 \cdot \vec{\vec{T}}_2 - \vec{v}_2 \cdot \vec{\vec{T}}_1) \cdot \hat{n}\, dS, \tag{5}$$

where Γ°_{ba} is the transmission coefficient in the absence of the flaw, Γ'_{ba} is the transmission coefficient in the presence of the flaw, and S_F is a closed surface containing the flaw with normal $\hat{n}$. $(\vec{v}_1, \vec{\vec{T}}_1)$ are the particle velocity and stress fields radiated by probe "a" in the <u>absence</u> of the flaw, when excited by an incident electrical power P. $(\vec{v}_2, \vec{\vec{T}}_2)$ are the corresponding fields radiated by probe "b" in the <u>presence</u> of the flaw, when that probe is used as a transmitter and is also illuminated by an incident electrical power P. The fact that $\Gamma_{ab} = \Gamma_{ba}$ allows one to interchange the roles of the transmitting and receiving probes, and taking the two probes to be identical provides reflection (pulse-echo) results.

It must be recognized that the total transmission coefficient is equal to

$$\Gamma_{ba} = \Gamma^e_{ba} + \Gamma^\circ_{ba} + \delta\Gamma_{ba} \tag{6}$$

where Γ^e_{ba} is the electrical transmission coefficient (which could be obtained by setting $\vec{B}_0 = 0$ so that no ultrasonic waves are generated.) A special application of the reciprocity relation allows one to estimate Γ°_{ba} when two separate transducers are used. Suppose we consider as a hypothetical flaw a planar crack of infinite extent, normal to the axis of the radiation pattern of probe "b" and separating the two probes. Then, if this hypothetical flaw were present, $\Gamma_{ba} = 0$ and $\Gamma^\circ_{ba} = -\delta\Gamma_{ba}$. The latter quantity can be obtained from the reciprocity relation. Taking S_F to coincide with the surface of the hypothetical flaw, it follows that $\vec{\vec{T}}_2 \cdot \hat{n} = 0$ because of the stress free boundary condition, and $\vec{v}_2 = 0$ on the shadow side of the flaw with respect to probe "b". In the spirit of a Kirchhoff approximation, one can assume $\vec{v}_2 \approx 2\vec{v}'_2$ on the illuminated surface of the crack with respect to probe "b", where $\vec{v}'_2$ is the particle velocity of the illuminating field. The latter conclusion follows from the doubling of displacement that occurs at a stress free surface upon reflection of normally incident waves. We thus conclude

$$\Gamma^\circ_{ba} \approx \frac{-1}{2P}\int \vec{v}'_2 \cdot \vec{\vec{T}}'_1 \cdot \hat{n}\, dS \tag{7}$$

where the integral is to be performed over the side illuminated by probe "b" (which we have shown could be either the actual transmitting or receiving probe).

In order to relate these results to standard voltage and current measurements in the laboraory, one needs to consider the simple equivalent circuit shown in Fig. 8a. There it is assumed that a source of open circuit peak voltage $2V_1$ has been matched to the transmission line of characteristic impedance Z_0. For this circuit, the available power (that would be delivered to a matched load) is $P = V_1^2/2Z_0$, the incident voltage is V_1, and the received voltage is observed across a matched load Z_0. From the definition of reflection and transmission coefficients and elementary circuit analysis, one concludes that

$$I_T = V_1(1 - \Gamma_{aa})/Z_0 \tag{8}$$

$$V_R = V_1 \Gamma_{ba} = V_0 Z_0/(Z_0 + Z_{bb}) \tag{9}$$

$$\Gamma_{aa} \approx (Z_{aa} - Z_0)/(Z_{aa} + Z_0) \tag{10}$$

Combination of Eqs. (4), and (8)-(10) shows that

$$Z_{ba} = \frac{\Gamma_{ba}(Z_0 + Z_{aa})(Z_0 + Z_{bb})}{2Z_0} \tag{11}$$

Recalling that $Z_0 = V_1^2/2P$, that $2V_1/(Z_0 + Z_{aa})$ is the current delivered to probe "a", and that $2V_1/(Z_0 + Z_{bb})$ is the current that would have been delivered to probe "b" had it been matched to the signal generator, leads to the conclusion

$$Z_{ba} = \frac{4P\Gamma_{ba}}{I_a I_b} \tag{12}$$

which directly relates the transmission coefficient that can be computed from Auld's relation to the laboratory transfer impedance. From this result and Eqs. (5) - (7), it immediately follows that the general expression for the transfer impedance is

$$Z_{ba} = Z_{ba}^e + Z_{ba}^o + \delta Z_{ba} \tag{13}$$

where Z_{ba}^e is the purely electrical response, the flaw free transmission is

$$Z_{ba}^o = -2 \int \tilde{v}_2^1 \cdot \tilde{T}_2^1 \cdot \hat{n} \, dS \tag{14}$$

and the flaw induced change is

$$\delta Z_{ba} = -\int_{S_F} (\tilde{v}_1^1 \cdot \tilde{T}_2 - \tilde{v}_2 \cdot \tilde{T}_1^1) \cdot \hat{n} \, dS \tag{15}$$

where the $\tilde{v}$ and $\tilde{T}$ are particle velocity and stress fields normalized by the drive current and a superscript "I" has been added where appropriate to denote the incident (flaw free) fields.

Simple relations can be written for $\delta\Gamma$ for a few special flaw cases. For a crack, one finds

$$\delta Z_{ba} = j\omega \int_{crack} \Delta \vec{u}_2 \cdot \vec{T}_2^! \cdot \hat{n} \, dS \tag{16}$$

where the integral is performed over the crack surface and $\Delta\vec{u}$ is the dynamic crack opening displacement. For a flaw small with respect to the beam profile,

$$\delta Z_{ba} = j\left(\frac{2\rho c_1^2 A}{f}\right)\hat{v}_1^! \hat{v}_2^! \tag{17}$$

5. RADIATION CHARACTERISTICS IN NONMAGNETIC METALS

The modeling of the radiation pattern of EMATs is important for two reasons. Given their rather low efficiencies, it is important to make estimates of expected signal levels before investing significant effort in application development. For example, transfer impedances can be measured in μ ohms, so that large drive signals will often be required to produce millivolt received signals. Careful attention in the design of an EMAT will maximize this level. This modeling can also play a very beneficial role in the utilization of an advantageous feature of EMATs, the ability to produce tailored radiation characteristics.

In analyzing EMAT radiation, two approaches present themselves as candidates. In the first, the fields are represented as angular spectra of plane waves. This approach is closely related to the propagation algorithms discussed by Langenberg in the context of imaging elsewhere in these lecture notes.[14] In the alternative, Green's function approach, the radiation is represented as the superposition of the responses of a distribution of point sources.[15]

In a full evaluation of the fields, the two approaches must give equivalent results, as will be discussed later. However, when simple relations are sought to develop intuition and guide transducer design, one or the other may lead to more useful approximations. Often, the plane wave analysis may be the most useful starting point to estimate near field response while Green's function approaches may be most useful in the spherically spreading far field.

5.1. Plane wave radiation

5.1.1. Normal radiation

Consider first a plane wave such as that which would be excited by the sheet coil shown in Fig. 6a. Application of Eq. (2) and the proper electromagnetic description of the eddy current distribution shows that the Lorentz forces are given by

$$f = \frac{nI(1+j)}{\delta}(B_{oz}\hat{a}_x - B_{ox}\hat{a}_z)e^{(1+j)z/\delta}. \tag{18}$$

where n is the number of turns per unit length of the EMAT coil. Using this as a body force in the isotropic wave equation for the half-space shown in Fig. 5, requiring that the stress vanish on the surface $z = 0$, and assuming a time harmonic variation of the form $e^{j\omega t}$, leads to the solution[16]

$$u = \frac{jT^e}{M[1 + j(\delta^2 k^2/2)]}[-\frac{e^{jkz}}{k} + \frac{(1+j)\delta}{2}e^{(1+j)z/\delta}], \qquad (19)$$

where $k = 2\pi/\lambda$ is the propagation constant of the ultrasonic wave, T^e is an effective surface stress and M is a modulus, the latter being given respectively by

$$T^e = nIB_{ox} \qquad (20a)$$

$$M = (\lambda + 2\mu) \qquad (20b)$$

when $B_{oz} = 0$ and longitudinal waves are generated, and

$$T^e = -nIB_{oz} \qquad (21a)$$

$$M = \mu \qquad (21b)$$

when $B_{ox} = 0$ and transverse waves are generated. Here λ and μ are the Lamé elastic constants. Note that the first term in brackets in the numerator of Eq. (19) describes the radiated wave while the second term gives local fields that do not propagate but are required to satisfy the boundary conditions.

An important physical implication of Eq. (19) is described by the term $(\delta^2 k^2/2) = (2\pi^2 \delta/\lambda)$ which appears in the factor in brackets in the denominator. When $\delta/\lambda \ll 1$, this term becomes small with respect to unity, showing that the amplitude of the generated wave is independent of electrical conductivity in the limit of high conductivity. Conversely, at elevated temperatures, resistivity and the skin depth rise and the factor in brackets describes the decrease in signal strength. In the former limit, the generated amplitude varies as $T^e/\omega\rho c$. Thus the displacement amplitude varies inversely as the acoustic impedance (ρc) of the medium, so that stronger signals are generated in a light material, such as aluminum, than in a heavy material, such as stainless steel.

The role of skin depth in determining the radiated amplitude has been verified by experiment, as illustrated in Fig. 9.[16] Here the variation of amplitude is shown as a function of temperature (controlling σ and thus δ) and frequency (controlling λ). The theory is substantiated by its excellent agreement with the experimental data.

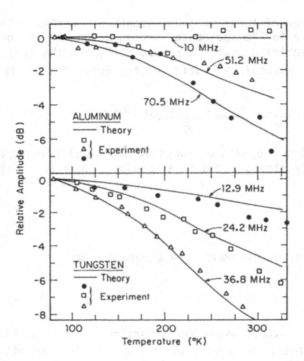

Fig. 9.   Variation or plane wave EMAT efficiency with skin depth, as
           controlled by temperature (after Ref. [16]).

5.1.2. Oblique radiation.

        Consider next the oblique radiation produced by the MC or PPM
EMATs. Elementary analysis shows that the stresses have the form

$$T = T^e(e^{j2\pi x/D} + e^{-j2\pi x/D})$$ (22)

where, for MC EMATs

$$\begin{pmatrix} T^e_{zz} \\ T^e_{xz} \end{pmatrix} \approx \frac{2I}{D}\left(\frac{\sin\pi a/D}{\pi a/D}\right)e^{-2\pi G/D}\begin{pmatrix} B_{ox} \\ -B_{oz} \end{pmatrix},$$ (23)

a is the width of the conductor, D is the coil period and G is the coil
lift-off as shown in Fig. 6b. Equation (23) is an approximation which
neglects the higher spatial frequencies, which are generally weaker in
amplitude, more attenuated with lift-off and not synchronous with the
radiated wavelength. The corresponding result for the PPM EMAT is

$$T^e_{yz} \approx \frac{-nIB_o}{(\pi/2)}e^{-2\pi G/D}$$ (24)

The factor $e^{-2\pi G/D}$ appearing in Eqs. (23) and (24) is of great practical importance since it describes an exponential weakening of the signals with lift-off. The implication is that the EMAT, while noncontact, must be close to the surface, with respect to a wavelength, for efficient operation. This exponential decay is a consequence of the fact that the magnetic fields excited by the coil are quasi-static in nature and hence satisfy Laplace's equation. Hence the real periodicity $e^{\pm j2\pi x/D}$ in the plane of the surface implies the complex decay $e^{-2\pi z/D}$ with lift-off.

The radiated displacement fields can be obtained by solving for the obliquely propagating plane waves radiated at an angle given by Eq. (3) by a stress of the form of Eq. (22). The solution is shown in normalized form in Fig. 10. There a normalized radiation resistance, as defined by

$$R_{L,S} = \left| \frac{\rho c_{L,S} \omega A_{L,S}}{T^e} \right|^2 \cos\theta \tag{25}$$

where A is the displacement amplitude of the radiated plane wave, is plotted as a function of the radiation angle. This is proportional to the power radiated per unit length of the transducer. Part (a) shows the angular variation of the longitudinal wave radiation and corresponds to physical expectations. For example, the radiation from a stress T^e_{xz} must vanish in the normal direction (0°) by symmetry and at grazing incidence (90°) because the longitudinal wave does not satisfy the stress free boundary conditions at the surface. The shear wave response, shown in part (b), is considerably more interesting. The behavior at 0° is dictated by symmetry. The structure near 30° is a consequence of the refracted longitudinal wave approaching the angle of critical refraction. For the SH wave, the rise as one approaches grazing retracted angles is a consequence of the constructive interference of the radiation from multiple sources. In the limit of $\theta = 90°$, the radiation from all EMAT elements (an unbounded number for the straight crested wave case) will add constructively because the SH wave satisfies the stress free boundary conditions exactly and hence can propagate along the surface without radiative loss.

An experimental confirmation of the theory is presented in Fig. 11. In the experiments, a pair of meander coils were attached on opposite sides of an aluminum plate, and the entire sample was placed in the gap of a large electromagnet. For each angle (determined by frequency in accordance with Eq. (3), the coils were translated to achieve maximum SV transmission, and the transfer impedance was measured for the magnetic field parallel and perpendicular to the metal surfaces. The ratio of those transfer impedances, equal to the ratio of the normalized radiation resistance, is plotted versus radiated angle and compared to theory in the figure. Agreement is seen to be excellent over two orders of magnitude.

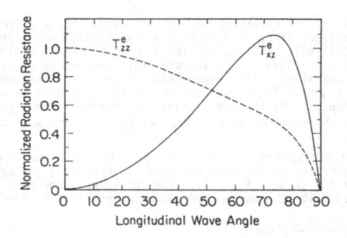

(a)

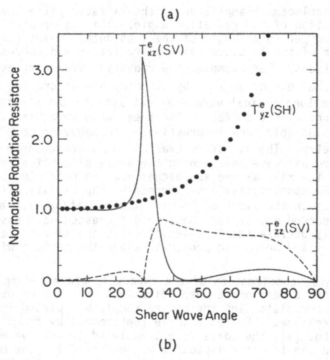

(b)

Fig. 10.   Normalized radiation resistance $R_{L,S}$, for plane wave generation
by MC and PPM EMATs in aluminum.   In each case, the curves are
labelled by the component of driving traction considered.   (a)
L-wave; (b) SV and SH waves (after Ref. [5])

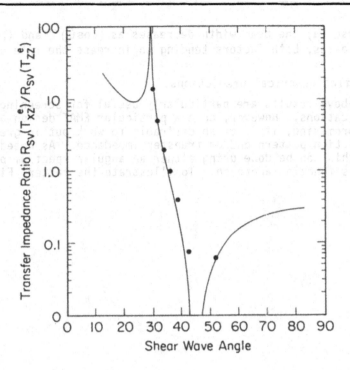

Fig. 11. Experimental test of ratio of SV wave transfer impedances for
          in-plane and normal tractions as a function of angle. (after
          Ref. [5]).

A practical implication of these results is that shear waves can be
radiated with particularly high efficiency when tuned such that the
associated longitudinal wave is near its critical angle.

5.2. Radiation of finite beams

5.2.1. Analytic estimates of maximum transfer impedance.

From the above expressions one can estimate the transfer
impedance when two bulk wave EMATs are placed in the near-field of one
another so that beam spread can be neglected. This involves combination
of Eq. (14) with various of the results of the previous section.
Although the detailed expressions are quite varied,[5] they can all be
placed in the uniform form.

$$|Z_{ba}| = \left(\frac{2}{\rho c}\right)\left(\frac{T^e}{I}\right)^2 RA \tag{26}$$

Thus the equivalent surface traction T^e and normalized radiation
resistance, R, are seen to be fundamental parameters in determining
transfer impedances. It should be emphasized, however, that Eq. (26) is
an upper bound for the signal transmitted between two EMATs. Any beam
spread will lower the observed signals. For a particular pair of EMAT's,
this will become more important as one scans to higher angles by lowering

frequency because (a) the beam width decreases as $(\cos\theta)^{-1}$ and (b) the wavelength increases, both factors tending to increase the rate of beam spread.

5.2.2.Detailed numerical predictions.

The above results are particularly useful for screening potential applications. However, once a particular EMAT design has been identified as promising, it is often desirable to work out in greater detail the radiation pattern and/or transfer impedance. As noted in the introduction, this can be done using either an angular spectrum of plane waves or Green's function approach. To illustrate the latter, Fig. 12

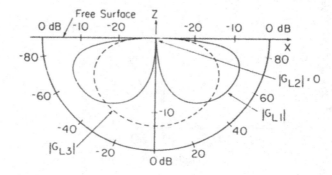

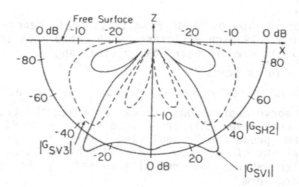

Fig. 12. Radiation patterns (Green's Function) of a line source. (top) L-wave; (bottom) SV and SH waves. In each case, the subscript is the radiated wave type followed by the direction of the excitation force (after Ref. [18]).

illustrates the radiation from line forces [17,18] directed in the $\hat{a}_1$, $\hat{a}_2$, and $\hat{a}_3$ directions. These can be seen to exhibit behaviors very similar to that of the normalized radiation resistance. The two can be directly related since stationary phase agreements can be employed to show that the far field Green's function is proportional to $A(\theta)\cos\theta/r^{1/2}$.

Examples of using such Green's function in detailed radiation pattern calculations are presented in Fig. 13[19].

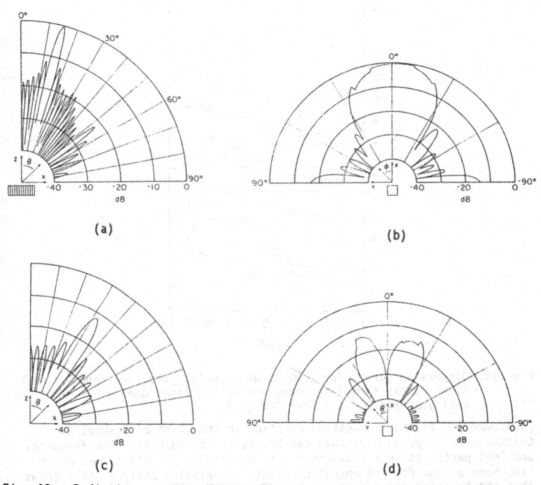

(a)

(b)

(c)

(d)

Fig. 13. Radiation pattern of MC EMAT having parameters L = 31.75 mm, D = 6.35 mm, G = 0.254 mm and f = 1.8 MHz: (a) angular dependence of SV main lobe in sagittal (x-y) plane; (b) cross-section of SV main lobe; (c) angular dependence of longitudinal wave radiation; (d) cross-section of SH radiation in plane of main SV lobe (after Ref. [19]).

5.3.   Radiation of Lamb waves

An important application of EMATs is in the excitation of Rayleigh waves on half-spaces or guided waves on plates, applications that take advantage of the coherent superposition of signals radiated from multiple elements. For example, it is well known that there is a family of Lamb waves, whose dispersion curves have the form illustrated in Fig. 14,

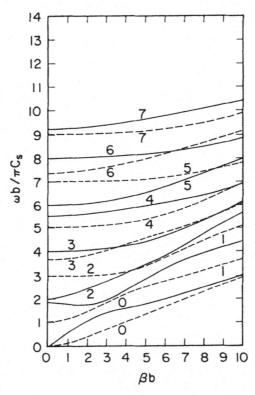

Fig. 14. Dispersion curves for Lamb modes in polycrystalline iron. Solid lines denote symmetric modes and dashed lines denote antisymmetric modes (after Ref [5]).

[13], where β is the propagation constant in the plane of a sheet of thickness b. A particular mode can be excited by adjusting the frequency and EMAT period so that they match the $(\omega = 2\pi f, \beta = 2\pi/D)$ values of the Lamb mode at the desired operation point. A detailed analysis then shows that the transfer impedances for the MC and PPM EMATs are given by

$$\text{MC:} \, |Z_{ba}| = \frac{\omega^2}{2} \, | \frac{T^e_{xz}}{IY^{1/2}_{M_x}} - e^{-j\phi} \frac{T^e_{zz}}{IY^{1/2}_{M_z}} |^2 L^2 W \qquad (27)$$

and

$$PPM: |Z_{ba}| = \frac{\omega^2}{2} | \frac{T_{yz}^e}{IY_{M_y}^{1/2}} |^2 L^2 W \tag{28}$$

In these formule, the properties of the Lamb waves enter through an admittance parameter, Y, which is defined as

$$P_M = \frac{1}{2} Y_{M_x} |U_{M_x}|^2 = \frac{1}{2} Y_{M_y} |U_{M_y}|^2 = \frac{1}{2} Y_{M_z} |U_{M_z}|^2 \tag{29}$$

where P_M is the acoustic power/unit length carried by the "M"-th mode having peak surface displacement amplitudes $U_{M_x}, U_{M_y}, U_{M_z}$. The physics described by this factor is that an EMAT producing a particular direction of surface traction will couple well to a mode having a large surface displacement (and hence small admittance) in that direction and will couple poorly to a mode having a small displacement in that direction. This is an important practical consideration, since the admittances have considerable structure as illustrated in Fig. 15 for the n = o and n = 1

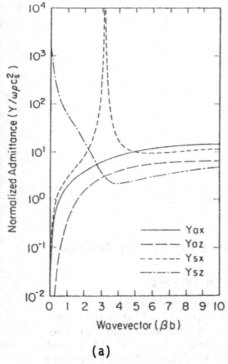

(a)

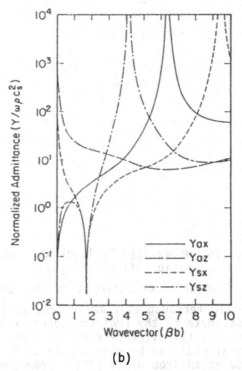

(b)

Fig. 15. Normalized admittances for Lamb modes. (a) n = o symmetric and antisymmetric modes; (b) n = 1 symmetric and antisymmetric modes (after Ref. [5]).

symmetric and antisymmetric Lamb modes. The behavior of Y_{M_y}, governing SH modes is somewhat simpler, being given by

$$Y_{M_y} = \rho \omega c_s^2 b [(\omega/c_s)^2 - (n\pi/b)^2]^{1/2, \epsilon_n} \qquad (30)$$

where $\epsilon_n = 1$ for $n = o$ and $\epsilon_n = 2$ for $n > o$.

For Rayleigh waves, the admittances are proportional to frequency and depend on Poisson's ratio as shown in Fig. 16.

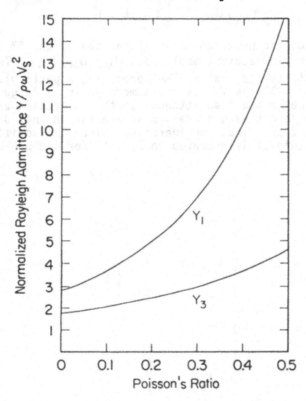

Fig. 16. Normalized Rayleigh wave admittance as a function of Poisson's ratio (after Ref. [5]).

6. Radiation characteristics in ferromagnetic materials

In ferromagnetic materials, additional magnetic mechanisms contribute to the EMAT transduction process. Figure 17 illustrates the experimental evidence for this comment by plotting the amplitude of the signal radiated between a pair of MC EMATs exciting a flexural (A_0) Lamb mode on an iron plate [20]. From the experimental data, it will be seen that the signal is approximately proportioneal to H_o for $H_o > 2 k O e$, but that peaks in the signal occur at lower fields denoting the presence of other mechanisms of transduction.

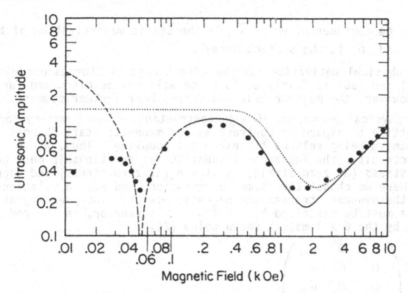

Fig. 17. Field dependence of the amplitude of the first antisymmetric
Lamb mode generated by a MC EMAT with a tangential bias field.
The frequency is 170 kHz, and the sample is a 1.27 cm thick,
hot-rolled Armco iron plate. Solid-broken line is a theoretical
prediction based on measurement of magnetostrictive properties,
and the dotted line is a theoretical prediction based on a first
principles model for grain rotation contributions to
magnetostriction (after Ref. [20]).

Such behavior can be understood in terms of magnetostrictive and
direct magnetic contributions to the transduction process [20-23]. In
such a view, the body force is composed of three terms

$$\vec{f} = \vec{f}_L + \vec{f}_M + \vec{f}_{MS} \tag{31}$$

which are given by

$$\vec{f}_L = \vec{J} \times \vec{B}_0 \tag{32}$$

$$\vec{f}_M = (\nabla \vec{B}) \cdot \vec{M}_0 \tag{33}$$

$$\vec{f}_{MS} = \nabla \cdot \overleftrightarrow{T}_{MS} \tag{34}$$

where $\overleftrightarrow{T}_{ms}$ is a magnetostructive stress tensor

$$\overleftrightarrow{T}_{MS} = -\overleftrightarrow{e} \cdot \vec{H} \tag{35}$$

and the e is a tensor of magnetic field dependent magnetostrictive
(piezomagnetic) coefficients of the material [20,21,24]. In this
formulation, a magnetic surface traction of the form

$$\hat{n}_0 \cdot \overleftrightarrow{T}_M = -\hat{n}_0 (\vec{B} \cdot \vec{M}_0) \tag{36}$$

must also be considered, where $\vec{M}_0$ is the static magnetization of the material and $\hat{n}_0$ is the surface normal.

The physical motivation for the direct magnetic forces described by Eqs. (33) and (36) is fairly obvious and will not be discussed further here. However, the magnetostrictive forces bear further comment.

The physical phenomena of magnetostriction is the tendency for a solid lattice to expand or contract as its magnetic state changes, with the mechanism being related to spin-orbit coupling. Thus, as in piezoelectricity, the fundamental constitutive relationship must be between stress (at zero strain) or strain (at zero stress) and magnetic state. Here we choose the former description, and seek a relationship between the dynamic stresses and magnetic fields. The piezomagnetic constants must be described by a fifth rank tensor or, in the reduced notation by the 6 x 3 matrix shown below [24]

$$e_{IJ} = \begin{pmatrix} 0 & 0 & e_{31} \\ 0 & 0 & e_{31} \\ 0 & 0 & e_{33} \\ 0 & e_{42} & 0 \\ e_{42} & 0 & 0 \\ 0 & 0 & 0 \end{pmatrix} \tag{37}$$

where the individual elements are functions of the bias magnetic field and, because of hysteresis, of the magnetic state that has been produced by the past history of the sample. In this particular case, the magnetic field has been assumed to be in the $\hat{a}_3$ direction. Symmetry arguments lead to the null elements and to the equality between e_{31} and e_{32} and between e_{51} and e_{42}, assuming that the material is isotropic in the demagnetized state. In many cases, magnetostriction conserves volume to a first approximation, under which condition $e_{31} \approx -\frac{1}{2} e_{33}$

In writing the elastic wave equation, one is concerned with the net force on a unit volume, which only exists for noniform stress. Hence, the divergence operator appears in Eq. (34) for the effective body force density. An interesting consequence occurs in magnetostrictive metals, for which the dynamic magnetic field might be described by the relation

$$H_x = H_s(x)e^{(1+j)z/\delta} \tag{38}$$

Consider the case in which the applied magnetic field is in the $\hat{a}_1$ direction. Then in matrix notation, the magnetostrictive stresses would be given by

$$T_I = \tilde{e}_{I1} H_x h(-z), I = 1,...,6 \tag{39}$$

where $\tilde{e}_{11}$, $\tilde{e}_{21} = \tilde{e}_{31}$ are now the non-vanishing elements of the piezomagnetic tensor (a rotated form of Eq. (37)) and h is the Heaviside step function. Application of Eq. (34) shows that

$$f_x = \tilde{e}_{11} \frac{\partial H_s(x)}{\partial x} e^{(1+j)z/\delta}, z < 0 \tag{40a}$$

$$f_y = 0 \tag{40b}$$

$$f_z = \tilde{e}_{31}[-\delta(z) + (1+j)/\delta]H_s(x)e^{(1+j)z/\delta}, z < 0 \tag{40c}$$

Under the common condition that $\delta/\lambda \ll 1$, one would next be tempted to integrate over z to obtain an effective surface traction as was done for the non-magnetic case. This would imply $T_{33}^e = 0$, a result that is inconsistent with a detailed analysis of the radiation pattern. This dilemma is resolved by a more careful analysis in which the depth dependence of the body forces are explicitly considered. After a series of operations involving representing the Green's function as a Taylor series in z, exploiting various consequences of the boundary conditions at the free surface, and integrating by parts, one concludes that the effective stresses are given by

$$T_{xz}^e = \int_{-\infty}^0 f_x dz + \left(\frac{\lambda}{\lambda + 2\mu}\right) \frac{\partial}{\partial x} \int_{-\infty}^0 f_z z dz + \ldots \tag{41a}$$

$$T_{yz}^e = \int_{-\infty}^0 f_x dz + \left(\frac{\lambda}{\lambda + 2\mu}\right) \frac{\partial}{\partial y} \int_{-\infty}^0 f_z z dz + \ldots \tag{41a}$$

$$T_{zz}^e = \int_{-\infty}^0 f_z dz + \frac{\partial}{\partial x} \int_{-\infty}^0 f_x z dz + \frac{\partial}{\partial y} \int_{-\infty}^0 f_y z dz + \ldots \tag{41c}$$

It can be seen that the important physical quantities are the moments of the force with respect to depth. For a given direction of force, changing the order of the moment changes the surface traction component. In the example considered, $\int f_z(z)dz = 0$ and both f_x and f_y contribute to the effective stress T_{xz}^e, a result which played a major role interpreting the experimental data in Fig. 17.

As in nonmangentic metals, the best predictions of radiation patterns occur when a detailed angular spectrum or Green's function analysis is employed. Figure 18 presents the results of such an analysis, as performed by Wilbrand, for a normal field MC EMAT radiating into a half-space. Part (a) shows the relative contributions of the various source components for a single element. In this case, the magnetostrictive contribution is not significant but the direct magnetic force considerable enhances the radiation at high angles. Part (b) shows the excellent agreement that is obtained between theory and experiment for the angular dependence of the radiation.

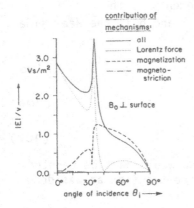

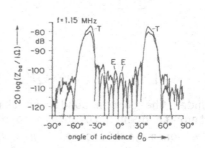

Fig. 18.   Angular dependence of SV transduction by MC EMAT with normal
           field in steel.   (a) Theoretical sensitivity pattern of single
           line source; (b) comparison of theory and experiment for
           apodized meander coil (after Ref. [23]).

        Examination of Eq. (37) reveals a component of the magnetostrictive
tensor which couples to shear stresses. This can be utilized by
orienting the bias field parallel to the major current carrying elements
of a meander coil. This structure is known as the transverse field EMAT.
Figure 19 shows the performance of such an EMAT. An interesting result
is that such an EMAT has no Lorentz force contribution (since $\vec{J} \times \vec{B}_0 = 0$,
as evidenced by the H_0^{-1} decrease in radiated amplitude at high fields.

7.   APPLICATIONS

        A number of EMAT applications have been described in a recent
article appearing in the ASNT Handbook on Ultrasonic Testing [6]. A few
of these are mentioned below to illustrate the range of applications of
those devices.

1.   Inspection of Moving Parts

        Because they do not require a couplant, EMATs are ideal for
inspecting materials which are moving. As a simple example, Fig. 20
presents data that was obtained while monitoring the thickness of
seamless steel tubing in an operating null. In this case, only the time
of arrival of a backsurface echo need to measured.

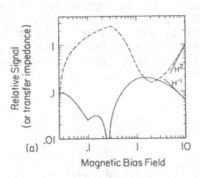

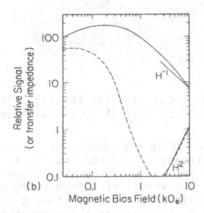

Fig. 19. Comparison of field dependence of SH generation by transverse
field EMAT (solid line) and flexural Lamb wave generation by MC
EMAT with tangential bias (broken line). Also indicated are
high-field asymptotes. (a) Iron; (b) nickel (after Ref. [25]).

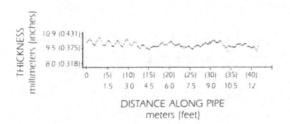

Fig. 20 Plot of a wall thickness profile produced by an automatic
thickness gage using an electromagnetic acoustic transducer on a
350 mm (14 in.) diameter, 9.5 mm (0.375 in.) wall seamless steel
tube at a production line speed of 0.9 m · s⁻¹ (3 ft · s⁻¹) in an
operating steel mill. (after Ref. [6]).

A more demanding application is the inspection of buried pipes.
Fig. 21 shows a photograph of a pipe inspection robot which is inserted
in a buried pipe and propelled by gas pressure. In this particular
system, ultrasonic Lamb waves are propagated around the circumference and
used to inspect for a variety of defect types. Operation at speeds up to
10 miles per hour is required.

Fig. 21.   An eight transducer robot used for testing a 750 mm (30 in.)
           diameter buried gas pipeline, 95 km (60 mi) long. (after Ref.
           [6]).

A third interesting application uses Rayleigh waves to inspect for
surface cracks in the wheel treads of railroad rails, as schematically
shown in Fig. 22.   Figure 23 shows a close-up photograph of the meander
coil EMAT as used to inspect the wheels of a German prototype high speed
train.   Figure 24 presents calibration results showing the response to a
saw cut as a function of its distance along the circumference from the
probe.   A full understanding of these signals requires one to recognize
that the EMATs are bidirectional, coupling to waves travelling in either
the clockwise or counterclockwise directions around the pipe.

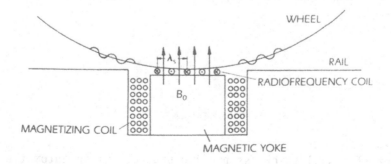

Fig. 22.   Schematic diagram of an electromagnetic acoustic transducer
           installed in a railroad rail for wheel tread testing. (after
           Ref. [6]).

Fig. 23. Detail of EMAT probes for wheel tests. (after Ref. [6]).

2.   High-Temperature Tests.

The facts that EMATs can operate with a lift-off and can be
constructed from temperature resistant materials allows EMATs to be used
to inspect materials at elevated temperature. Figure 25 shows a device
designed to measure the time of flight of longitudinal waves propagating
through a red-hot steel billet at temperature in excess of 1,100°C
(2,000°F). This was developed as part of a program to infer the internal
temperature profile in a continuously cast steel billet from ultrasonic
velocity measurements.

3.   Testing In Vacuum

A novel application of EMATs is inspection in vacuum. Figure 25
illustrates one such system, with the meander coil EMATs visible on the
right-hand side. Use of this system allows electron beam welds to be
inspected as they are made. Any necessary repairs can be made in-situ,
without breaking the vacuum, at a considerable savings in time.

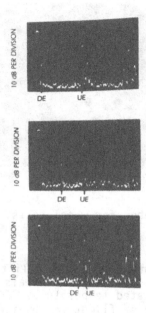

Fig. 24.   Display of tests for a 1 mm (0.04 in.) deep saw cut in the
center of railway wheel tread (see the following figure) for
distances from transmitter and receiver to discontinuity of:
(a) x = 155 mm (6 in.), (b) x = 720 mm (28 in.) and (c) x =
1,250 mm (49 in.). (after Ref. [6]).

Fig. 25.   Portable system for measuring the ultrasonic transit time
           through steel billets emerging from a continuous casting
           furnace at temperatures over 1,100 °C (2,000 °F);
           electromagnetic acoustic transducers are shown on top of the
           transport cases, positioned on opposite sides of a 100 mm (4
           in.) square billet. (after Ref. [6]).

Fig. 26. Mechanical scanning system and cylindrical test object inside a
          vacuum chamber; the electron beam weld is circumferential and
          positioned between two angle beam, focusing EMATs; the
          transducers scan downward while the cylinder is rotated to give
          100 percent coverage of the weld zone. (after Ref. [6]).

REFERENCES

1. Silk, M. G.: Ultrasonic Transducer for Nondestructive Testing, Adam
    Hilger, Bristol, 1984.

2. Adler, L.: Lecture notes in this volume, 1992.

3. Dobbs, E. R.: Electromagnetic Generation of Ultrasonic Waves, in:
    Physical Acoustics, Vol. 10, (W. P. Mason and R. N. Thurston, Eds.),
    Academic Press, New York, 1976, 127-193.

4. Frost, H. M.: Electromagnetic - Ultrasound Transducers: Principles,
    Practice and Applications, in: Physical Acoustics, Vol. 14, (W. P.
    Mason and R. N. Thirston. Eds.), Academic Press, New York, 1979,
    179-276.

5. Thompson, R. B.: Physical Principles of Measurements with EMAT
    Transducers, in: Physical Acoustics, Vol. 19 (R. N. Thurston and A.
    D. Pierce, Eds.) Academic Press, New York, 1990, 157-200.

6. Alers, G. A., G. Huebschen, B. W. Maxfield, W. Repplinger, J. Salzburger, R. B. Thompson, and A. Willbrand: Electromagnetic Acoustic Transducers in: Nondestructive Testing Handbook, American Society for Nondestructive Testing, Columbus, OH., 1991, 326-240.

7. Scruby, C.B. and L. E. Drain: Laser Ultrasonics, Adam Hilger, Bristol, 1990.

8. Wagner, J. W.: Optical Detection of Ultrasound, in: Physical Acoustics, Vol. 19 (R. N. Thurston and A. D. Pierce, Eds.) Academic Press, New York, 1990, 201-266.

9. Scruby, C. B.: Lecture notes in this volume, 1992.

10. Thompson, R. B.: A Model for the Electromagnetic Generation and Detection of Rayleigh and Lamb Waves, in: IEEE Trans. on Sonics and Ultrason. SU-20, 1973, 340-346.

11. Moran, T. J. and R. M. Panos: in: Appl. Phys. 47, 1976, 2225-2227.

12. Vasile, C. F., and R. B. Thompson: Excitation of Horizontally Polarized Shear Elastic Waves by Electromagnetic Transducers with Periodic Permanent Magnets, in: J. Appl. Phys. 50, 1979, 2583-2588.

13. Auld, B. A.: General Electromechanical Reciprocity Relations Applied to the Calculation of Elastic Wave Scattering Coefficients, in: Wave Motion 1, 1979, 3-10.

14. Langenberg, K. J. Lecture notes in this volume, 1992.

15. Achenbach, J. D. Lecture notes in this volume, 1992.

16. Gaerttner, M. R., W. D. Wallace and B. W. Maxfield: Experiments Relating to the Theory of Magnetic Direct Generation of Ultrasound in Metals, in: Phys. Rev. 184, 1969, 702-704.

17. Miller, F., and H. Pursey: The Field and Radiation Impedance of Mechanical Radiation on the Free Surface of a Semi-infinite Isotropic Solid, in: Proc. Roy. Soc. London A-223, 1954, 521-541.

18. Thompson, R. B.: The Relationship Between Radiating Body Forces and Equivalent Surface Stress: Analysis and Application to EMAT Design, in: J. Nondestr. Eval., 1, 1980, 79-85.

19. Pardee, W. J., and R. B. Thompson: Half-space Radiation by EMATs, in: J. Nondestr. Eval., 1, 1980, 157-181.

20. Thompson, R. B.: A Model for the Electromagnetic Generation of Ultrasonic Guided Waves in Ferromagnetic Metal Polycrystals, in: IEEE Trans. on Sonics and Ultrason., SU-25 1978, 7-15.

21. Thompson, R. B.: Mechanisms of Electromagnetic Generation and Detection of Lamb Waves in Iron-Nickel Alloy Polycrystals, in: J. Appl. Phys., 48 (1977), 4942-4950.

22. Wilbrand, A.: EMUS - Probes for Bulk Waves and Rayleigh Waves. Model for Sound Field and Efficiency Calculations, in: New Procedures in Nondestructive Testing, (P. Höller, Ed.) Springer-Verlag, Berlin, 1983, 71-80.

23. Wilbrand, A.: Quantitative Modeling and Experimental Analysis of the Physical Properties of Electromagnetic - Ultrasonic Transducers, in: Review of Progress in Quantitative Nondestructive Evaluation, Vol. 7A (D. O. Thompson and D. E. Chimenti, Eds.), Plenum Press, New York, 1987, 671-680.

24. Berlincourt, D. A., D. R. Curran, and H. Jaffe: Piezoelectric and Piezomagnetic Materials and Their Function in Transduction, in: Physical Acoustics, Vol. 1A, (W. P. Mason, ed.), Academic Press, New York, 169-270.

25. Thompson, R. B.: Generation of Horizontally Polarized Shear Waves in Ferromagnetic Materials Using Magnetostrictively Coupled, Meander-Coil Electromagnetic Transducers, in: Appl. Phys. Lett., 34 (1979), 175-177.

MEASUREMENT MODELS FOR QUANTITATIVE ULTRASONICS

J.D. Achenbach

Northwestern University, Evanston, IL, USA

ABSTRACT

This lecture emphasizes the importance of measurement models. Two examples are discussed both from a theoretical and an experimental point of view. These are backscattering of body waves from a surface-breaking crack and reflection and transmission of surface waves by a surface-breaking crack. Measurement models also play a useful role in the training of neural networks, as discussed by the example of backscattering from a surface-breaking crack.

7.1 MEASUREMENT MODELS

Over the last decade one of the most significant advances in nondestructive evaluation has been the evolution of NDE from a conglomeration of empirical techniques to a well defined field of interdisciplinary science and engineering. In the course of this development it has become well recognized that a fundamental approach to NDE must be based on quantitative models of the measurement processes of the various inspection techniques. A model's principal purpose is to predict, from first principles, the measurement system's response to specific anomalies in a given material or structure, (e.g. cracks, voids, distributed damage, corrosion, deviations in material properties from specifications, and others). Thus, a measurement model includes the configuration of probe and component being inspected as well as a description of the generation, propagation, and reception of the interrogating energy. In the ultrasonic case, this description requires computations of the transducer radiation pattern, refraction of the beam at the parts' surface, the beam profile and the propagation characteristics in the host material including effects of material anisotropy, attenuation, diffraction losses, etc. Detailed modeling of the field-flaw interactions which generate the measurement system's response function are also included, as well as information of material and other conditions that produce noise and add an uncertainty to the measurement results. A well constructed measurement model should be able to predict specific instrumental responses to anomalies in complex materials and structures as well as to "standard" flaws placed in various calibration blocks.

A number of measurement models have been formulated in the past several years for different inspection techniques. For practical applications, the challenge lies in making approximations that permit the computations to be tractable while retaining sufficient accuracy so that the engineering applications are not compromised. The status of models for ultrasonics, eddy current methods and radiographic techniques has recently been discussed by Gray et al. [7.1].

The availability of a measurement model has many benefits. Numerical results based on a reliable model are very helpful in the design and optimization of efficient testing configurations. A good model is also indispensable in the interpretation of experimental data and the recognition of characteristic signal features. The relative ease of parametrical studies based on a measurement model facilitates an assessment of the requirement for the development of an inverse technique based on quantitative data. Last but not least a measurement model whose accuracy has been tested by comparison with experimental data provides a practical way of generating a training set for a neural network or a knowledge base for an expert system.

In this lecture we will concentrate on an essential component of a measurement model for quantitative ultrasonics, namely, the modeling of the interaction of ultrasound with a defect.

7.2 CRACK-SCATTERING THEORY

The formulation of the scattering problem of transient ultrasound by a crack can be reduced to a system of singular boundary integral equations (BIE) for the crack-opening displacement. These equations can be solved numerically by a direct time-stepping approach in conjunction with a boundary element discretization (BEM) over the crack surface. An alternate approach, which presently seems to be preferred and which was discussed in lecture 4, first disposes of the time dependence by the use of a Fourier integral. The resulting problem in the frequency domain only requires a boundary element discretization over the crack surface. A Fast-Fourier Transform can subsequently be employed to construct the time-domain solution. Naturally it may be necessary to calculate frequency-domain results for a fairly large range of frequencies, which may require a considerable investment in computer time. Nevertheless the frequency domain approach has the advantage of being straight forward and for elastodynamic crack problems it is, so far, the most frequently used method. In a paper by Kitahara, Hirose and Achenbach [7.2] the direct-time approach has been discussed. The frequency-domain approach has been discussed in detail by Li and Achenbach [7.3].

A special class of applications concerns a crack which breaks a free surface or is located in the vicinity of a free surface. Since elastodynamic Green's functions for bounded bodies are exceedingly complicated, it is generally advantageous to use full space Green's functions. The use of such Green's functions implies, however, that the system of boundary integral equations includes equations over the external surface(s) of the body. Examples of surface-breaking and near surface cracks in an elastic half-space have been discussed by Zhang and Achenbach [7.4].

A catalogue of two-dimensional problems of interaction of ultrasound with cracks is shown in Fig. 7.1. Some of the corresponding three-dimensional problems have also been solved generally at the expense of considerably greater computer time, but without fundamental difficulties.

SCATTERING BY CRACKS: TWO-DIMENSIONAL CONFIGURATION

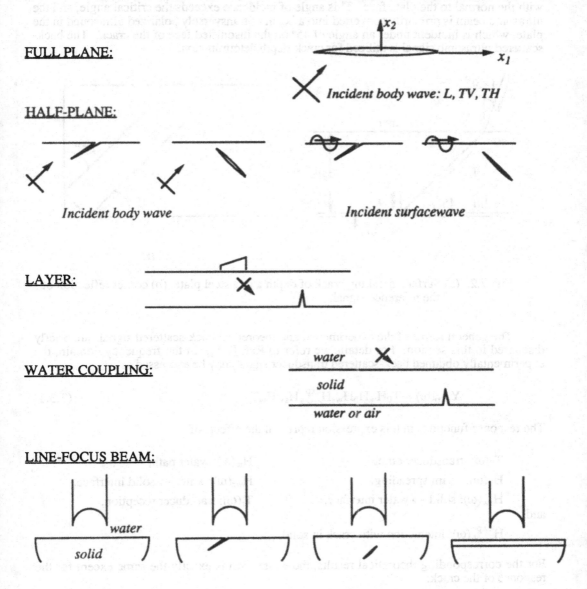

Fig. 7.1. Catalogue of solved scattering problems.

7.3 BACKSCATTERING FROM A SURFACE-BREAKING CRACK

A surface-breaking crack of depth a in a steel plate of thickness h, immersed in a water bath, is shown in Fig. 7.2. Ultrasound is generated by an immersed piezoelectric transducer. The angle of incidence on the insonified top face of the plate is taken to be 18.9° with the normal to the plate face. This angle of incidence exceeds the critical angle, and the ultrasonic beam is primarily converted into a beam of transversely polarized ultrasound in the plate, which is incident under an angle of 45° on the insonified face of the crack. The back-scattered ultrasonic signal is utilized for crack-depth determination.

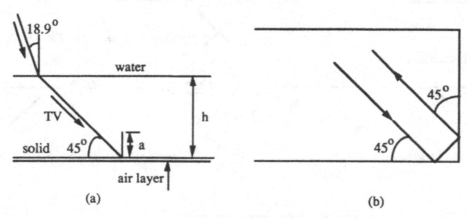

(a)   (b)

Fig. 7.2. (a) Surface breaking crack of depth a in a steel plate, (b) corner reflection of the reference signal.

The general forms of the experimental and theoretical back-scattered signals are briefly discussed in this section. For details we refer to Ref. [7.5]. In the frequency domain, the experimentally obtained back-scattered transducer signal may be expressed as

$$Y_{exp}(\omega) = T_0 H_w H_b H_{ws} H_{crack}^{exp} H_{sw} H_w T_r. \tag{7.3.1}$$

The response functions in this expression represent the effects of

$T_0(\omega)$: transducer output, $H_w(\omega)$: water path,

$H_b(\omega)$: beam spreading, $H_{ws}(\omega)$: water $\rightarrow$ solid interface,

$H_{sw}(\omega)$: solid $\rightarrow$ water interface, $T_r(\omega)$: transducer reception,

and

$H_{crack}^{exp}(\omega)$: interaction with crack in solid.

For the corresponding theoretical results, the expression is exactly the same except for the response of the crack:

$$Y_{theory}(\omega) = T_0 H_w H_b H_{ws} H_{crack}^{BEM} H_{sw} H_w T_r. \tag{7.3.2}$$

In equation (7.3.2) H_{crack}^{BEM} represents the interaction with the crack of the incident wave as calculated by the boundary element method (BEM). The BEM calculation is based on two-dimensional elastodynamic theory for an elastic body with a surface-breaking crack.

To uncouple the theoretical signal in Eq.(7.3.2) from the response functions, the signal for a corner reflection is introduced as the reference signal, see Fig. 7.2. For the same transducer angle, the same water paths, and the same specimen but with a rectangular corner, this reference corner signal can be written as

$$X_{ref}(\omega) = T_0 H_w H_b H_{ws} H_{cor}(\omega) H_{sw} H_w T_r, \tag{7.3.3}$$

where $H_{cor}(\omega)$ represents the corner reflections in the solid. For a solid-air interface, the term $H_{cor}(\omega)$ can be expressed in simple form as shown in Ref.[7.5]. The formal deconvolution of the theoretical signal of Eq.(7.3.2) by the reference signal of Eq.(7.3.3) yields

$$\frac{Y_{theory}(\omega)}{X_{ref}(\omega)} = \frac{H_{crack}^{BEM}(\omega)}{H_{cor}(\omega)}. \tag{7.3.4}$$

Thus, the theoretical signal can be expressed as

$$Y_{theory}(\omega) = \frac{X_{ref}(\omega)}{H_{cor}(\omega)} H_{crack}^{BEM}(\omega), \tag{7.3.5}$$

where the term X_{ref}/H_{cor} accounts for the beam paths in the water and across the solid-water interface. The theoretical signal in Eq. (7.3.5) is a convolved signal of the water path, X_{ref}/H_{cor}, and the elastodynamic interaction, $H_{crack}^{BEM}(\omega)$ with a crack in the solid. We call the signal $Y_{theory}(\omega)$, of Eq. (7.3.5) the theoretical signal, which may be directly compared with the experimental signal, $Y_{exp}(\omega)$, of Eq.(7.3.1). The numerical calculations of the elastodynamic interaction term, $H_{crack}^{BEM}(\omega)$, have been discussed in detail by Zhang and Achenbach [7.4].

In recent work, [7.5] and [7.6], a neural network has been developed to determine the depths of a surface-breaking crack from backscattering data. In the neural network strategy of Ref. [7.6], the theoretical signals of Eq. (7.3.5) are used for the training of the network. In the theoretical analysis, the boundary element calculation is carried out in the frequency domain to evaluate the interaction term $H_{crack}^{BEM}(\omega)$. The numerical result is subsequently convolved with the term X_{ref}/H_{cor} to obtain $Y_{theory}(\omega)$. Next the time domain signal is generated by the use of the FFT algorithm. The back scattered signals have been calculated for the cases of perfect mathematical cracks and notches of 0.5mm width. The calculated time and frequency domain signals are calibrated by comparison with experimental data and they are then used as synthetic data to train the network.

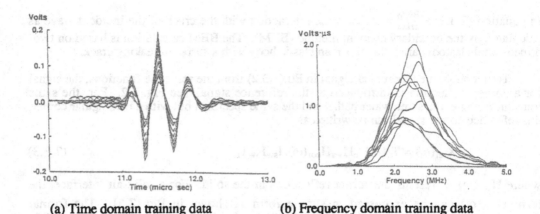

| (a) Time domain training data | (b) Frequency domain training data |

Fig. 7.3. Theoretical training signals for ten notch depths ranging from 0.6mm to 2.4mm with 0.2mm increments.

For network training, synthetic data were obtained for a total of ten crack depth ranging from 0.6mm to 2.4mm, with equal increments of 0.2mm. The time and frequency domain training data for notches are shown in Fig. 7.3. The network training has been discussed in Ref. [7.5]. The center frequency of the transducer is 2.25MHz. For the experimental measurements the time domain signal is acquired and the frequency domain signal is obtained by the use of the FFT algorithm. The network processes the time and frequency domain signals for each set of experimental data and produces information on the depth of the crack.

Table 1

Network performance for experimental data inputs

		Output unit									
MM		0.60	0.80	1.00	1.20	1.40	1.60	1.80	2.00	2.20	2.40
1.05	Time		0.22	0.97							
	Freq.				0.99						
1.49	Time					0.52	0.45				
	Freq.					0.29	0.10				
2.19	Time									0.97	
	Freq.									0.98	

Table 1 summarizes the network performance when the experimental backscattered data for cracks of depth 1.05mm, 1.49mm and 2.19mm are entered into the network as input data for crack-depth characterization. The outputs from the response units are listed in the table. The first row of the table lists the crack depths for the synthetic data used to train the network. Three sets of two rows each list the response numbers for the three sets of experimental input data. The labels time and frequency indicate that the output numbers in the labeled rows were obtained from time domain and frequency domain data, respectively. Thus the response numbers 0.97 and 0.98 for the time domain and frequency domain inputs of the 2.19mm crack show that for both sets of data the network indicates a crack depth of 2.2mm. The response numbers for the frequency domain data are also quite conclusive for the 1.49mm crack. For this case the time domain results may be interpreted as indicating a crack of depth in-between 1.4 and 1.6mm. On the other hand the frequency domain data suggest a crack depth quite close to 1.6mm. For the 1.05mm case the time domain data suggest a crack depth close to 1mm, where the frequency domain data indicate a crack depth close to 1.2mm.

The number of experimental data available for network testing was too small to allow unambiguous conclusion on the network's performance. It can, however, be stated that at least for the time domain data, the network produces numbers that indicate each depth within an acceptable error range of the actual crack depth.

7.4 CRACK-DEPTH MEASUREMENTS BY SURFACE WAVE TECHNIQUE

When a surface containing a surface-breaking crack is directly accessible for the generation of ultrasound, the use of surface waves is a preferred technique for depth determination of the crack. In principle, the crack depth can be determined by an accurate measurement at a fixed frequency or over a range of frequencies of either or both the reflection, R, and transmission, T, coefficients, and by subsequent comparison of these measurements with theoretical results for R and/or T as function of crack depth.

Analytical and numerical investigations of the reflection and transmission coefficients for normal as well as oblique incidence have been presented in [7.7]-[7.9]. Here we consider the generation and detection of surface waves by contact transducers. Various experimental studies using contact transducers have been presented in [7.7]-[7.12]. A major problem in the experimental work has been the uncoupling of the measured results from the response functions of the contact transducers, including the coupling between the transducers and specimen. To account for the coupling between transducers and specimen, various calibration procedures have been employed, such as a parallel reference specimen [7.10], preliminary calibration on a flaw-free specimen [7.11] and the use of calibration coefficients [7.12].

In a recent paper by Achenbach et al. [7.13] the depth of a surface-breaking crack is determined by measuring the absolute values of the ratio of the transmission and reflection coefficients, $|T/R|$, instead of $|T|$ and $|R|$ separately. This measurement can be performed in a self-calibrating manner using a configuration of commercial surface wave transducers. In Ref. [7.13], application of the technique is presented for both normal and oblique surface wave incidence on a surface-breaking crack. Here we will discuss normal incidence only.

For normal incidence two surface wave transducers are used to measure R/T. The two transducers are coupled to the specimen by a thin layer of oil, and they are placed at different sides of the crack, as shown in Fig. 7.4. Transducer 1 produces the signal, and receives the reflected signal with voltage

$$V_{11} = A_1 D_{10} R_c D_{01} S_1, \tag{7.4.1}$$

where

A_1 = response function of transducer 1, including the transmission from the transducer to the specimen.

D_{10} = response function for transmission over the distance from 1 to 0, including attenuation and diffraction,

R_c = reflection coefficient of the crack,

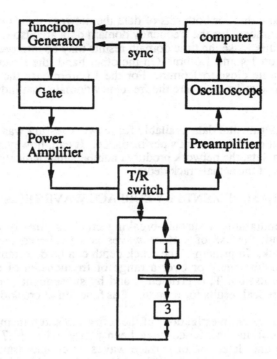

Fig. 7.4. Schematic of the experiment for for normal incidence.

and D_{01} and S_1 are defined analogously to D_{10} and A_1, respectively.

Transducer 3, placed on the other side of the crack receives the transmitted signal with voltage

$$V_{13} = A_1 D_{10} T_c D_{03} S_3,\qquad\qquad(7.4.2)$$

where

T_c = transmission coefficient of the crack,
D_{03} = response function for transmission along the distance from 0 to 3 and,
S_3 = response function of transducer 3, including transmission from the specimen to the transducer.

In the conventional technique it would be attempted to obtain the reflection and transmission coefficients directly from (7.4.1) and (7.4.2). This is , however, very difficult, primarily because the coupling between the transducers and the specimen is unpredictable and difficult to reproduce from one measurement to another. Hence there is no guarantee that the incident wave (response function A_1) is the same for each measurement nor that the reflected and transmitted waves would be measured with the same response functions (S_1 and S_3, respectively) by the receiving transducer. Since it is awkward and time consuming to calibrate the complete set-up for each measurement, a configuration of transducers suitable for a self-calibrating measurement technique has been proposed in Ref. [7.14].

By the technique of Ref. [7.14], transducer 3 is also fired, and V_{33} and V_{31} are measured. Next by considering the ratio $V_{13}V_{31}/V_{11}V_{33}$ it easily follows that

$$\left|\frac{T}{R}\right| = \left|\frac{V_{13} \cdot V_{31}}{V_{11} \cdot V_{33}}\right|^{1/2} . \tag{7.4.3}$$

The experimental configuration is shown in Fig. 7.4. To maintain exactly the same conditions of electrical signal generation and amplification, the two transducers are connected parallel. Since the measured voltage amplitudes for the transmitted signals, V_{13} and V_{31}, are overlapping, the signals used in equation (7.4.3) are taken as half the total measured

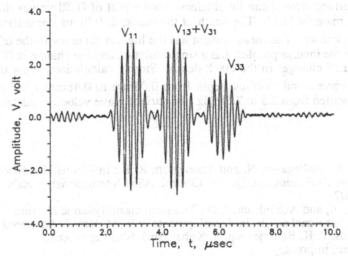

Fig. 7.5. Waveforms for normal incidence of a surface wave pulse on a surface-breaking crack.

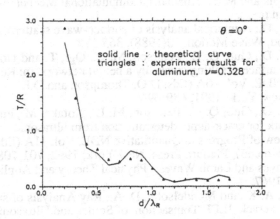

Fig. 7.6. Comparison of the theoretical and experimental values of |T/R| for a surface-breaking crack.

amplitude. To prevent overlap of the signals V_{11} and V_{33} the transducers should be installed at different distances from the crack. The wave forms acquired are shown in Fig. 7.5. As shown in Fig. 7.6, a comparison between the experimental results with the theoretical values obtained by Angel and Achenbach [7.9] shows very good agreement.

The general case of oblique incidence has been discussed in Ref. [7.14].

The self-calibrating technique can be efficiently used to determine the depth of a surface-breaking crack. For this purpose the frequency of the generated surface waves should be varied and $|T/R|$ should be measured at different frequencies. Then, d/λ_R, where λ_R is the wavelength of surface waves, can be obtained from a plot of $|T/R|$ versus d/λ_R calculated for the selected material [7.9]. The depth of the crack, d, follows immediately since the wavelength, λ_R is known. The measurement has the highest accuracy in the d/λ_R range from 0.2 to 0.4 where the inverse problem has a single solution and big change in $|T/R|$ correspond to relatively small changes in the crack depth. Simple calculations show that d/λ_R values in this range correspond to crack depth from 0.08mm to 0.48mm when the surface wave frequency is varied from 2.5 to 7.5MHz (the surface wave velocity was taken as V_R =3000m/s).

References

7.1 Gray, J. B., Gray, T. A., Nakagawa, N. and Thompson, R. B.: in Metals Handbook, Vol. 17: Nondestructive Evaluation and Quality Control, ASM International, Metals Park, OH, 1989, P.702.

7.2 Kitahara, M., Hirose, S., and Achenbach, J. D.: Transient elastodynamic fracture mechanics of three-dimensional solids, in: Developments in BEM, Vol. 7, Advanced Dynamic Analysis (Eds. P. K. Banerjee and S. Kobayashi), Elsevier Science Publishers, Amsterdam, in press.

7.3 Li, Z. L. and Achenbach, J. D.: BEM computations of elastodynamic fields containing internal, near-surface and surface-breaking cracks, in: Advances in BEM for Fracture Mechanics (Eds. C. A. Brebbia and M. H. Aliabadi), Computational Mechanics Publication, Southampton, in press.

7.4 Zhang, C. and Achenbach, J. D.: Numerical analysis of surface-wave scattering by the boundary element method, Wave Motion, 10(1988), 365-374.

7.5 Kitahara, M., Achenbach, J. D., Guo, Q. C., Peterson, M. L., Ogi, T. and Notake, M.: Depth determination of surface-breaking cracks by a neural network, in: Review of Progress in Quantitative NDE, Vol. 10A (Eds. D.O. Thompson and D.E. Chimenti), Plenum Press, New York, 1991, 689-696.

7.6 Kitahara, M., Achenbach, J. D., Guo, Q. C., Peterson, M. L., Notake, M., and Takadoya, M.: Neural network for crack-depth determination from ultrasonic backscattering data, in: Review of Progress in Quantitative NDE, Vol. 11A (Eds. D.O.Thompson and D.E. Chimenti), Plenum Press, New York, 1992, 701-708.

7.7 Viktorov, I. A.: Rayleigh Waves and Lamb Waves - Physical Theory and Application, Plenum Press, New York, 1967.

7.8 Achenbach, J. D., Gautesen, A. K. and Mendelsohn, D. A., Ray Analysis of surface-wave interaction with an edge crack, IEEE Transaction of Sonics and Ultrasonics, SU-27(1980), 124-129.

7.9 Angel, Y. C., and Achenbach, J. D.: Reflection and transmission of obliquely incident Rayleigh waves by a surface-breaking crack, J. Acoust. Soc. Am., 75(1984),

313-319.
7.10 Vu, B. Q. and Kinra, V.K.: Diffraction of Rayleigh waves on a half-space. I. Normal edge crack, J. Acoust. Soc. Am., 77(1985), 1425-1430.
7.11 Dong, R. and Adler, L.: Measurements of reflection and transmission coefficients of Rayleigh waves from cracks, J. Acoust. Soc. Am., 76(1984), 1761-1773.
7.12 Resch, M. T., Nelson, D. V., Yuee, H. H. and Ramusat, G. F.: A surface acoustic wave technique for monitoring the growth behavior of small surface fatigue cracks, J. of NDE, 5(1985), 1-7.
7.13 Achenbach, J. D., Komsky, I., Lee, Y. C., and Angel, Y. C.: Self-calibrating ultrasonic technique for R/T and T/R measurements, in: Review of Progress in Quantitative NDE, Vol. 11A (Eds. D.O. Thompson and D.E. Chimenti), Plenum Press, New York, 1992, 1035-1042.

NUMERICAL TECHNIQUES

P.P. Delsanto
Politecnico di Torino, Torino, Italy

ABSTRACT

After a brief introduction on the application of finite elements and finite
difference equation techniques in the field of ultrasonic characterization of
materials, a simulation method, based on the use of highly parallel computers,
is discussed. The versatility and efficiency of this novel numerical approach
make it very suitable for important applications, such as the analysis of Epstein
layers, composite materials and acoustic tomography.

1. INTRODUCTION

Although the analytical solutions of the ultrasonic wave equation are well
known, the application of numerical methods becomes necessary, whenever

non elementary boundary conditions are applied. Among the most widely used numerical techniques are the methods of finite elements (FE), finite difference equations (FDE), computer simulation and tomography.

There exists a vast body of literature devoted to FE techniques. For a very comprehensive and up-to-date bibliography on the subject we refer to [1]. An overview of FE analysis and its applications to the modeling of ultrasonic NDE phenomena has been presented recently by Lord and collaborators [2]. FE methods are also used as tools for computer simulation [3].

FDE's have also been widely used to study the propagation of ultrasonic waves. In fact it is quite natural to transform, for a numerical analysis, a partial differential equation into a FDE. However, most of the applications are due to geophysicists, who have developed FDE method as a tool for the quantitative interpretation of seismograms [4-10]. "Mutatis mutandis" much of their work can also be adopted for NDE purposes (e.g. the ultrasonic characterization of materials and structures). Some of the basic concepts for the development of FDE techniques are reviewed in the next Section.

FDE's can also be used for computer simulation, since they are apt to reduce a problem from a global to a local level. The advent of highly parallel computation and the anticipation of its wide diffusion and projected progress [11] give a special urgency to the need of developing efficient simulation techniques, which exploit this novel computing architecture. In fact, parallel computers, which control millions of independent processors, are the natural tools for simulating the behaviour of a material at a local level, since a correspondence can be created between the processors and spatial "cells" in the material.

In Sec. 2 we describe a numerical approach, which has been devised to take full advantage of parallel computing. Finally in Sec. 3 we discuss some of the possible applications (e.g. to layered media, Epstein layers and tomography) and show some illustrative results.

2. FINITE DIFFERENCE EQUATIONS

Finite difference equations have been an important topic of mathematical physics long before the advent of sequential (von Neumann) computers [12]. However, with the availability of computers they have become a very useful tool of numerical analysis. As such, they are well covered in several textbooks, e.g. [13-17]. They are also treated in monographies on the numerical solution of partial differential equations [18], some of which are specifically devoted to applications in acoustics and ultrasonics [19-20].

There are also, as we have noted in the Introduction, many research articles, devoted to the development of more sophisticated FDE techniques. The goal is often to improve the convergence, so that fewer grid points are needed for the calculation. In fact, for a sequential computer, the computing time is roughly proportional to the number of points in the grid, which is used to reduce the differential equation into a FDE. This economizing is, however, much less of consequence in the case of parallel computers, since each grid point is associated to a different processor (if their number is sufficiently large). Since all the processors are working simultaneously and independently, the computer time is (almost) independent on the number of grid points (provided that the problem is managed in such a way that each processor is assigned an equivalent amount of work).

Therefore, we limit ourselves, in the following, to give a brief overview of the basic concepts of the FDE formalism, neglecting all those improvements and other features, which become less relevant in the switch from sequential to parallel computing. Let us assume that we study the propagation of ultrasonic waves in a homogeneous material plate. The elastodynamic equation for the particle displacements w_k is [21-2]

$$\partial_l(S_{klmn}\partial_n w_m) = \rho\ddot{w}_k \tag{1}$$

where ρ is the density and S in the stiffness tensor. If the wave propagates through the plate, normal to its surface, the problem becomes one-dimensional, with a single displacement w, representing the entire plane wavefront:

$$S w'' = \rho\ddot{w} \tag{2}$$

In order to transform Eq. (2) into a FDE, we discretize time and space. We define an elementary time unit δ, so that the continuous time t becomes $t\delta$, with $t = 0, 1, 2...$. We also divide the propagation path into "cells" of length ε, and label the nodepoints between them with indices i, with $i = o$ and $i = N$ representing the two plate surfaces.

We then write the time and space derivatives for each time $t > o$ and for each nodepoint i (except $i = o$ and $i = N$), keeping only the first order terms in δ and ε :

$$\ddot{w}_{i,t} = (w_{i,t+1} + w_{i,t-1} - 2w_{i,t})/\delta^2$$

$$\tag{3}$$

$$w''_{i,t} = (w_{i+1,t} + w_{i-1,t} - w_{i,t})/\varepsilon^2$$

Substituting into Eq. (2) we obtain

$$w_{i,t+1} = c(w_{i+1,t} + w_{i-1,t}) + 2(1 - C)w_{i,t} - w_{i,t-1} \tag{4}$$

where

$$c = v_L/v_c \tag{5}$$

$$v_L = \sqrt{S/\rho} \tag{6}$$

and

$$v_c = \varepsilon/\delta \tag{7}$$

In Eqs. (5-7) v_L is the longitudinal velocity and v_c is the "cell" or "characteristic" velocity. Their ratio c can be considered as a free parameter (since both ε and δ can be chosen at will, provided they are sufficiently small to justify the approximations). However, from the theory of FDE's [13], it follows that the best choice is $c = 1$, since then the regions of determination (i.e. all the values of $w_{i',t'}$ with $t' < t$ which contribute to $w_{i,t}$) of both the differential and finite difference equations coincide. If $c = 1$, Eq. (4) becomes simply

$$w_{i,t+1} = w_{i+1,t} + w_{i-1,t} - w_{i,t-1} \tag{8}$$

Eq. (8) allows for a very simple numerical solution of Eq. (2), since, once the displacements at the times $t = o$ and $t = 1$ are specified, it allows to calculate the displacements at all successive times. Eq. (8) is also very suitable for parallel computing, since, if each nodepoint is associated to a different processor , all the processors calculate the "new" displacements simultaneously and indipendenty. Also, the amount of work is the same for each processor, which is important , as we noted previously, in order to obtain a perfect "speed-up" [10]. Finally, the computation of Eq. (8) requires a minimal amount of memory, which is also important since, "ceteris paribus", when the number of processors increases, the amount of memory allocated to each of them must decrease of the same factor.

Eqs.(4) or (8) can be easily generalized to two-or three-dimensional problems. E.g. in two dimensions (or in three dimensions, when there is complete symmetry with respect to one of the coordinates) the same FDE formalism yields two iteration equations for the components u and v of the displacements:

$$u_{i,j,t+1} = \alpha(u_{i-1,j,t} + u_{i+1,j,t}) + \beta(u_{i,j-1,t} + u_{i,j+1,t})$$

$$+\gamma(v_{i-1,j-1,t} + v_{i-1,j+1,t} + v_{i+1,j-1,t} + v_{i+1,j+1,t}) - 2\xi u_{i,j,t} - u_{i,j,t-1}$$

$$\tag{9}$$

$$v_{i,j,t+1} = \beta(v_{i-1,j,t} + v_{i+1,j,t}) + \alpha(v_{i,j-1,t} + v_{i,j+1,t})$$

$$+\gamma(u_{i-1,j-1,t} + u_{i-1,j+1,t} + u_{i+1,j-1,t} + u_{i+1,j+1,t}) - 2\xi v_{i,j,t} - u_{i,j,t-1}$$

where

$$\alpha = (v_L/v_C)^2$$

$$\beta = (v_T/v_C)^2 \tag{10}$$

$$\gamma = (\alpha - \beta)/4$$

$$\xi = \alpha + \beta - 1$$

and v_T is the transversal (shear) velocity in the material.

3. THE CONNECTION MACHINE APPROACH

The FDE formalism, presented in the previous Section, can be very conveniently exploited in conjunction with the use of a massively parallel computer , such as the Connection Machine (CM) [24]. The main idea is that, since all the CM processors are independent of each other, also the material "cells" can be independent, i.e. they can have all different physical properties. In other words, if we want to study an arbitrarily complex material specimen, we divide it into a sufficiently large number of cells. The physical properties are kept constant within each cell, but can vary from cell to cell, in order to match as precisily as possible with the physical properties of the real specimen. E.g. a multilayered specimen can be represented by one or more cells (all of equal physical properties for each layer).

Within each layer one can use the iteration equations obtained in the previous Section. At the interfaces between layers, however, one must match both the displacements and the stresses for continuity. This can be done analytically, by making use of the FDE formalism [25], or it can be inferred heuristically, in a much simpler way [26], by considering the transmission coefficients in the forward (t) and backward (t') directions at the interface:

$$t_I = 1 + r_I = 2/(1 + \xi)$$

$$\text{(11)}$$

$$t'_I = 1 - r_I = 2\xi/(1 + \xi)$$

where

$$r_I = (1 - \xi)/(1 + \xi) \qquad \text{(12)}$$

is the reflection coefficient and

$$\xi = Z_2/Z_1 \qquad \text{(13)}$$

is the ratio between the impedances of the two layers.

Eq.(8) can then be generalized to all the nodepoints of the multilayer:

$$w_{i,t+1} = t'_i w_{i+1,t} + t_i w_{i-1,t} - w_{i,t-1} \qquad (14)$$

since in all the nodepoints within a layer $t_i = t'_i = 1$ and at the surfaces t_i and t'_i give the correct transmission amplitudes.

The CM approach can also be used for attenuative media by considering complex values both for the displacements and the velocities [27]. Within a single layer, the effect of absorption can be included, with a simple heuristic argument, by considering the cell attenuation factor [23]

$$q = \exp(-\alpha\varepsilon) \qquad (15)$$

where α is the attenuation factor. Then Eq. (8) becomes

$$w_{i,t+1} = q(w_{i+1,t} + w_{i-1,t}) - q^2 w_{i,t-1} \qquad (16)$$

In fact it takes one step of cell length ε to propagate (at the velocity v_c) from the nodepoints $i \pm 1$ to the nodepoint i, but it takes two steps (2ε) to return to the nodepoint i after two time steps δ (via the neighbors $i \pm 1$).

To conclude, by putting together the effect of the interfaces and of the attenuation, we obtain a completely general iteration equation, in which $w_{i,t+1}$ is computed as a linear combination of $w_{i\pm1,t}$ and $w_{i,t-1}$. This iteration equation allows to solve any one-dimensional problem, i.e. it can be applied to study normal propagation through any multi-layered material.

Likewise, any two- or three-dimensional wave propagation problem can be reduced to an iteration equation in which the displacement components at any grid point P at the time $t+1$ are obtained as linear combinations of the displacements in P and in the neighboring grid points at the times t and $t-1$. Other propagation mechanisms, such as the kissing bond in thin films [28], may also be included through an "ad hoc" analysis of the local interaction or, if it is more convenient, through the application of the FDE formalism.

4. APPLICATIONS

The CM approach, which has been briefly described in the previous Section, can been applied to a variety of problems. As a first example of application of the method , at the one-dimensional level, we show in Fig. 1 the propagation of a gaussian pulse through a simulated piston damper ring (with layers of polyethylene, rubber and steel) in water, without attenuation, vs. the time (in abscissa, in units of δ). The horizontal lines represent different depths within the various layers. We notice that there is little reflection at the interfaces between water and polyethylene and between polyethylene and rubber, due to the small change in impedance. Since steel has a much larger impedance than rubber, we observe a large reflection at the rubber-steel interface.

It is important to notice that, due to the "speed- up" of parallel computing, plots like Fig.1 require only a few seconds of computing time. They yield not only the reflection and transmission rates, but also the amplitude of all the reflected and transmitted pulses at each time step and depth.

Fig.2 shows that the CM approach can also be used to treat Epstein layers (EL), which are defined as media whose physical properties vary continuously along one direction, but are constant in the other two directions [29]. We

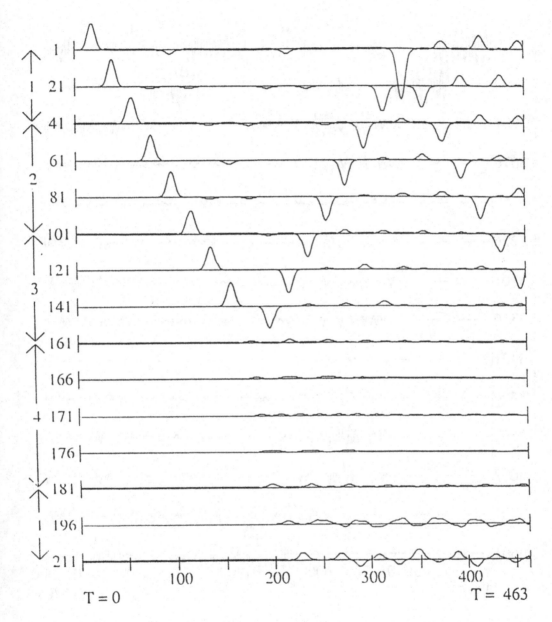

Fig.1. Propagation of a gaussian pulse through a piston in water.
Layers: 1= water, 2= Polyethylene, 3= rubber, 4= steel.

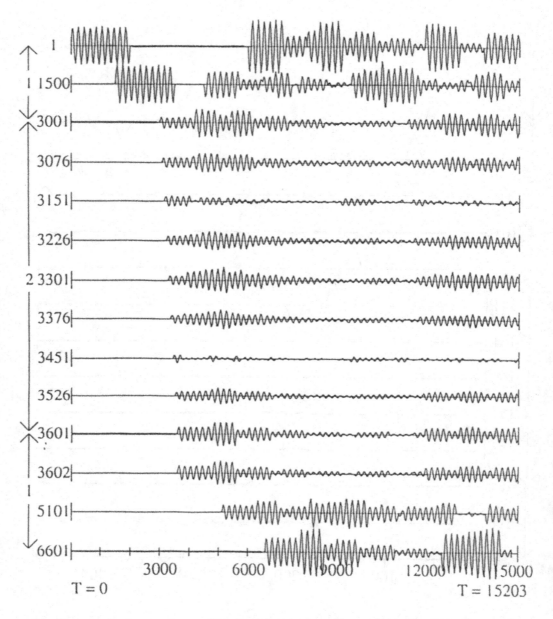

Fig.2 Propagation of a multiple sine pulse through an Epstein layer in water.

consider an EL whose impedance varies by a factor 2 between the top and bottom surface. By dividing it into a sufficiently large number of discrete layers, with linearly varying impendances, the EL can be treated with a good approximation as a multilayered material. In Fig. 2 the EL is divided into 600 discrete layers of thickness $\lambda/200$. A multiple sine pulse is used as source wave. One can notice the pattern of destructive and constructive interference, developing inside the EL, due to the interference between the incoming and reflected pulse. No reflections are observed from the discrete layers making up the EL, which appears as a single homogeneous layer.

Epstein layers are very important, since there are many examples of media that can be approximated as EL's : e.g. the earth's core and atmosphere, the sea, any material plate having a stress (or texture) gradient, etc. Another important application of the CM approach at the one-dimensional level is given by composite materials, which in some cases can be treated as bilayers.

Several examples of application of the CM approach at the two-dimensional level are presented in ref. [23]. It is remarkable that, in the transition from one to two (or three) dimensions, the amount of computer time does not appreciably increase. In fact only the number of grid points (and therefore of processors) increases, but since all processors work simultaneously, the computing time is not affected. Here a very important application is acoustic tomography [30], which, for the characterization of structural materials, may be more suitable than x-rays tomography. In fact, for the NDE inspection of industrial materials, the ultrasonic propagation velocity and signal amplitude are the relevant quantities. However, the bending of rays in nonhomogeneous media and the directional dependence of the propagation velocity in anisotropic materials have so far presented insurmountable difficulties for a general treatment. The efficiency of the CM approach and its speed, due to the tremendous computer time savings provided by parallel processing, give us the expectation that a practical method may be found for acoustic tomography in anisotropic inhomogeneous media, without prohibitively extensive calculations or overly restrictive assumptions [31].

REFERENCES

1.   Noor, A.K., Appl. Mech. Rev. 44, 307 (1991).
2.   Lord, W., Ludwig, R. and Z. You, J. of Nondestr. Ev. 9, 129 (1990).
3.   Harumi, K. and M. Uchida, J. of Nondestr. Ev. 9, 81 (1990).
4.   Ungar A. and A. Ilan, J. Geophys. 43, 33 (1970).
5.   Rusanov, V.V., J. Comp. Phys. 5, 507 (1970).
6.   Boore, D.M., in: Methods in Computational Phys. (Ed.B.A. Bolt), Academic Press 1972, Vol. 11, 1.
7.   Alford, R.M., Kelly, K.R. and D.M. Boore, Geophys. 39, 834 (1974).
8.   Huang, C.H. and J.H. Cushman, J. Comp. Phys. 40, 376 (1981).
9.   Virieux, J., Geophys. 49, 1033 (1984).
10.   Terki-Hassaine, O. and E.L. Leiss, Int. J. Supercomp. Appl. 2, 49 (1988).
11.   Denning, P.J. and W.F. Tichy, Science 250, 1222 (1990).
12.   Courant, R., Friedrichs, K. and H. Lewy, Math. Ann. 100, 32-74 (1928).
13.   Hildebrand, F.B.: Methods of Applied Mathematics, Prentice Hall 1958.
14.   Carnahan, B., Luther, H.A., Wilkes J.O.: Applied Numerical Methods, J. Wiley 1969.
15.   Ames, W.F. : Numerical Methods for Partial Differential Equations, Academic Press. 1977.
16.   Press, W.H., Flannery, B.P., Teukolsky, S.A. and W.T. Vetterling: Numerical Recipes, Cambridge Univ. Press. 1986.
17.   Hildebrand, F.B.: Finite-Difference Equations and Simulation, Prentice Hall 1968.

18. Noye, J., in: Numerical Solutions of Partial Differential Equations, North-Holland 1982.

19. Siegmann, W.L., Lee, D. and G. Botseas, in: Computational Acoustics, Elsevier Sci. 91-109 , (1988).

20. Mikens, R.E., ibid., p. 387-93.

21. Eringen A.C. and E.S. Suhubi: Elastodynamics, Academic Press., 1974, vol. 1.

22. Achenbach, J.D. : Wave Propagation in Elasic Solids, North Holland, 1973.

23. Delsanto, P.P., Chaskelis, H.H., Mignogna, R.B., Whitcombe, T.V. and R.S. Schechter, to appear in the Proc. of the 18th "Annual Review of Progress in Quantitative NDE, Brunswick, July 1991.

24. Hillis, D.W.: The Connection Machine, The MIT Press, 1985.

25. Delsanto, P.P., Whitcombe, T., Chaskelis, H.H. and R.B. Mignogna, to appear in: Wave Motion, 1992.

26. Delsanto, P.P., H.H. Chaskelis, Whitcombe T. and R.B. Mignogna, in: Nondestructive Characterization of Materials (Eds. C.O. Ruud, R.E. Green and J.F. Bussiere), Plenum Press, 1991, Vol.4.

27. Delsanto, P.P., Whitcombe, T., Batra, N.K., Chaskelis, H.H. and R.B. Mignogna, in : Review of Progress in QNDE (Eds. D.O. Thompson and D.E. Chimenti), Plenum Press, Vol. 10A, 38, 241 (1991).

28. Chaskelis, H.H. and A.V. Clark, Mat. Eval. 38, 20 (1980).

29. Brekhovskikh, L.M.: Waves in Layered Media, Academic Press, 1980.

30. Devaney, A.I., in : Inverse Problems of Acoustic and Elastic Waves (Eds. F. Santosa, Y.H. Pao, W.W. Symes and C. Holland), SIAM 1984, 250.

31. Mignogna, R.B., Kline, R.A., Chaskelis, H.H., Schechter, R.S. and P.P. Delsanto: Work in progress.

SPECTROSCOPIC EVALUATION OF LAYERED SUBSTRATES

L. Adler

The Ohio State University, Columbus, OH, USA

ABSTRACT

In this paper an ultrasonic method to evaluate bond quality between a layer of substrate is suggested. The method uses guided modes in the layered substrate structure, rather than the conventional bulk waves. It is demonstrated that the interface quality between the layer and the substrate can be monitored by the dispersion behavior of the lowest mode's phase velocity. Both dissimilar and similar material combinations of layer and substrate with various bonding conditions are investigated.

1 INTRODUCTION

Generally a layered structure can sustain an infinite number of guided modes the so-called generalized Lamb mode, which is usually strongly dispersive. i.e. their velocity changes with frequency. As for thin layers, when the layer thickness is small compared to the wavelength of the interrogating ultrasonic wave, there is only one principal mode of practical interest, the so called dispersive Rayleigh mode (modified Rayleigh mode). At very low frequencies the thin layer has a negligible effect on the surface wave propagation and the principal mode behaves like a simple Rayleigh wave on the free surface of the substrate. At very high frequencies the substrate has less and less effect on the layer and the principal mode degenerates into the simple Rayleigh mode on the free surface of the layer material. In between the effect of the layer on the substrate can be readily investigated from the frequency dependent surface wave velocity. [1].As will be discussed in the following section, it has been demonstrated that the spectral behavior of the modified Rayleigh mode can play an important role in the evaluation of bond quality in layered substrates [2,3].

2 THEORY

The problem of wave propagation in a thin layer on half space with rigid boundary conditions has been addressed by several investigators [1]. It can be shown that for an isotropic case and when the wavelength is much larger than the layer thickness, the expression of the characteristic equation is identical to the classical Rayleigh waves at the surface on the half space. For kh=0 where k is the wave number and h is the layer thickness, this equation becomes:

$$[(V_R^2/V_{t1}) - 2]^2 - 4[1 - (V_R/V_{t1})^2]^{1/2}[1 - (V_R/Vl_{l1})^2 = 0 \qquad (1)$$

where V_R = Rayleigh wave velocity,
  V_{l1} = longitudinal wave velocity in the medium (half-space substrate),
and
  V_{t1} = shear wave velocity in the medium (half-space substrate).

Consequently, one of the solutions of the wave equation for a layered half-space degenerates the Rayleigh wave solution on an unlayered half-space when kh = 0.

It has been [1] shown that the initial slope of the dispersion curve of this mode is positive if

$$4[1 - (V^{t2}/V_{l2})^2] V_{t2}^2 > V_R^2[1 + \{1-(V_R/V_{l1})^2/1-(V_R/V_{t1})^2\}] \qquad (2)$$

where V_{l2} and V_{t2} are the longitudinal and shear wave velocities in the layer medium. This equation ca be simplified to [1]:

$$V_{t2}/V_{t1} > \{[1 - (V_{t1}/V_{l1})]^2/[1 - (V_{t2}/V_{l2})^2]\}^{1/2} \tag{3}$$

For real materials, Poisson's ratio is positive and smaller than 0.5. As a consequence, the longitudinal wave velocity is higher than $\sqrt{2}$ times the shear velocity and the extreme limits of the right hand side of eq.3 are $\sqrt{2}$ and $1/\sqrt{2}$, which means

$$V_{t2} = \sqrt{2}\, V_{t1} \text{ and } V_{t1} = \sqrt{2}\, V_{t2}. \tag{4}$$

In two regions the situation is quite obvious:

if $V_{t2} > \sqrt{2}\, V_{t1}$ it is stiffening
if $V_{t2} < V_{t1}/\sqrt{2}$ it is loading   (5)

"Loading" means that the initial slope is negative and the wave velocity begins to decrease as function of frequency below the Rayleigh velocity of the substrate. This is illustrated on figure 1 where a zinc oxide layer loads the silicon substrate.

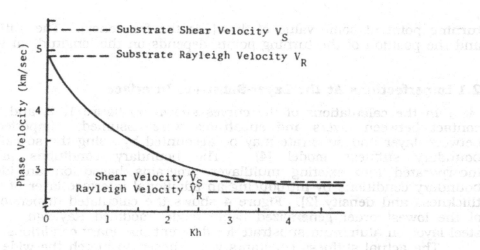

Figure 1: Dispersion curve for zinc oxide on silicon ($V_S < V_s$). The layer loads the substrate. (see ref. 1)

"Stiffening" means that the initial slope of the dispersion curve is positive and that the wave velocity begins to increase above the Rayleigh velocity of the substrate. This is shown on figure 2 where a silicon layer stiffens the zinc oxide substrate.

An interesting case of practical interest is when a thin layer of steel is bonded to aluminum or a thin layer of aluminum is bonded to a steel substrate. On figure 3a, and b. the dispersion curves are plotted for the cases [2]. The behavior of the lowest curve is interesting because it has a

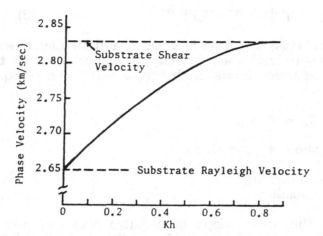

Figure 2: Dispersion curve for silicon on zinc oxide ($V_s > V_s$). The layer stiffens the substrate. (See ref. 1)

turning point at some value of fh. [f is the frequency]. The initial shape and the position of the turning points depends on the longitudinal velocities.

2.1 Imperfections At the Layer-Substrate Interface

In the calculations of the curves shown on figure 1, 2, and 3 a rigid contact between layers and substrates were assumed. Imperfect bonds between layer and substrate may be accounted by using the so called finite boundary stiffness model [4]. The boundary conditions are easily incorporated into existing multilayer programs based on an ideal rigid boundary conditions by introducing an additional interface layer of negligible thickness and density [2]. Figure 4 shows the calculated dispersion curves of the lowest order generalized Lamb mode [modified Rayleigh mode] of a steel layer on aluminum substrate for different boundary conditions.

The actual stiffness constants were chosen to match the wide range of experimental data [to be described later] obtained on inertia friction welded specimen. As can be expected at very low frequencies the interface looks perfect and the dispersion curves approximate the rigid bond case. As the frequency increases, the interface looks more and more loose and the dispersion curves approximate the completely delaminated free plate case. In between, the phase velocity seems to be sensitive to the boundary stiffness i.e. bond quality. Another example for imperfect layered-substrate is illustrated on figure 5. In this case both the layer and the substrate were selected to be silicon in order to emphasize the effect of the interface imperfections. The results are similar as in figure 4. At low frequencies and at high frequencies the phase velocity approaches the perfect bond and the free plate result respectively. In between the phase velocity has a minimum which is characteristic of the bond quality.

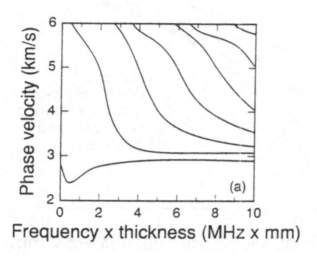

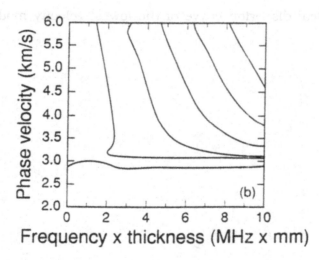

Figure 3: Theoretical dispersion curves for various guided modes. (a) Steel layer on aluminum substrate. (b) Aluminum layer on a steel substrate.

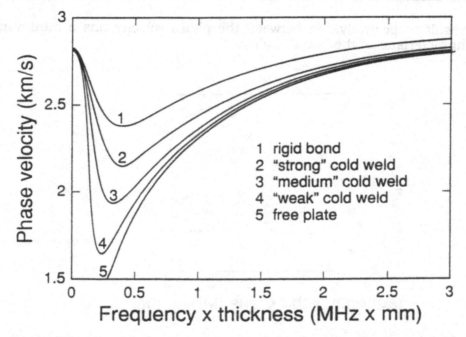

Figure 4: Theoretical disperion curve of the lowest-velocity mode for various bond qualities.

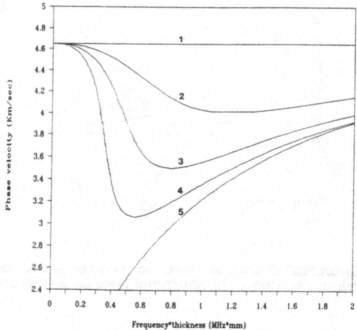

Figure 5: Calculated dispersion curves of the lowest order mode on an imperfect layered substrate: Similar Materials

3 EXPERIMENT

3.1 Sample Preparation

In this section the inertia friction welded aluminum-steel samples and the diffusions bonded silicon-silicon samples will be briefly described.

3.1.1 Inertia friction welded aluminum-steel samples

The aluminum-steel samples were prepared by standard inertia-friction solid-state welding process. The process and its results are described in details elsewhere [5]. The cylindrical samples were 75mm in diameter and 100mm in length before machining one of the materials, in order to obtain a thin layer (thickness < 400μm). The quality of the bond obtained using this process depends on various parameters: pressure, rotational speed, etc. In general, the quality of the bond is not uniform on the cylindrical sample. this quality decreases from the outside to the center, which corresponds to the axis of rotation and, consequently, to a zero relative speed.

3.1.2 Silicon Layer on Silicon Substrate

The layered substrate specimens were prepared by solid-state diffusion process. The quality of the bond was dependent on various factors but the variable in this study was the size (or periodicity) of the mesh. By intervening a mesh layer of a certain size, specimen of different percentage of unbonded area could be prepared. the thickness of the meshed intervening layer was only a very small fraction of the thickness of the top layer, and both dimensions were smaller than the ultrasonic wavelength used. Therefore, this intervening layer can be treated as an induced imperfect interfaces between layer and substrate.

For the reason of simplicity, both the layer and the substrate used in this study were made of the same material. Both the silicon layer and substrate were in mirror finish conditions. The silicon layer is 50mm in diameter and 0.283 mm in thickness. The meshed gold layer has the overall dimension of 25mm x25mm, and the size of mesh varies from 25 to 50 micron, and the thickness is only few microns. The components of a layered substrate are shown in Figure 6a with the side view of a bonded specimen which is shown in Figure 6b.

3.2 Experimental Techniques

As usual for Lamb waves, the dispersion curves are plotted, which is velocity versus the product fh. Two techniques have been used in order to measure the variations of the phase velocity versus frequency. The first one (figure 7a) is based on the existence of a minimum in the reflection coefficient when the Snell-Descartes law is satisfied for one of the waves generated in the sample, that is when

$$\sin\Theta = V_w/V(f) \qquad (6)$$

$$Sin\Theta = V_w/V(f) \tag{6}$$

where V_w is the velocity of sound in the surrounding water.

Components of the Specimen

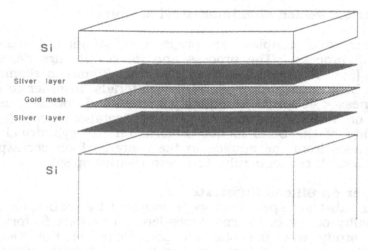

Side View of the Bonded Specimen

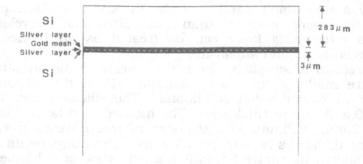

Figure 6:   (a)   The components of a diffusion bonded silicon layer on a silicon substrate. (b) Side view of the bonded sample.

An ultrasonic tone burst is reflected from the sample. The incidence Θ_i and reflection Θ_R angles are varied simultaneously in such a way that they remain equal to one another.   For each center frequency of the toneburst one measures the angles Θ_i where a minimum reflection occurs (Figure 7b). From the angles, using relation(6), one obtains the velocities V_i of waves generated at this frequency.   The center frequency of the tone burst is changed and another measurement is carried out. from each measurement

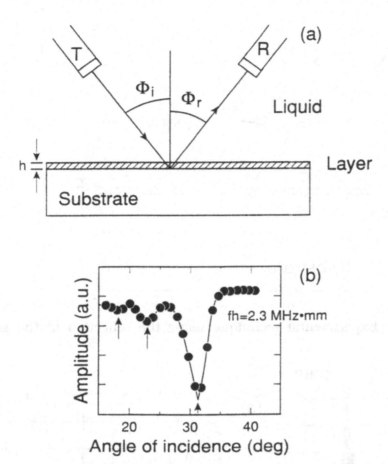

Figure 7: (a) Experimental technique using the minimum of the reflection coefficient. (b) Angular dependence of the reflected amplitude.

one obtains a set of points situated on the same vertical line in the dispersion curve.

The second experimental setup is shown in Figure 8a. It uses a single transducer broadband pulse technique. The wave is reflected by the right-hand corner of the sample, and because of the leaking property of this wave, part of its energy goes back towards the transducer. A typical example of the signal received is plotted in Figure 8b. The received signal is gated out and spectrum analyzed. The obtained frequency spectrum (figure 8c) exhibits maxima at frequencies f_i. The velocity V_i of the generated waves is given by [1]. One plots the dispersion curves as the points corresponding to frequencies f_i on the horizontal line V_i. Then by changing the angle of incidence Θ_i and do the same for various angles of incidence.

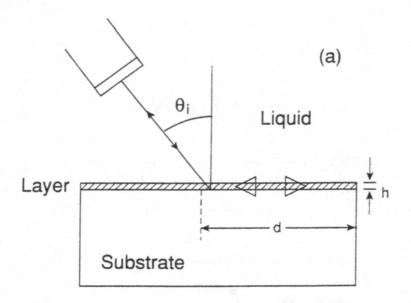

Figure 8: (a) Experimental technique using the reflection at the sample's edge.

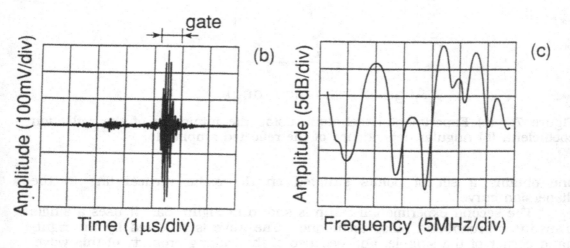

Figure 8: (b) Typical time signal obtained with the configurationof (a).
(c) Frequency spectrum of the previous signal [(b)].

Both techniques were used in this study to obtain the dispersion curves for bonded layer-substrate structures. Frequencies used in the experiments ranged form 0.4 MHz to 15 MHz.

4 COMPARISON BETWEEN THEORY AND EXPERIMENTS

4.1 Steel Layer on Aluminum

Figure 9 shows the comparison between experimental data and theoretical prediction. The dispersion curve of the generalized Lamb waves is for the case of steel layer on aluminum substrate for "good" bonding conditions. The significance of this result is the excellent agreement between theory and experiment for the lowest mode, especially for the low frequency region.

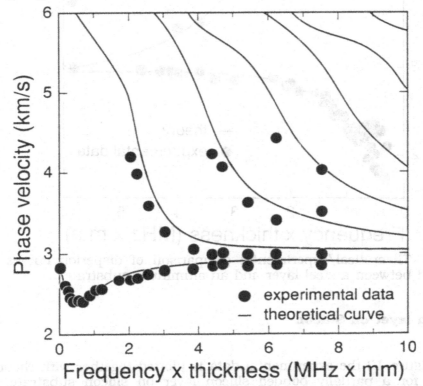

Figure 9: Theoretical/experimental comparison of dispersion curves for a "good" bond between a steel layer and an aluminum substrate.

For the sample with the "weak" bond, experimental data is plotted in Figure 10. the behavior of the lowest mode is quite different than for the "good" bond case. On figure 11 the lowest order measured generalized Lamb mode results are plotted together with theoretical results obtained from the two extreme boundary conditions. As can be seen the "good" bond corresponds to rigid boundary conditions and the "very weak" bond to free plate. For partial bond quality the data points are scattered between these two limits which may be related to theoretical results shown in fig. 4.

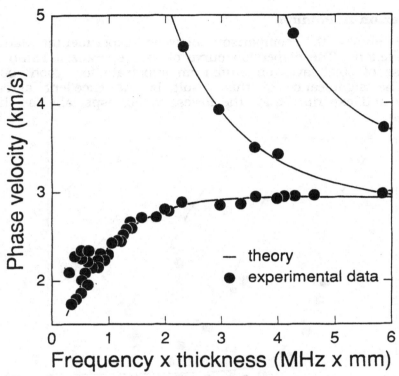

Figure 10: Theoretical/experimental comparison of disperion curves for a "weak" bond between a steel layer and an aluminum substrate.

4.2 Silicon layer on Silicon

On Figure 12 the experimental data is plotted together with theoretical predictions for a partially bonded silicon layer on silicon substrate. The dispersion curve is for lowest order generalized Lamb mode. The theoretical curve was calculated by using normal and transverse stiffness constants with values of $S_n = 1.22 \times 10^{14} N/m^3$ and $S_t = 9.6 \times 10^{13} N/m^3$. Results of a worse partial bond case are plotted on figure 13. In this case the normal and transverse stiffness constants are $S_n = 5.5 \times 10^{13} N/m^3$ and $S_t = 4.35 \times 10^{13} N/m^3$. As predicted by the analytical model, the shift of the minimum in the dispersion curve toward the lower frequencies indicate a less perfect bond between layer and substrate. These types of measurements can be used to obtain quantitative information about the bond quality of layered substrate systems.

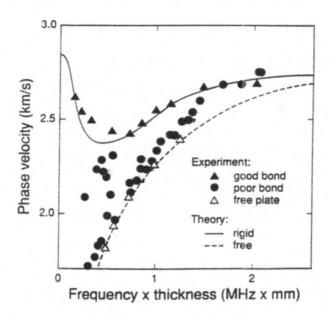

Figure 11: Experimental dispersion curve of the lowest-velocity mode for various bond qualities.

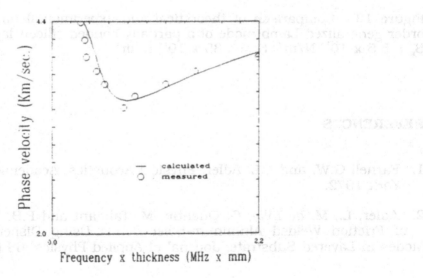

Figure 12: Comparison of theoretical and experimental data for the lowest order generalized Lamb mode of a partially bonded silicon layered substrate. $S_n = 1.22 \times 10^{14}$ N/m^3 $S_t = 9.6 \times 10^{13}$ N/m^3

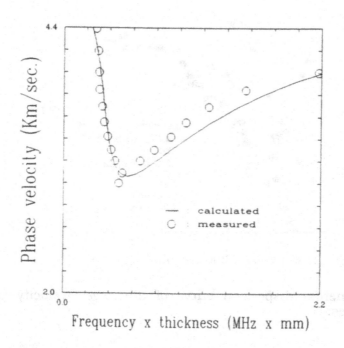

Figure 13: Comparison of theoretical and experimental data for the lowest order generalized Lamb mode of a partially bonded silicon layered substrate. $S_n = 5.5 \times 10^{13}$ N/m^3 $S_t = 4.35 \times 10^{13}$ N/m^3

REFERENCES

1. Farnell G.W. and E.L. Adler: Physical Acoustics, Academic Press, New York 1972.

2. Adler, L., M. de Billy, G. Quentin, M. Talmant and P.B. Nagy: Evaluation of Friction Welded Aluminum-Steel Bonds Using Dispersive Guided Modes in Layered Substrate, Journal of Applied Physics, 68 (1990), 6072.

3.   Ko, R., P.B. Nagy and L. Adler: Experimental Study of Interface properties Between Layer and Substrate, Review of Progress in Quantitative NDE, 11B (1992), 1967.

4. Baik, J.M. and R.B. Thompson: Ultrasonic Scattering from Imperfect Interface a Quasi-Static Model, Journal of NonDestructive Evaluation, 4 (1985), 177.

5. Nagy, P.B. and L. Adler: Ultrasonic Nondestructive Evaluation of Solid-State Bonds: Inertia and Friction Welds, Journal of Nondestructive Evaluation, 7 (1988), 199.

DETERMINATION OF THE MECHANICAL PROPERTIES OF COMPOSITES

R.B. Thompson
Iowa State University, Ames, IA, USA

ABSTRACT

Ultrasonic techniques to measure mechanical properties of composites are described. After a general introduction, detailed discussions are presented of two examples, measurements of anisotropic elastic constants and of porosity. The anisotropic elastic constants are inferred from the velocities of ultrasonic waves propagating in various directions. Porosity information is contained in attenuation or backscattered noise data, both a consequence of the scattering of ultrasound by the pores. Results of a variety of recent application studies are discussed.

1. INTRODUCTION

Modern composites involve a great diversity of materials. Included are the broadly used graphite-reinforced epoxy composites as well as more specialized materials such as metal-matrix, ceramic-matrix, and carbon-carbon composites. Each material system can, in turn, incorporate a variety of morphologies, involving unidirectional, woven, braided, knitted and stitched and filament wound fibers; particle reinforcements; and multiple-plies. These are designed to enhance the material's response in tension, compression or shear loads.

Because of the complex processing steps involved, it is possible for nominally identical materials to exhibit a range of mechanical properties. One of the roles of quantitative ultrasonics is to provide the information needed to ensure that appropriate properties have been achieved. Examples of properties of concern include anisotropic elastic moduli, porosity, integrity of fiber-matrix and inter-ply interfaces, fiber-resin ratios, degree of cure, residual stress, distributed damage, and discrete flaws [1]. The recovery of such information is complicated by the fact that the propagation of ultrasonic waves is influenced by the anisotropy of the media and that the intrinsic heterogeneity of the composites can lead to such complications as pulse distortion (frequency dependent velocity and attenuation) and low signal-noise ratio due to high attenuation and internal scattering.

In this set of notes, primary attenuation will be given to the problem of elastic constant and porosity measurements.

2. WAVE PROPAGATION IN ANISOTROPIC MEDIA

The first step in the determination of anisotropic elastic constants, or in the interpretation of a variety of other measurements, is the understanding of the influence of the material anisotropy on the propagation of ultrasonic waves. For any direction of propagation, in isotropic materials, there are three distinct plane waves which can propagate. These are a longitudinal wave, polarized in the direction of propagation, and a pair of shear waves, polarized perpendicular to the direction of propagation. In anisotropic materials, the situation is considerably more complex. For any given propagation direction, (defined as a normal to phase fronts) there are again three plane waves which can propagate, but these need not be polarized purely parallel or perpendicular to the propagation direction. Moreover, if a finite beam is excited by a transducer, the energy will not necessarily flow along the propagation direction but may "skew" in some other direction.

The mathematics describing this behavior is well known [2,3]. When one assumes a plane wave solution of the form

$$u_k = p_k \exp[i\omega(S_r x_r - t)] \tag{1}$$

one finds from the wave equation and constitutive relations that

$$(c_{ijkl} S_j S_l - \rho \delta_{ik}) p_k = 0. \tag{2}$$

Here $\vec{u}$ is the displacement vector, $\vec{p}$ is the polarization vector, $\vec{S}$ is the slowness vector, $\overset{\leftrightarrow}{c}$ is the elastic stiffness tensor, ρ is the density, and δ_{ik} is the Kronecker delta. For a given propagation direction $\hat{S}$, the slowness $|S|$ (reciprocal of phase velocity) is obtained by requiring the determinant of the left hand matrix of Eq. (1) to vanish. The direction of the polarization then equals that of the eigenvector $\vec{p}$.

Figure 1 schematically illustrates these results for the case of a zinc single crystal [2]. Parts (i) and (ii) respectively show the phase velocity and its inverse, the slowness, in the 1-3 plane. The group velocity $\overline{v}_g$ (speed and direction of energy flow) can be obtained by a graphical construction. The direction of the group velocity is equal to that of the normal to the slowness surface. Its magnitude is determined by the relation

$$v_g / v_p = (\cos \Psi)^{-1} \tag{3}$$

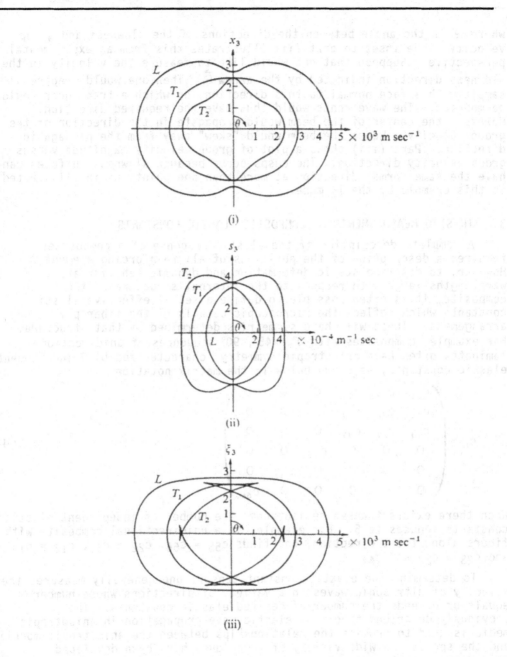

Fig. 1.   Zonal sections of (i) velocity, (ii) slowness and (iii) wave
           surfaces for zinc. Rotation about the â₃ direction in each
           case would produce the corresponding surface. The sketch
           beside (ii) indicates an experimental implementation. (after
           Ref. [2]).

where Ψ is the angle between the directions of the slowness and group
velocity. The inset to part (ii) illustrates this from an experimental
perspective. Suppose that one would like to measure the velocity in the
slowness direction indicated by the arrow $\vec{S}$. Then one would prepare a
sample with a face normal to this direction, on which a transducer would
be mounted. The wavefronts would thus have the required direction.
However, the center of the beam would propagate in the direction of the
group velocity. Hence the beam would "skew" away from the propagation
direction. Part (iii) shows a plot of group velocity magnitude versus
group velocity direction. The cusps occur because slowness surfaces can
have the same normal direction at more than one point, as in illustrated
in this example by the T_2 mode.

3.   IN-SITU MEASUREMENTS OF COMPOSITE ELASTIC CONSTANTS

A complete description of the elastic response of a composites
requires a description of the positions of all reinforcing elements.
However, to describe static deformation and dynamic behavior at
wavelengths large with respect to the internal structure of the
composite, it is often possible to define a set of effective elastic
constants which reflect the microscopic details of the fiber ply
arrangement. These will have symmetries determined by that structure.
For example, composites with $0°$, $\pm 45°$, $90°$ sequences of unidirectional
laminates often have orthotropic symmetry, characterized by 9 independent
elastic constants, as shown below in the matrix notation

$$c_{IJ} = \begin{pmatrix} c_{11} & c_{12} & c_{13} & 0 & 0 & 0 \\ c_{12} & c_{22} & c_{23} & 0 & 0 & 0 \\ c_{13} & c_{23} & c_{33} & 0 & 0 & 0 \\ 0 & 0 & 0 & c_{44} & 0 & 0 \\ 0 & 0 & 0 & 0 & c_{55} & 0 \\ 0 & 0 & 0 & 0 & 0 & c_{66} \end{pmatrix}. \tag{4}$$

When there exists transverse isotropy, the number of independent elastic
constants reduces to 5. For example, in a unidirectional composite with
fibers along the 1 direction, one finds $c_{55} = c_{66}$, $c_{22} = c_{33}$, $c_{12} = c_{13}$
and $c_{23} = c_{22} - 2 c_{44}$.

To determine the elastic constant tensor, one generally measures the
velocity of ultrasonic waves in a variety of directions whose number
equals or exceeds the number of desired elastic constants. The
previously described theory of elastic wave propagation in anisotropic
media is used to provide the relationships between the anisotropic moduli
and the speeds. A wide variety of techniques have been developed,
depending on whether one has access to one or two sides of the sample and
the type of wave used in the measurement. Here, examples selected from
results presented in recent conferences will be reviewed, as described in
the proceedings. Many have, or will be, written up in greater detail in
the reviewed literature, and the reader is encouraged to look there for
more detail.

3.1.Two sided access

The determination of elastic constants is probably simplest when one has access to both sides of the sample. In the most common approach, one measures the velocities of bulk waves propagating at various directions through the sample [4-7]. In some cases, transducers are placed on both sides of the sample. In others, a reflector is placed on one side so that the wave passes through the sample twice before being detected in a pulse-echo mode. Figure 2 illustrates two configurations that use piezoelectric transducers in an immersion mode [4,5].

For purposes of error reduction, it is often desirable to obtain more data than the number of unknown elastic constants. It is then possible to determine the constants by a least square fitting procedure wherein the elastic constants are varied to minimize the difference between calculated and observed wave speeds. These procedures have the advantage are reducing the effects of noise on individual data points. As an example, Table I presents a set of predicted elastic constants, along with statistical estimates of errors, on two unidirectional composites [6]. The data were obtained with a point-source, point-receiver configuration, as distinct from those illustrated in Fig. 2. Note the large uncertainties in the prediction of c_{12}. This is typical of the ultrasonic techniques, which predict some elastic constants with much more accuracy than others.

It is sometimes more convenient to measure the characteristics of guided waves propagating in the composite plate rather than bulk waves propagating through its thickness. Such techniques involve the theory of Lamb waves in anisotropic plates [3]. An example may be found in Ref. [8].

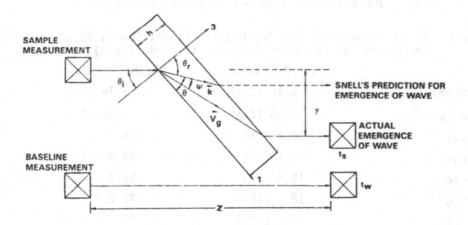

Fig. 2.   Determination of anisotropic elastic constants with two-sided access.
(a) Single transmission method (after Ref. [4]).

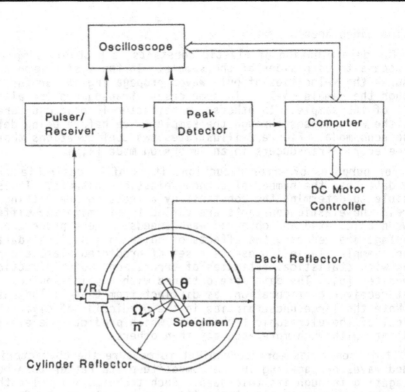

Fig. 2.   Determination of anisotropic elastic constants with two-sided
           access.
           (b) Double transmission method (after Ref. [5]).

Table I.   Elastic Constants for Two Unidirectional Composite as
           Determined by Two Sided Access. (after Ref. [6]).

	Glass/Polyester	Graphite/epoxy
h(mm)	6.38 (0.1%)	3.76 (6%)
ρ (g/cm^3)	1.90 (0.5%)	1.56 (6%)
c_{11}	56.4 (2%)	140 ---
c_{33}	42.1 (7%)	90.8 ---
c_{44}	12.8 (4%)	22.7 ---
c_{66}	18.6 (2%)	83.2 ---
c_{12}	10.2 (18%)	69.0 ---
c_{23}	16.5 (5%)	45.4 ---

all c_{ij} in GPA.
fibers in 1-direction, sample normal in 3-direction

3.2.Single-sided access

The large size of composite panels or other geometrical complexities may render single-sided access difficult in practical situations. A variety of techniques have also been developed for this purpose. The apparatus shown in Fig. 2b can also be used, since the cylindrical reflector allows a double reflection mode of operation. This can be used to infer moduli from the reflection properties of bulk or Lamb waves [9]. Figures 3a and 3b show geometries which excite Rayleigh and surface skimming longitudinal waves for single-sided measurements [10]. Table II presents values of the elastic constants measured by this technique at 0.5 MHz, as compared with values obtained from ultrasonic measurements made at various frequencies on cubes cut from the sample. Two interesting points can be noted. First, the frequency dependence of the data obtained on the cubic sample shows a dispersion that marks the beginning of a breakdown of the effective elastic constant notion discussed previously. Second, the varying leverage of the data on different elastic constants, as noted in the discussion of Table I, is manifest here by the inability to obtain values for two of the constants.

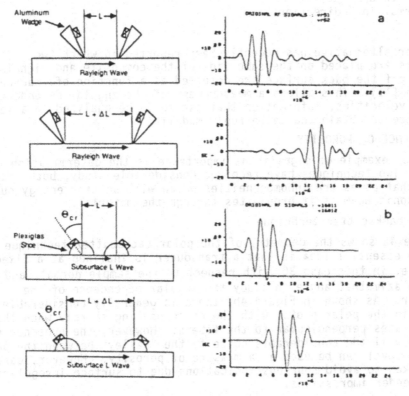

Fig. 3.   Determination of anisotropic elastic constants with one-sided access: Techniques and typical signals.
           (a) Surface waves (after REf. [10]).
           (b) Sub-surface longitudinal waves (after Ref. [10]).

Table II. Elastic Constants for Orthotropic Composite as Determined by
One-side Access at 0.5 MHz and Compared to Measurements Made at
Various Frequencies on Cube Shaped Samples. (after Ref. [10])

	One Sided Technique	0 MHz	.1 MHz	1.0 MHz	2.0 MHz
c11	97.4	102.6	101.5	91.5	79.5
c22	27.2	38.3	37.4	29.2	20.4
c33	11.3	10	10	10	10
c44	3.4	3.1	3.1	3.1	3.1
c55	4.5	4.2	4.2	4.2	4.2
c66	***	19.8	20	20	20
c12	***	26.7	24.8	19.3	11.5
c13	25.5	11.9	11.8	12.2	12.6
c23	12.6	5.2	5.6	5.7	4.4

all moduli in GPA
sample normal in 3-direction

Yet another alternative has recently been reported in which two
transducers are placed on the same side of the composite and signals
reflected off the back surface are detected as a function of transducer
separation [11]. This allows a measurement of the angular dependence of
bulk wave velocities, information that can again be analyzed in a least
square sense to obtain the anisotropic moduli.

4.   INFERENCE OF POROSITY

Another example of significant importance is the determination of
porosity. Two techniques have received considerable study, both based on
the idea that pores are inhomogeneities which will scatter energy out of
the ultrasonic beam as it propagates through the material.

4.1 Polar backscatter technique

Figure 4a shows the geometry of the polar backscatter technique
[12]. The essential idea is that a transducer is inclined at a fixed
polar angle, in this case 30° with respect to the sample normal, and then
scanned in azimuthal angle to study the angular dependence of the
backscatter. As shown in Figure 4b, there is generally considerable
structure in the polar plots, with the noise peaking sharply when the
beam propagates perpendicular to the fibers. However, the presence of
anisotropy will add considerable noise to the "valley" between the peaks.
This noise level can be used as a measure of porosity. However, care
must be taken to avoid spurious indications due to surface irregularities
such as bleeder impressions.

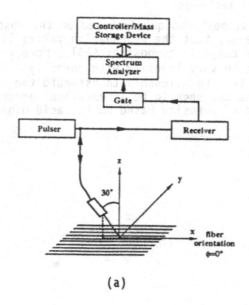

(a)

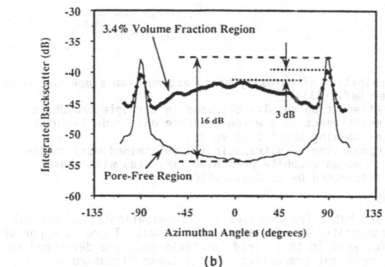

(b)

Fig. 4. Determination of porosity using polar backscattering technique (after Ref. [12]).
(a) Experimental configuration
(b) Typical data

4.2 Attenuation slope technique

The attenuation slope technique is based on the observation, illustrated in Figure 5a, that the attenuation varies linearly with frequency over a wide range in composites [13]. Moreover, this slope, dα/df, has been found to vary linearly with porosity. Figure 5b shows the results of using this relationship to estimate the porosity. The ultrasonic predictions are seen to be in excellent agreement with independent estimates of porosity based on the acid digestion technique.

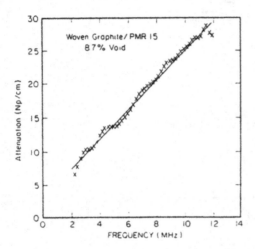 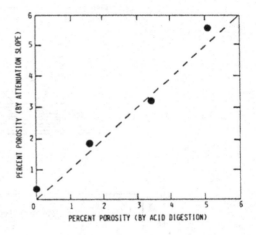

Fig. 5.   Determination of porosity using attenuation slope technique
           (after Ref. [13]).
           (a) Attenuation results obtained in a single broadband
                measurement on a woven graphite polyimide laminate
                containing about 8.7% voids.
           (b) Comparison of ultrasonically determined void content
                in woven graphite polyimide laminates with those
                determined by acid digestion.

5.   CONCLUSIONS

The above are but a few examples of determination of mechanical properties of composites by ultrasonic techniques. There is opportunity for much fruitful work in this field, as techniques are developed to estimate other important properties such as those discussed in the introduction.

6.   REFERENCES

 1. Sachse, W., Towards a Quantitative Ultrasonic NDE of Thick
    Composites, in: Review of Progress in Quantitative Nondestructive
    Evaluation, Vol. 10B, (D. O. Thompson and D. E. Chimenti, Eds.),
    Plenum Press, New York, 1991, 1575-1582.

 2. Musgrave, M. J. P., Crystal Acoustics, Holden-Day, San Francisco,
    1970.

3. Auld, B. A., Acoustic Fields and Waves in Solids, John-Wiley & Sons, New York, 1973.

4. Pearson, L. H. and Murri, W. J., Measurement of Ultrasonic Wavespeeds in Off Axis Directions of Composite Materials, in: Review of Progress in Quantitative Nondestructive Evaluation, Vol. 6B (D. Thompson and D. E. Chimenti, Eds.) Plenum Press, New York, 1987, 1093-1101.

5. Rokhlin, S. I., and Wang, W. Ultrasonic Evaluation of In-Plane and Out-of-Plane Elastic Properties of Composite Materials, in: Review of Progress in Quantitative Nondestructive Evaluation, Vol. 8B (D. O. Thompson and D. E. Chimenti, Eds.), Plenum Press, New York, 1989, 1489-1496.

6. Castagnede, B. and Sachse, W. Optimized Determination of Elastic Constants of Anisotropic Solids from Wavespeed Measurements, in: Review of Progress in Quantitative Nondestructive Evaluation, Vol. 8B, (D. O. Thompson and D. E. Chimenti, Eds.), Plenum Press, New York, 1989, 1855-1862.

7. Mignogna, R. B. Ultrasonic Determination of Elastic Constants from Oblique Angles of Incidence in Non-Symmetry Planes, in: Review of Progress in Quantitative Nondestructive Evaluation, Vol. 9B (D. O. Thompson and D. E. Chimenti, Eds.), Plenum Press, New York, 1990, 1565-1572. Also Determination of Elastic Constants of Anisotropic Materials From Oblique Angle Ultrasonic Measurements I: Analysis in: Vol. 10B, 1991, 1669-1675.

8. Rokhlin, S. I., Wu, C. Y., and Wang, L. Application of Coupled Ultrasonic Plate Modes for Elastic Constant Reconstruction of Anisotropic Composites, in: Review of Progress in Quantitative Nondestructive Evaluation, Vol. 9B (D. O. Thompson and D. E. Chimenti, Eds.) Plenum Press, New York, 1990, 1403-1410.

9. Rokhlin. S. I. and Chimenti, D. E., Reconstruction of Elastic Constants from Ultrasonic Reflectivity Data in a Fluid Coupled Composite Plate, in: Review of Progress in Quantitative Nondestructive Evaluation, Vol. 9B (D. O. Thompson and D. E. Chimenti, Eds.) Plenum Press, New York, 1990, 1411-1418.

10. Rose, J. L., Huang, Y., Ditri, J. J., Dandekar, D. P., and Chou, S. S., One Sided Inspection for Elastic Constant Determination of Advanced Materials, in: Review of Progress in Quantitative Nondestructive Evaluation, Vol. 10B, (D. O. Thompson and D. E. Chimenti, Eds.) Plenum Press, New York, 1991, 1469-1476.

11. Minachi, A. and Hsu, D. K., Determination of Elastic Constants Using Acousto-Ultrasonic Technique, in: Review of Progress in Quantitative Nondestructive Evaluation, Vol. 11B (D. O. Thompson and D. E. Chimenti, Eds.) Plenum Press, New York, in press.

12. Handley, S. M., Miller, J. G., and Madaras, E. I., Effects of Bleeder Cloth Impressions on the Use of Polar Backscatter to Detect

Porosity, in: Review of Progress in Quantitative Nondestructive
Evaluation, Vol. 8B (D. O. Thompson and D. E. Chimenti, Eds.) Plenum
Press, New York, 1989, 1581-1588.

13. Hsu, D. K., Ultrasonic Measurements of Porosity in Woven Graphite
Polyimide Composites, in: Review of Progress in Quantitative
Nondestructive Evaluation, Vol. 7B (D. O. Thompson and D. E.
Chimenti, Eds.) Plenum Press, New York, 1988, 1063-1068.

CHARACTERIZATION OF POROUS MATERIALS BY ULTRASONIC SPECTROSCOPY

L. Adler

The Ohio State University, Columbus, OH, USA

ABSTRACT

The problem of determining pore size and pore concentration in cast and weld material as well as in composites are of technological importance. In this paper analytical and experimental work is presented to correlate ultrasonic attenuation due to porosity in a solid body. From the spectral behavior of the attenuation coefficient both pore size and pore concentration ca be determined. The method is applicable if the total pore volume is less than 6% and if grain scattering is negligible.

1 INTRODUCTION

Porosity in structural materials limits their ultimate strength and hence their utility. Consequently nondestructive methods of characterizing the amount and nature of the porosity are important for this and other reasons. The number of pores as well as their distribution in size, shape and orientation are quantities which an ideal characterization method would determine. ultrasonic methods are particularly good candidates since they are sensitive to these parameters. One of the fundamental measurements in ultrasonics is the attenuation of an incident beam or pulse. The spectral behavior of the ultrasonic attenuation is sensitive to the volume fraction and average pore size.

Recently several studies which discuss the porosity induced ultrasonic attenuation have appeared [1-4]. Porosity assessment by ultrasonic attenuation measurement involves two principal problems: first, how to relate the porosity induced ultrasonic attenuation to porosity parameters such as average pore radius and volume fraction, and second, how to separate the sought porosity induced attenuation from other components contributing to the actually measured total attenuation. In this paper these questions will be addressed and ultrasonic porosity assessment will be described in aluminum cast material.

2 THEORY

2.1 Ultrasonic Scattering in Porous Materials

Scattering of ultrasound from porosity will be described in terms of the following model. Consider an isotropic elastic space which is uniform and everywhere homogeneous except between two parallel planes. Between the two planes (the first at $z=0$ and the second at $z=z_0$) there is assumed to be a spatially uniform distribution of spherical voids (pores). All of the pores are assumed to have the same radius, a_p. The validity of this latter approximation will be discussed in the text.

An idealized scattering experiment is modeled as follows. A longitudinally polarized plane wave is incident along the z-axis which is normal to the planes bounding the porous region. A portion of the incident wave is transmitted coherently. Fundamental measurements are made to determine the reflection and transmission. Clearly both depend on the frequency dependent attenuation of the incident pulse as it propagates through the porous region. Figure 1 shows a schematic of the idealized experimental geometry.

The description of the scattering and attenuation is greatly simplified in the limit that the volume fraction of pores c becomes small.[4] This allows multiple scattering to be ignored. The scattering amplitude becomes just the coherent sum of the amplitudes generated at each pore. The volume fractions measured in our experiments varied between 0% and 6% and c << 1.

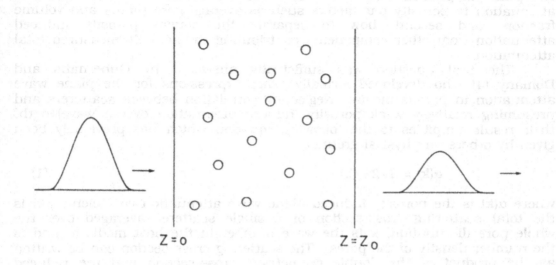

Figure 1: Schematic representation of the idealized experimental situation.

The attenuation $\alpha(k)$ describes the power scattered out of the main beam as a function of the frequency. On the other hand, the power scattered by each pore is described by its total cross section $\gamma(k)$. In the small volume fraction limit, one may thus expect the attenuation to be proportional to the scattering cross section. Here k is the magnitude of the longitudinal wave vector where $k = 2\pi f/V_L$; f is the frequency and V_L denotes the longitudinal velocity.

The attenuation, reflection, and transmission can be understood qualitatively by considering the low and high frequency limits. For long wavelengths, $ka_p << 1$, the power scattered per pore vanishes as frequency to the fourth power $[\alpha(k) = k^4]$. Consequently, the sample is relatively transparent and little ultrasound is reflected. At high frequencies $(ka_p >> 1)$ scattering can be described by ray concepts and the geometric cross-section of the flaw. For $ka_p >> 1$, the attenuation, reflection, and transmission become relatively independent of frequency, and the sample becomes more opaque compared to lower frequencies.

The transition from long to short-wavelength behavior typically occurs when $ka_p = 1$. Since the transition is relatively abrupt, measurements of the transmission or reflection can be used to determine the pore size. Further, as is discussed later, the overall strength of the signal can then be used to find the volume fraction of pores.

2.1 Porosity Induced Attenuation

Porosity assessment by ultrasonic attenuation measurement involves two principal problems: first, how to relate the porosity induced ultrasonic attenuation to porosity parameters such as average pore radius and volume fraction, and second, how to separate the sought porosity induced attenuation from other components contributing the actually measured total attenuation.

The first question was sufficiently answered by Gubernatis and Domany [4] who developed formally exact expressions for the plane wave attenuation in porous media. Neglecting correlation between scatterers and presuming relatively weak porosity induced attenuation over a wavelength, their result simplifies to the following equation which has previously been given by others on physical grounds:

$$\alpha(k) = 1/2n\gamma(k) \qquad (1)$$

where $\alpha(k)$ is the porosity induced plane wave attenuation coefficient, $\gamma(k)$ is the total scattering cross-section of a single scatterer averaged over the whole pore distribution, k is the wave number in the host medium, and is the number density of the pores. The scattering cross-section can be written as the product of the double geometrical cross-section and the reduced scattering cross-section $\Gamma(ka_p)$

$$\delta(k) = 2a^2{}_p\pi\Gamma(ka_p) \qquad (2)$$

where a_p denotes the pore radius. The reduced scattering cross-section $\Gamma(ka_p)$ depends on the host medium only through its Poisson ratio, which makes it possible to determine the pore radius a_p from the shape of the attenuation coefficient versus frequency curve. Subsequently, the pore density can be calculated from the scattering induced attenuation by using Eq. 1. Gubernatis and Domany[4] showed that this simple data reduction technique yields accurate porosity parameters for spherical voids of peaked size distribution and less than 5% volume fraction, and a recent comprehensive experimental study [3] confirmed these predictions.

The second problem mentioned is how to measure the porosity induced attenuation coefficient $\alpha(k)$. The total measured attenuation loss L from the porous sample as determined by comparing the received front surface frequency spectrum A_f to the backwall echo frequency spectrum A_b, shown schematically in Figure 2. The total attenuation contains many factors:

$$L = Ln\ (A_f/A_b) = L_{IMP} + L_{DIFF} + L_{GRAIN} + L_{SURF} + L_p \qquad (3)$$

L_{IMP} and L_{DIFF} are due to the double transmission loss at the liquid-solid interface and to ultrasonic beam spread as it propagates back and forth inside the sample, respectively. These two factors can be readily calculated from known parameters or completely eliminated by comparing the backwall echo from the porous sample to one from a similar porosity-free sample. The loss due to grain scattering L_{GRAIN} depends on the anisotropic elastic properties of the host material and the grain size as well. There seems to be no feasible correction for this term, and the ultrasonic porosity assessment techniques are limited to applications where the grain scattering induced attenuation is negligible with respect to the porosity induced component L_p. Attenuation loss due to rough surface L_{SURF} will be described in a later section. From Eq 3, the only remaining term is the attenuation loss due to porosity L_p which relates the porosity induced attenuation to the ultrasonically measured attenuation.

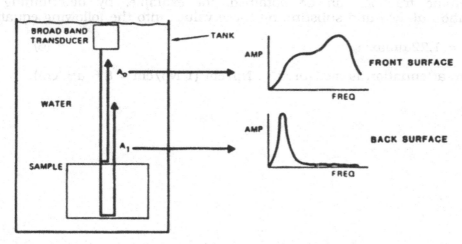

Figure 2:   Schematic diagram of the ultrasonic attenuation measurement.

2.2 Characterizing the Pore Distribution

The volume fraction and pore radius for porosity can be determined once the porosity induced attenuation is measured. The porosity induced attenuation coefficient $\alpha(k)$ is determined by dividing the porosity induced attenuation by twice the sample thickness d, as shown by eq. 4.

$$\alpha(k) = (L_p/2d) \qquad\qquad (4)$$

The resultant attenuation coefficient was plotted as a function of frequency as shown graphically in Figure 3a.

Determination of the pore size and volume fraction is suggested by the work of Gubernatis and Domany[4]. It is approximate and suitable when the pore size distribution is peaked. To make the method as concrete as possible, the theoretical frequency dependent attenuation is shown for the A357 alloy in Figure 3a. A volume fraction of 2% and a pore radius of

150µm are chosen. Figure 3b shows the result of dividing the attenuation coefficient by the frequency. There is a single large peak near $ka_p = 1$. The occurrence of this peak is characteristic of isotropic materials. The location of the peak determines the pore size. First one measures the value max at which the peak occurs. Then the size is determined from Figure 3b. For the example case of Al A357 alloy, the figure indicates that $a_p = 1.05/max$.

In principle, any measurement of the attenuation determines the volume fraction once the pore radius is known. Combining Eqs. 1 and 2 and the definition of the volume fraction, one finds

$$c = 4/3[\alpha(k)a_p/\Gamma(ka_p)] \tag{5}$$

The volume fraction can be obtained, for example, by determining the attenuation $\alpha(max)$ and substituting these values into the following equation:

$$c = 1.22\alpha(max)a_p \tag{6}$$

Here the attenuation is measured in Np/cm (1 Np/cm = 8.6 dB/cm).

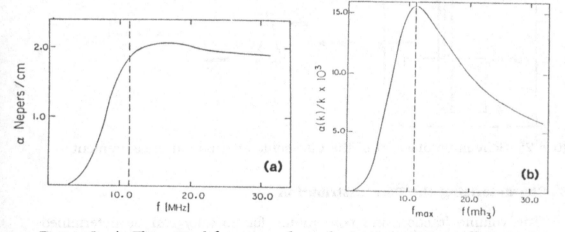

Figure 3: a) Theoretical frequency-dependent attenuation coefficient.
        b) Theoretical frequency-dependent attenuation coefficient divided by frequency.

3 ULTRASONIC POROSITY ASSESSMENT

In this section the attenuation measurements made on the cast aluminum alloy samples are discussed. The measurements are based on the use of an ultrasonic spectroscopy system which is first described. After the experimental procedures are described, the results of the porosity measurements are given.

3.1 Ultrasonic Spectroscopy System

The ultrasonic spectroscopy system, which is schematically displayed in Figure 4, is assembled around a PDP 11/34A minicomputer. A broad bandwidth ultrasonic pulse is produced by exciting an untuned ceramic transducer with a fast rise-time, high voltage pulse (the transducer bandwidth is from 2 to 20 MHz). Reflected signals are received by the same transducer (pulse-echo configuration). The electrical pulse generated by the received waveform is filtered and amplified. The time domain signal can either be fed to a conventional spectrum analyzer or it can be sampled and converted into digital data to be processed by a computer. For processing by the spectrum analyzer, a stepless gate is used to select apportion of the received signal. The receiver output, as well as the gated waveform, is displayed on an analog spectrum analyzer.

For conversion to digital data a high-speed transient recorder is used to store the signal amplitude at discrete times in its digital memory. The computer controls the acquisition of the ultrasonic pulse data and then transfers the digitally represented signal from the recorder to the minicomputer memory. The rf signal data may be permanently recorded on floppy disks. Processing of the ultrasonic signal is performed on the minicomputer and includes the following operations: gating, auto correlation, averaging, Fourier Transform (Fast Fourier Transform procedure), deconvolution, and plotting. Plots in the time domain, as well as in the frequency domain, are displayed on a graphics terminal and accompanying hard copy unit.

3.2 Attenuation measurements

The attenuation of the ultrasonic waves in a porous sample is given as

$$\alpha(k) = 1/(z-z_1) \ln [A(k,z)/A(k,z_1)] \qquad (7)$$

Here $A(k,z)$ represents the amplitude at the front surface of the sample (z), and $A(k,z_1)$ represents the amplitude at the back surface of the sample (z_1). These amplitudes are corrected for beam divergence using diffraction corrections for $A(k,z)$ and $A(k,z_1)$. Corrections are also made for the liquid-solid interface losses. The attenuation coefficient $\alpha(k)$ is then obtained by dividing the front surface spectrum by the backwall echo spectrum as shown in Figure 5. The resultant attenuation coefficient versus frequency curve is also shown in Figure 5.

Determination of the pore radius was made by dividing the frequency dependent attenuation coefficient by its frequency, thus approximating a single peak near $ka_p = 1$. The value of f_p determined from the $\alpha(k)/f$ graph was then substituted into the following equation

$$a_p = 1.5 \times V_L/(2\pi \times f_p) \qquad (8)$$

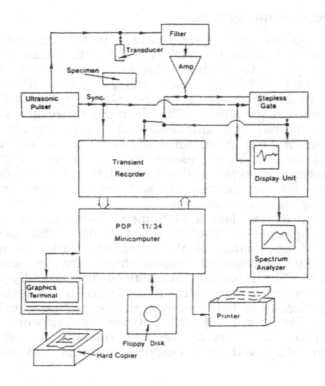

Figure 4: Schematic diagram of the ultrasonic spectroscopy system.

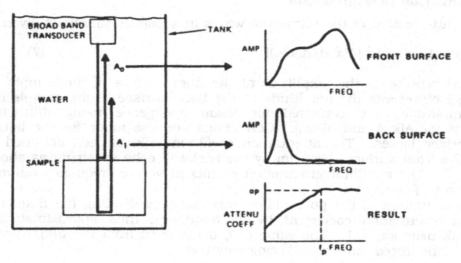

Figure 5: Illustration of the ultrasonic porosity assessment technique.

where V_L is the longitudinal velocity of each sample. The volume fraction of porosity was calculated by substituting the value of a_p obtained from Eq. 9. and $\alpha(k_P)$, the value of the attenuation coefficient located at f_p into eq. 10.

$$c = 1.22 \times \alpha(k_P) \times a_p \tag{9}$$

3.3 Comparison of Theory and Experiment

Finally we would like to present some results which compare theoretical prediction of the attenuation coefficient versus frequency with experimental results Figures 6,7, and 8 are sample #1820, #1520, and #1920, respectively. Theoretical estimates of the frequency dependent attenuation were obtained from Eq. 2 and by calculating the reduced scattering cross-section $\Gamma(ka_p)$ for A357 aluminum alloy. The agreement between theory and experiment is reasonably good for each case. The unusual low-frequency behavior below 2 MHz of the experimental data is due to the lack of transducer response in this region.

Other similar ultrasonic results for the different samples are listed on Table 1 together with estimates for the volume fraction of porosity based on weight-density measurements. The results are encouraging for most cases. that is, the ultrasonic method agrees well with the porosity determinations based on density measurements in most cases. An exception, for example, is sample #1820 where the error is over 30%. Detailed studies indicate that the pores are not spherical in the sample, and they are somewhat oriented, too. Therefore, the anisotropic shape of the flaws plays a major role in the measured porosity induced attenuation.

3.4 Rough Surface Corrections

We have investigated the feasibility of correcting for the surface roughness induced attenuation component by subtracting its estimated value from the measured attenuation. The surface roughness induced attenuation usually cannot be measured directly because of the lack of a porosity free sample of the same surface quality as the piece under investigation. On the other hand, the surface roughness induced attenuation can be calculated easily by taking advantage of the strict relation between the front and backwall attenuations. We have shown that the surface roughness attenuates the reflected and transmitted components in a similar way, and their attenuation ratio is independent of frequency, rms roughness, or even surface profile. Obviously, the same is true for their difference, e.g. for aluminum immersed in water; the sought surface roughness induced attenuation of the whole sample is negative and about 2/3 of the readily measured surface roughness induced attenuation of the frontwall echo.

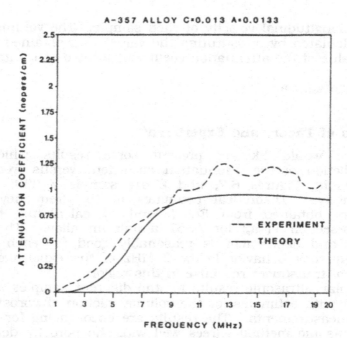

Figure 6: Sample 1520.

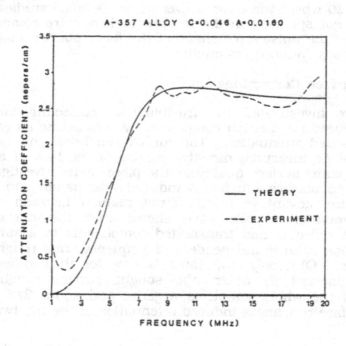

Figure 7: Sample 1920

TABLE 1: Summary of ultrasonic results of cast samples.

Sample	Attenuation Coefficient (N$_p$/cm)	Frequency (MHz)	K$_o$ (1/cm)	Pore Radius (µm)	Porosity(%) Exp.	Density
Al 013	0.39	15	142	70	0.24	...
Al 1011	0.22	12	114	87	0.17	0
Al 1210	0.20	8	76	131	0.23	0.22
Al 1410H	1.02	13	123	80	0.74	1.3
Al 1410L	0.20	10	95	105	0.19	1.3
Al 1510	1.12	9	85	206	2.41	2.18
Al 1810	1.00	10	95	105	1.42	0.9
Al 1820	1.25	12	114	145	1.19	1.05
					1.57	
					1.57	
Al 1830	1.20	11.5	109	132	1.33	1.2
Al 1850	1.00	11	104	136	1.33	1.08
Al 1920	2.85	8	76	160	4.5	4.6
Al 0000	0.1				0	0.16
Al (Air cast)	1.00	4	36	528	5.8	5.0

Results were obtained by using a smooth front reference signal as a bias of comparison for porosity assessment of rough surface samples. To illustrate this point, aluminum samples having a surface roughness of 20µm rms were ultrasonically measured and compared with a smooth surface sample. The received front and backwall amplitudes are showing Figure 8, respectively.

The attenuation coefficient was then calculated and plotted as a function of frequency for the following combinations: (1) smooth front/smooth backwall, (2a) rough front/rough backwall, and (2b) smooth front/rough backwall, as shown in Figure 9. For each combination, the volume fraction of porosity was calculated and is listed in Table 2. By disregarding the rough front surface and comparing the smooth surface reference to the rough backwall, the measured volume fraction of porosity for the rough surface sample was properly estimated.

The porosity assessment results were improved by using a smooth frontwall reference echo, however, a slight overestimation of the volume fraction of porosity occurs using this technique. It is our opinion that some overestimation of porosity is more favorable than underestimation and, for ease of analysis, the latter method should be used. To avoid either overestimation or underestimation of the volume fraction of porosity, a more accurate correction for surface roughness was considered. In order to do this, we can take advantage of the fact that the surface roughness induced attenuation of the frontwall echo is three times higher than that of the backwall echo (this is a material constant independent of frequency, surface profile, or rms roughness). As opposed to the backwall echo, the surface roughness induced attenuation of the frontwall echo can be measured easily, and the sought correction L_{SURF} will be minus 2/3 of the measured attenuation.

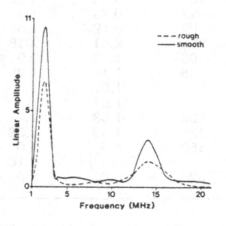 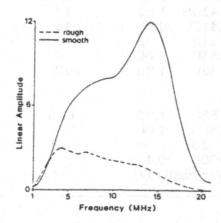

Figure 8: Amplitude spectra of front surface and backwall echoes.

TABLE 2: Porosity assessment results for rough surface.

	Pore Radius	Volume Fraction
1. Smooth Surface	250 μm	4.1%
2. Rough Surface		
a. Rough Reference	280 μm	2.8%
b. Smooth Reference	250 μm	4.3%

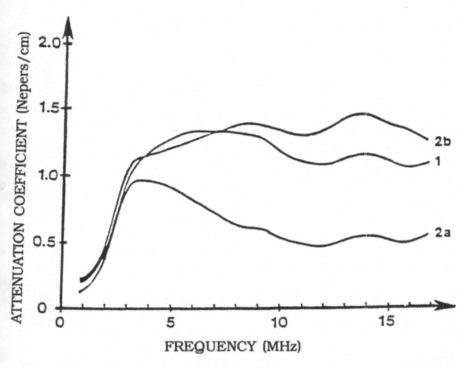

Figure 9: Attenuation coefficient spectrum for rough surface.

REFERENCES

1. Thompson, D.O., S.J. Wormley, J.H. Rose, and R.B. Thompson: Elastic Wave Scattering from Multiple Voids (porosity),Review of Progress in Quantitative Nondestructive Evaluation, 2A (1983), 867.

2. Varadan, V.K., V.V. Varadan, and L. Adler: Frequency Dependent Properties of Materials Containing a Distribution of Pores and/or Inclusions, in: Nondestructive Methods for Material Property Determination, Plenum Press, New York, 1984, 179.

3. Adler, L., J.H. Rose, and C. Mobley: Ultrasonic Method to Determine Gas Porosity in Aluminum Alloy Castings: Theory and Experiment, Journal of Applied Physics, 59 (1986), 336.

4. Gubernatis, J.E. and E. Domany: Effects of Microstructure on the Speed of Elastic Waves: Formal Theory and Simple Approximations, Review of Progress in Quantitative Nondestructive Evaluation, 2A (1983).

NUMERICAL WAVE ANALYSIS FOR THE LINE FOCUS ACOUSTIC MICROSCOPE: A MEASUREMENT MODEL

J.D. Achenbach, V.S. Ahn and J.O. Kim
Northwestern University, Evanston, IL, USA

J.G. Harris
University of Illinois, Urbana, IL, USA

ABSTRACT

A combined numerical and analytical model for the voltage measured by a line focus acoustic microscope as it scans the surface of a specimen is described. Particular emphasis is placed upon integration of the various parts into the total measurement model. Specimens having surface-breaking and subsurface cracks are considered as are ones coated by a thin film. The scanning acoustic microscope holds out the promise of quantitative measurements of very localized near surface properties, and of surface and near surface defects whose lengths are of the the order of a few microns.

12.1 INTRODUCTION

The present lecture reviews the work of the authors on developing a model of the voltage recorded by a scanning acoustic microscope, having a line focus (almost cylindrical) lens, as it scans a specimen. Specimens containing surface-breaking and subsurface cracks and ones coated by a thin film are considered. This paper is a synopsis of [12.1], [12.2] and [12.3]. Among other papers in this direction are those by Somekh, et al. [12.4], Li et al. [12.5] and, Rebinsky and Harris [12.6]. The first two papers use an approach roughly analogous to that used for Fourier optics, while the later two emphasize working with asymptotic approximations to the wavefields in real space.

12.2 THE MICROSCOPE

Figure 12.1 shows an overview of the acoustic microscope. The transducer radiates a radio frequency pulse down the buffer rod. At the lower end an approximately cylindrical lens focuses it to a line (really a diffraction limited strip). The lens is usually coated with a quarter wavelength matching layer to minimize reflections. The lens is coupled to a solid specimen by a fluid layer, usually water. The focal line can be placed at, above or below the surface of the specimen. What makes the device of real interest is that,

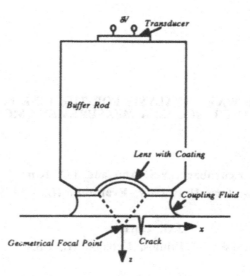

Fig. 12.1, The scanning acoustic microscope showing its principle features. Note that the origin of the coordinate system is placed at the geometrical focal point.

in addition to a reflected wave, a leaky surface wave is excited at the surface. This wave propagates along the specimen's surface and continuously radiates back into the fluid. For a surface without a coating the surface wave is a leaky Rayleigh wave. The lens collects both the reflected and leaky surface waves and the two signals propagate up the buffer rod and strike the transducer. The transducer *adds the two signals, both magnitude and phase, to produce a change in the voltage* caused by the presence of the specimen. While multiply reflected signals are present, the delay introduced by the buffer rod usually allows one to separate the initially returning signals from the multiply reflected ones. By scanning the focused beam in the horizontal and vertical directions, a signal commonly referred to as the acoustic (material) signature is recorded. This signal is really a record of the interference between the beam reflected from the surface of the specimen and the leaky surface wave.

The acoustic microscope typically operates from 100 MHz to 1 or 2 GHz, though instruments working at lower frequencies are available and ones operating at higher frequencies are being developed. In practise the penetration of the focused wave into the specimen is quite limited, on the order of a longitudinal wavelength, so that the device is used to measure surface and near surface properties. However, because the leaky surface wave participates in the measurement process, the microscope is particularly sensitive to features such as surface-breaking cracks, and regions of delamination between thin films and their substrates. Such measurements are of importance. For example, the initial macroscopic manifestation of metal fatigue is the appearance of clusters of micron-sized, surface-breaking cracks. Ceramic components are often coated to improve their resistance to wear and fracture so that their reliability depends upon the quality of the coating and its adhesion to the substrate. The scanning acoustic microscope holds out the promise of making detailed measurements of such features, even though their lengths are on the order of microns. However, a measurement model is needed to make quantitative interpretations of these measurements.

12.3 THE MODEL

To relate the change in voltage to the mechanical wavefields scattered at the surface of the specimen, an electromechanical reciprocity identity is used. Because this relation plays an essential role in this work its derivation is briefly reviewed, following closely that given by Auld [12.7]. The starting point is the following identity

$$(-i\omega u_{k1}\sigma_{jk2} + i\omega u_{k2}\sigma_{jk1} + e_{jkm}E_{k1}H_{m2} - e_{jkm}E_{k2}H_{m1}),_j = 0 \qquad (12.1)$$

which can be derived from the equations that describe a piezoelectric solid and which is valid in a region free of sources. In Eq.(12.1) the subscripts 1 and 2 indicate two different solutions to the piezoelectric equations, u_i is a component of the particle displacement, σ_{ij} is a component of the stress, E_i is a component the electric field and H_i is a component of the magnetic field. The term e_{ijk} is the alternating tensor. This equation is next integrated over a surface that surrounds the transducer, the buffer rod and the lens (Fig.12.1). The electromagnetic field is assumed to be zero everywhere except where the surface of integration cuts the coaxial cable coming from the terminals of the transducer and the elastodynamic field is assumed to be zero everywhere except over a surface that lies at or below the surface of the lens. Field 1 is excited if the specimen is not present and field 2 is excited in the presence of the specimen. Following Auld [12.7], the integration over the cross section of the coaxial cable is found to be proportional to the change in voltage caused by the specimen. The change in voltage δV is then given by

$$\delta V = C\int_{S_1} (u_i^{in}t_i - t_i^{in}u_i) \, dS \qquad (12.2)$$

Here u_i^{in} and t_i^{in} are the components of the incident particle displacement and traction, u_i and t_i are those of the total particle displacement and traction, and C is a normalizing constant. The contour of integration S_1 is chosen to be the surface of the specimen.

To calculate the incident wavefield at the specimen's surface a model of the propagation and subsequent focusing of the incident wave through the buffer rod, lens and coupling fluid has been developed. It is described in Section II of [12.1]. The initial assumption is that the transducer radiates a beam that at the aperture of the lens is approximately Gaussian in profile.and has a harmonic time-dependence. This assumption can only be checked by comparing the model's predictions with measurements (however, the model can readily be changed to accommodate alternative beam profiles). The lens' coating is modeled with a layer of distributed masses and springs. The wavefield incident to the surface of the specimen is calculated by formulating the problem as a system of integral equations and solving them using the boundary element method. The mass-spring model of the coating facilitates the formulation and numerical solution of these equations. The solution technique will be illustrated later, in connection with scattering from a specimen with a surface-breaking crack, so that the details of this calculation are omitted. From this calculation the wavefields incident on the specimen, u_i^{in} and t_i^{in}, are now known. The figures in [12.1] show that, using parameter values discussed below, this model produces a sharply focused acoustic beam in a fluid. It remains to calculate the wavefields scattered from the specimen.

The various parameters used for the microscope are explained in [12.1]. The more important are the following. The microscope is designed to operate at 225 MHz. At this frequency the normalized focal length f/λ_w is approximately 9.94×10, where the focal length f is measured from the aperture of the lens and λ_w is the wavelength of sound in water. The normalized aperture half-width b/λ_w is 1.32×10^2. The coupling fluid is water whose density and wavespeed are 9.98×10^2 kg/m^3 and 1.48×10^3 m/s, respectively.

12.4 ACOUSTIC SIGNATURES

12.4.1 Defect free surface

The focused wavefield is incident to the surface of the specimen. Calculating the wavefields reflected from the surface of a defect-free specimen using the boundary element method is straightforward. Again we refer to [12.1] for the details. Knowing the incident and scattered wavefields we use Eq. (12.2) to calculate the acoustic signatures. Figure 12.2 shows the theoretical $|\delta V|$ (solid line) plotted against z/λ_w. The term z/λ_w defines the distance that the geometrical focal point lies above (+) or below (-) the surface of a defect-free specimen. Also shown is a measured $|\delta V|$ (dashed line). The material is fused quartz. Its density, longitudinal wavespeed and transverse wavespeed are 2.2×10^3 kg/m^3, 5.9×10^3 m/s and 3.7×10^3 m/s. Note that we have plotted the two curves so that the locations of the secondary peaks coincide. We believe that the locations of the largest peaks do not quite match because the focal points of the microscope and our model do not quite coincide. The good agreement, though no explicit account of attenuation has been made, indicates that the model is satisfactory.

12.4.2 Surface-breaking crack

Next we summarize how the wavefield scattered from a surface having a surface-breaking crack is calculated. Figure 12.3 shows a blown-up view of the lens and specimen. Along contour S_1, the surface of the specimen, we impose the continuity conditions

$$u_n = u_n^f, \quad t_n = -t_n^f, \quad t_s = 0 \tag{12.3}$$

where u_n^f and t_n^f are the normal components of particle displacement and traction in the fluid, u_n and t_n are the corresponding normal components in the solid, and t_s is the tangential component of traction in the solid. Note that $t_n^f = p$, where p is the pressure in

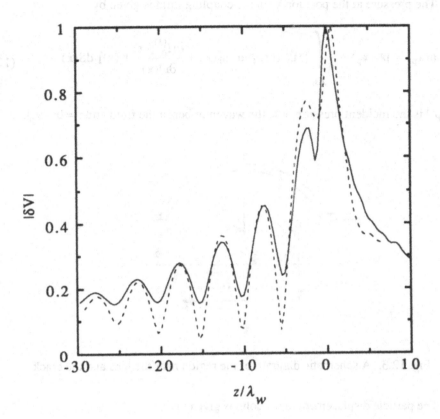

Fig. 12.2, The magnitude of δV as a function of z/λ_w for fused quartz. The solid curve
indicates the numerical result and dashed one the measured result.

the fluid. Though it is not clear whether or not the fluid penetrates the crack, we assume
that there is no penetration and therefore that the crack faces, labelled S_2 in Fig. 12.3, are
free of traction. That is,

$$t_n = t_s = 0 \qquad\qquad\qquad (12.4)$$

where t_n and t_s are the normal and tangential components of traction. The particle
displacement on the left crack face is designated by u_k^-, and that on the right face by u_k^+.

The pressure at the position x_p in the coupling fluid is given by

$$p(x_p) = p^{in}(x_p) + \frac{i}{4} \int_{S_1} [H_0^{(1)}(kr) \rho \omega^2 u_n^f(x) + \frac{\partial H_0^{(1)}(kr)}{\partial n^f(x)} t_n^f(x)] \, dS(x) \qquad (12.5)$$

where p^{in} is the incident pressure, k is the wavenumber in the fluid and $r = |x - x_p|$.

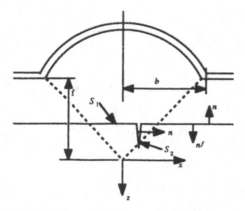

Fig. 12.3, A schematic diagram of the region near the lens and the crack.

The particle displacement in the solid is given by

$$u_i(x_p) = \int_S [u_{ij}^G(x,x_p) \, t_j(x) - \sigma_{ijk}^G \, u_j(x)n_k(x)] \, dS(x) \qquad (12.6)$$

where u_j and t_j are the components of the (total) particle displacement and traction in the solid specimen, u_{ij}^G and σ_{ijk}^G are the components of the particle displacement and stress for the Green's tensor, given in Appendix B of [12.1], and $S = S_1 + S_2$. These two integral representations in combination with the boundary and continuity conditions on the contours given by Eqs.(12.3) and (12.4) become a closed system of integral equations when x_p approaches points on the contours S_1 and S_2. The system is solved by the boundary element method. Special care is required to deal with the hypersingular integrals. The details are given in [12.2].

Throughout the numerical work a constant shape function is used for each element. The integration of the regular terms is performed using the Gaussian quadrature formula for constant elements. The singular terms are integrated analytically by expanding the kernels in series. The incident wavefield is focused onto a narrow region of the surface S_1 so that the region that must be discretized is small. Along S_1 from 8 to 10 elements per Rayleigh wavelength were used and the size of the region discretized was determined by trial and error. Along the surface S_2 greater care was needed. Again by trial and error it was found that from 60 to 70 elements per Rayleigh wavelength were needed, but that no special element was needed to take account of the crack tip. A greater number of elements did not change in a significant way the scattered wavefields.

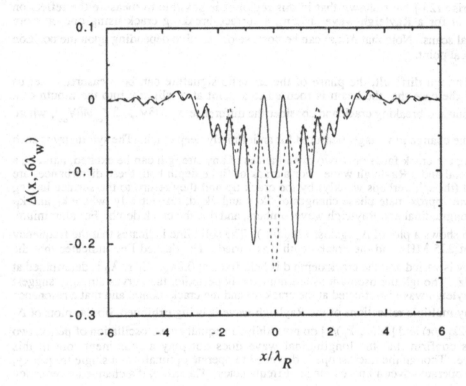

Fig. 12.4, Horizontal scans for a specimen of aluminium. The focal point is $6\lambda_w$ below the surface. The solid line indicates a crack λ_R deep and the dashed line indicates one $0.3\lambda_R$ deep.

Figure 12.4 shows a horizontal scan for an aluminium specimen, when the focal point is $6\lambda_w$ below the surface, for vertical surface-breaking cracks $0.3\lambda_R$ (dashed line) and λ_R (solid line) deep. The parameter λ_R is the Rayleigh wavelength. The density, longitudinal wavespeed and transverse wavespeed of aluminium are 2.7×10^3 kg/m^3, 6.32 $\times 10^3$ m/s and 3.13×10^3 m/s, respectively. In this figure the difference $\Delta(x,z)$ (for $z/\lambda_w =$

6) between the magnitude of the signature with a crack, $|\delta V|_{cr}$, and that of the signature without a crack, $|\delta V|_{nc}$, namely, $\Delta(x,z) = |\delta V|_{cr} - |\delta V|_{nc}$, is plotted against x/λ_R. Substantial differences for the two crack depths are noted. Of interest are the broad and deep minima at the crack's location indicating that a vertical discontinuity is strongly indicated. Away from the crack the periodicity of the curve is approximately $\lambda_R/2$ indicating that a standing wave pattern formed by incident and reflected leaky Rayleigh waves has been set up. Rebinsky and Harris [12.6] have shown that in this region it is possible to measure the reflection coefficient for a Rayleigh wave striking a surface-breaking crack using one or more horizontal scans. Note that $\Delta(x,z)$ can be positive or negative depending upon the position of the focal point.

Though difficult, the phase of the acoustic signature can be measured. Let us consider the case that the beam is focused to a point at or slightly into the mouth of a vertical surface-breaking crack and consider the difference $\Delta_d = |\delta V_{cr} - \delta V_\infty|/|\delta V_\infty|$, where δV_∞ is the change in voltage measured for an infinitely deep crack. The symmetry is such that along the crack faces only two disturbances of any strength can be excited, namely, a longitudinal and a Rayleigh wave. For a crack of finite depth both these disturbances are reflected (though perhaps weakly) by the crack tip and they return to the surface having undergone approximate phase changes of $2k_L d$ and $2k_R d$, respectively, where k_L and k_R are the longitudinal and Rayleigh wavenumbers, and d is the crack depth. For aluminium, Fig. 12.5 shows a plot of Δ_d against $(2k_R d/\pi)$. The solid line indicates that the frequency is held at 225 MHz and the crack depth d is varied. The dashed line indicates that the frequency is varied and the crack depth d is held fixed at $0.8\lambda_R$, where λ_R is determined at 225 MHz. Though the oscillations are not exactly periodic, the curves strongly suggest that a Rayleigh wave is reflected at the crack tip and the crack mouth, and that a resonance caused by multiple reflections of the Rayleigh wave is being exhibited. Similar plots of Δ_d against $(2k_L d/\pi)$ and $[2d(k_R-k_L)/\pi]$ do not exhibit an equally clear oscillation of period two and thus confirm that the longitudinal wave does not play a prominent role in this resonance. Though the microscope is designed to operate optimally at a single frequency, it can be operated over a limited range of frequencies. Therefore, if a change in frequency Δf causes the Δ_d curve to shift from peak to peak, d is given by

$$d = c_R/(2\,\Delta f) \tag{12.8}$$

where c_R is the Rayleigh wavespeed. Note that the accuracy of Eq.(12.8) assumes that clearly detectable surface waves are excited on the crack faces. This is the case only for relatively smooth crack faces that are not in contact.

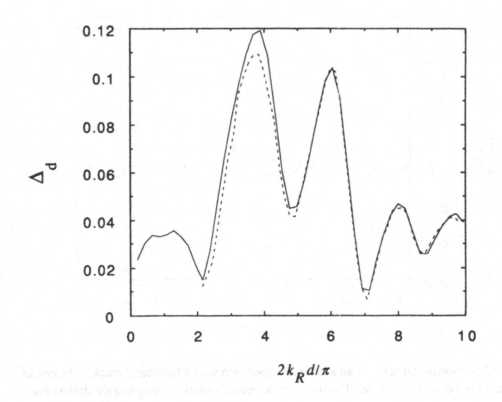

Fig. 12.5 A plot for aluminium of Δ_d against $(2k_Rd/\pi)$ as k_R and d are varied.

12.4.3 Subsurface cracks

Figures 12.6 and 12.7 illustrate horizontal scans, with the focal point at the surface, for aluminium specimens containing subsurface cracks. The calculation of the scattered wavefields was done using the boundary element method as illustrated in the previous section. Figure 12.6 illustrates $\Delta(x,0)$ for horizontal subsurface cracks of lengths λ_R (solid line) and $1.5\lambda_R$ (dashed line). The cracks are $0.1\lambda_R$ below the surface. The difference $\Delta(x,0)$ indicates the wavefield reflected from the upper crack face. Thus when the lens is over the crack the geometrically reflected wavefield is quite large and the curve for $\Delta(x,0)$ shows a clear image of the subsurface crack. Note that the width of the large disturbance in the $\Delta(x,0)$ curve is approximately equal to the crack length. The oscillations caused by the leaky Rayleigh waves are small because the reflection of the Rayleigh wave by the horizontal crack is relatively small. Still, a standing wave pattern showing a periodicity of

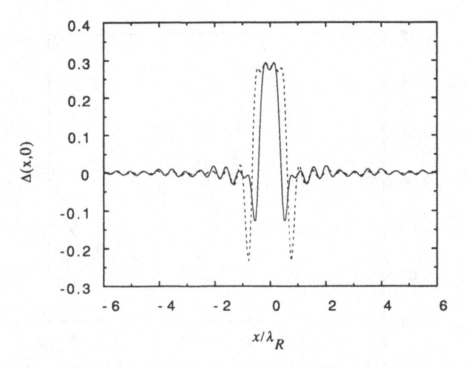

Fig. 12.6, Horizontal scan for an aluminium specimen with a horizontal crack. The cracks are $0.1\lambda_R$ below the surface. The solid line indicates a crack λ_R long and the dashed line indicates one $0.5\lambda_R$ long.

$\lambda_R/2$ is evident.. Figure 12.7 illustrates $\Delta(x,0)$ for vertical subsurface cracks of lengths λ_R (solid line) and $0.5\lambda_R$ (dashed line). The upper tip is $0.1\lambda_R$ below the surface. Near the crack one notices the strong and complicated nature of the signal, while for $|x/\lambda_R|$ somewhat greater than one, an interference pattern is set up by reflected and transmitted Rayleigh waves.

12.4.4 Specimens coated with thin films

The acoustic signatures for specimens coated with thin films are of interest because of their important industrial applications. The film modifies the nature of the leaky surface wave that participates in the formation of the acoustic signature. The presence of the film allows more than one leaky surface wave to participate. The transverse wavespeed of a solid thin film is usually greater than that of the coupling fluid so that any waves that it guides will leak into the fluid and hence be collected by the lens. However, the film can

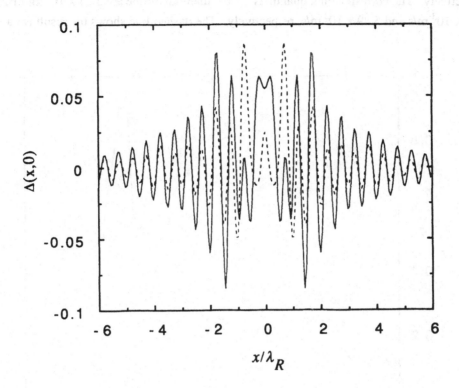

Fig. 12.7 Horizontal scan for an aluminium specimen with a vertical crack. The upper crack tips are $0.1\lambda_R$ below the surface. The solid line indicates a crack λ_R long and the dashed line indicates one $0.5\lambda_R$ long.

essentially either mass-load the substrate, lowering the speed of the lowest guided mode below the Rayleigh wavespeed of the substrate, or it can stiffen the substrate, raising the speed of the lowest guided mode above the Rayleigh wavespeed of the substrate.Moreover, a film that mass-loads the substrate can usually guide one detectable higher mode, known as a Sezawa wave, so that two leaky surface waves can be present. The formulation and calculation, using the boundary element method, of the scattered wavefields is given in [3].

Figure 12.8 shows a vertical scan, similar to that in Fig. 12.2, of a specimen made from steel coated with a titanium-nitride film. The density, longitudinal wavespeed and transverse wavespeed for steel are 7.9×10^3 kg/m^3, 5.84×10^3 m/s and 3.13×10^3 m/s, respectively. The corresponding quantities for the titanium nitride are 5.21×10^3 kg/m^3, 9.58×10^3 m/s and 5.89×10^3 m/s, respectively. The dashed line shows the result for a

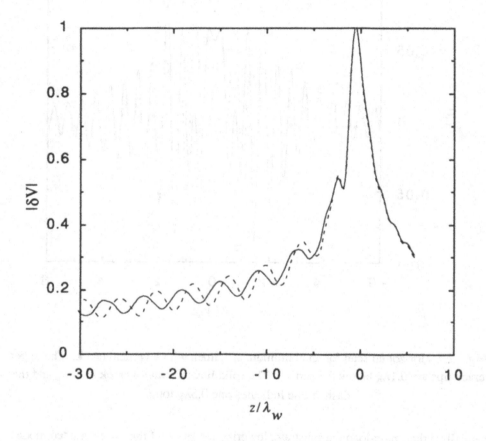

Fig. 12.8, A vertical scan for a titanium-nitride layer on a steel substrate. Dashed line indicates a 0.5μm thick coating. Solid line indicates a 1μm thick coating.

coating 0.5 μm thick and the solid line for one 1.0 μm thick. The film stiffens the substrate so that only one mode, the leaky Rayleigh mode, is excited, and its wavespeed increases above that of the substrate as the coating thickness increases. It can be shown that the speed of propagation in the coating c is related to the spacing between the peaks of the vertical scan Δz at a given frequency f by

$$c \approx (c_w \, f \, \Delta z)^{1/2}. \tag{12.9}$$

From this one may infer information about the thickness and mechanical properties of the coating.

Calculated vertical scans have also been compared with experimental curves. Examples are given in [12.3].

12.5 SUMMARY

This paper has described the synthesis of an integrated measurement model of the scanning acoustic microscope, using both computational and analytical methods. It has been shown that the model gives results that agree with and facilitate quantitative interpretation of measurements. Examples show that the results of this paper may be expected to aid in the measurement of the depth of a surface-breaking crack, the width of a subsurface crack and the properties of a thin film or coating.

REFERENCES

12.1. Achenbach, J.D., V.S. Ahn and J.G. Harris: Wave analysis of the acoustic material signature for the line focus microscope, IEEE Ultrason. Ferroelect. Freq. Contr. 34, (1991) 380-387.

12.2. Ahn, V.S., J.G. Harris and J.D. Achenbach: Numerical analysis of he acoustic signature of a surface-breaking crack, IEEE Ultrason. Ferroelect. Freq. Contr. 39 (1991) 112-118.

12.3. Ahn, V.S., J.D. Achenbach, Z.L. Li and J.O. Kim: Numerical modeling of the V(z) curve for a thin-layer/substrate configuration, Res. Nondestr. Eval. 3 (1991) 183-200.

12.4. Somekh, M.G., H.L. Bertoni, G.A.D. Briggs: A two-dimensional imaging theory of surface discontinuities with the scanning acoustic microscope, Proc. R. Soc. Lond. A 401 (1985) 29-51.

12.5. Li, Z.L., J.D. Achenbach and J.O. Kim: Effect of surface discontinuities on V(z) and $V(z,x_o)$ for the line focus acoustic microscope, Wave Motion 14 (1991) 187-203.,

12.6. Rebinsky, D.A. and J.G. Harris: An asymptotic calculation of the acoustic material signature of a cracked surface for the line focus scanning acoustic microscope, Proc. R. Soc. Lond. A 436 (1992) 251-265.

12.7. Auld, B.A. General electromechanical reciprocity relations applied to the calculation of elastic wave scattering coefficients, Wave Motion 1 (1979) 3-10.

GENERATION OF ULTRASOUND BY LASER

C.B. Scruby

National NDT Centre, Oxfordshire, UK

1.   INTRODUCTION

Various physical processes may take place when a solid surface is illuminated by a laser, some of which generate ultrasound. At lower incident powers these include heating, the generation of thermal waves, elastic waves (ultrasound), and in materials such as semiconductors, electric currents may be caused to flow. At higher powers, material may be ablated from the surface, and a plasma formed, while in the sample there may be melting, plastic deformation and even cracks formed.

In this lecture I shall only consider laser power regimes that are suitable for nondestructive evaluation. Most attention will be concentrated on the results of the localised heating produced by laser radiation which in turn generates the thermoelastic stresses and strain that act nondestructively as an ultrasonic source. Generation of ultrasound by ablation is included because the damage this causes is acceptable for some NDE applications.

2.   ABSORPTION OF LASER RADIATION

When the coherent light from a laser is incident on a solid sample, some of the energy is absorbed by mechanisms that depend on the nature of the sample and the frequency of the radiation, while the remainder is reflected or scattered from the surface. I first assume that the sample is too thick for any transmission to occur, and also that the intensity of the radiation is too low for ablation or damage.

If the specimen is a non-reflecting insulator, then the radiation is progressively attenuated by various types of absorption process as it penetrates into the sample. The intensity as a function of depth decays exponentially, and a characteristic absorption depth can be defined as the reciprocal of the absorption coefficient (γ).

If, however, the radiation is incident at the surface of a conductor, some energy is absorbed by eddy current resistive losses and some reflected within a thin surface layer known as the skin depth (δ). In common metals δ is in the range 1 - 10 nm for visible light, and is inversely proportional to the square root of the optical frequency. The absorption of light by a clean, polished metal surface is inversely proportional to δ.

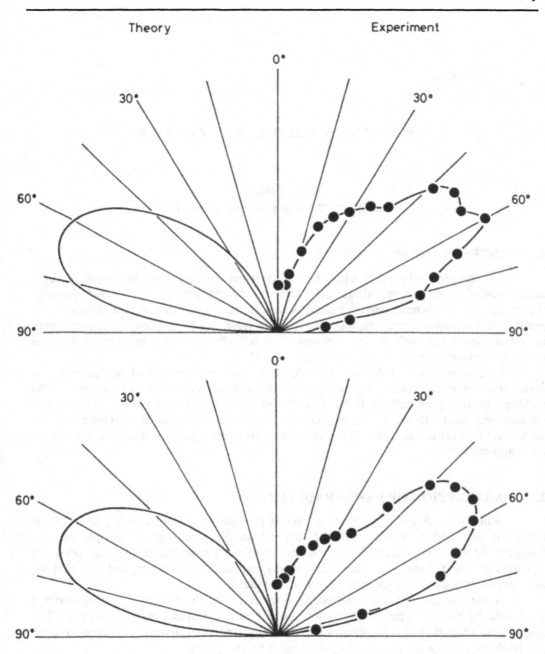

Figure 1
Comparison of experimental data [3] for 1 mm thermoelastic source and 1MHz
piezoelectric receiver on 50 mm hemisphere with calculation [2].

$$u(\theta) \approx \frac{2k^2 \cos\theta \left(k^2 - 2\sin^2\theta\right) F}{4\pi\rho c_1^2 r \left[\left(k^2 - 2\sin^2\theta\right)^2 + 4\sin^2\theta \left(1 - \sin^2\theta\right)^{1/2} \left(k^2 - \sin^2\theta\right)^{1/2}\right]} \tag{5}$$

where F is the ablative force. When compared with experimental data [3] in Figure 2, it takes the form of a broad lobe perpendicular to the surface, which is vastly different from the thermoelastic source. The shear directivity comprises a number of oriented lobes with different amplitude and phase from thermoelastic. Directivity patterns of constrained surface sources are more difficult to calculate. Experiments indicate that they mostly resemble ablation.

8. BULK ULTRASONIC WAVEFORMS

The majority of measurements have been carried out with the source and receiver directly opposite one another on a parallel-sided plate, a standard arrangement for many ultrasonic applications (Figure 3a [4]). Apart from a small positive initial pulse, the whole waveform is negative, implying that the surface at the receiver is depressed inwards. Intuitively, a rise of the surface would be expected from a centre of expansion, and this is indeed the case for all orientations except the epicentre. This waveform arises because the plane stress boundary conditions cancel the farfield compression (P) wavefield (cf Figure 1) which is a positive pulse, leaving only near-field terms.

Figure 3b shows a waveform calculated using the Green's function formalism, but neglecting thermal conductivity. This waveform was then convolved with a Gaussian broadening function (Figure 3c) to simulate the effect of both laser pulse width and limited bandwidth in the recording system. The magnitude of the P-wave step is given by $8B\delta V/(\pi\mu r^2 k^5)$. The agreement with experiment is indeed very good, with the exception of the small positive pulse at the commencement of the waveform. Various workers have confirmed that this small pulse is at least partially due to thermal conduction into the bulk of the specimen [2].

Experimental data [5] confirm that at higher power densities the ultrasonic waveforms and hence the acoustic sources are very different. The power density is raised at constant incident energy by using a lens to vary the beam diameter. Below $\sim 20\ \text{MW cm}^{-2}$ the waveform does not change significantly. Between 20 - 100 MW cm^{-2} the initial P-wave pulse increases in magnitude, while the shear wave step decreases (suggestive of a change of state). Above 100 MW cm^{-2} there is a huge increase in P-wave magnitude as strong ablation and plasma take place. In terms of waveforms (Figure 4) the P-wave arrival first becomes dominated by a pulse rather than a step of increasing size (a - c), and then later (d - f) it is modified into a step. The air-breakdown source [6] and under certain circumstances the constrained source, generate waveforms rather similar to metallic ablation. Successful calculations have also be made for the ablation regime, modelling the source as a combination of a reactive force normal to the surface and thermoelastic stresses.

$3\alpha/\delta C$ increases with temperature in both aluminium and steel, raising the ultrasonic generation efficiency. However, as the temperature is raised it becomes more difficult to remain within this regime, and energetically easier for melting and ablation to occur.

5.   CHANGES OF STATE

As incident power densities and pulse energies are raised, effects other than heating occur. Surface temperatures become high enough in both metals and insulators for melting, vaporisation and plasma formation.

Laser-induced melting of the surface is detrimental to the generation of ultrasound unlike laser welding applications, for it absorbs energy and is likely to be accompanied by permanent changes to material properties on cooling. With Q-switched lasers, the power regime in which melting occurs without vaporisation is fortunately too narrow to be of much practical concern. For common metals melting tends to occur when the incident power density $\sim 10^6$ - 10^7 W cm^{-2} for Q-switched pulses.

Vaporisation of the specimen surface is relatively easy to produce in metals and insulators with Q-switched lasers at power densities above $\sim 10^7$ W cm^{-2}, ie modest focusing of most pulsed laser beams. The physical processes that take place in this regime are extremely complex. In simple terms, as the incident optical power increases, the surface temperature rises until the boiling point of the material is reached, and some material is vaporised, ionised and a spark or plasma formed. With Q-switched pulses, the rate of heat dissipation is too fast for appreciable thermal conductivity to take place, so that the main factors controlling vaporisation are specific and latent heats.

Vaporised material is ejected perpendicularly from the surface - a process known as ablation - producing a net stress in reaction against the specimen surface, which can be simply calculated from the rate of change of momentum. For an incident power density of 10^8 W cm^{-2} the ablative stress in aluminium or steel is estimated to be ~ 50 MPa, ie of the same order of magnitude (but in a different orientation) as the stresses generated thermoelastically. Because of the extremely high rate of change of power input during nanosecond pulses, additional phenomena must be considered. First the plasma exerts a high pressure on the surface which in turn suppresses vaporisation by raising the boiling point well above its normal value. The plasma absorbs light from the laser pulse and becomes extremely hot. It expands to produces an impulsive reaction on the surface, and also radiates heat back to the surface, maintaining a high temperature for some time after the incident laser pulse power has started to fall.

If the ambient surface temperature is raised, ablation should become more efficient for two main reasons. Decreased reflectivity should improve the coupling of energy into the surface, and less energy is required to raise the surface temperature up to the boiling point. NDE applications should therefore be more efficient than at elevated than room temperatures in both regimes.

6.   SURFACE CONSTRAINTS

The main benefit of laser generated ultrasound is that it can be used without contact with the specimen, couplant or surface constraint.   There are nevertheless some requirements for laser generation at a surface constrained either by a transparent solid layer or a liquid.   The effects of surface modification or constraint can be very complicated, since optical absorption, conversion to ultrasonic energy and ultrasonic propagation are all affected by boundary conditions. Roughness and coating the surface with a thin solid layer of, for instance, paint or rust increase optical absorption and may also be accompanied by ablation of the layer.   Both these effects enhance the ultrasonic wave field. Covering the surface with a transparent solid such as glass or liquid changes the boundary conditions modifying and enhancing the source directivity.   In the latter case expansion and/or vaporisation of the liquid cause further amplification. Finally, constraining a thin layer of liquid between a transparent solid and the sample also generates an enhanced source because of gross thermal expansion in the liquid.

7.   LASER-ULTRASONIC RADIATION PATTERNS

Radiation (directivity) patterns can be deduced for the main types of laser induced surface source by a number of methods [2].

7.1 Thermoelastic source

The radiation pattern for compression waves, $u(\theta)$, from a point source is given by:

$$u(\theta) \approx \frac{3\alpha B \sin 2\theta \sin \theta \left(k^2 - \sin^2 \theta\right)^{\frac{1}{2}} \frac{\partial}{\partial t}(\delta E)}{\pi(\lambda + 2\mu)\rho C c_1 r \left[\left(k^2 - 2\sin^2 \theta\right)^2 + 4\sin^2 \theta \left(1 - \sin^2 \theta\right)^{\frac{1}{2}} \left(k^2 - \sin^2 \theta\right)^{\frac{1}{2}}\right]} \tag{4}$$

where k is the ratio of the compression and shear velocities (c_1/c_2). As Figure 1 shows, the pattern consists of a broad lobe at $\sim 60°$ to the normal with rotational symmetry. This is somewhat unusual compared with more common ultrasonic sources and arises because the dominant tractions are parallel to the surface as a result of the boundary conditions. The shear wave directivity consists of a number of lobes which change phase at the critical angle ($\sim 30°$) and at 45°. Experimental data [3] do not show the predicted zero at $\theta = 0$ in the compression wave radiation. This could be due to various factors: source and receiver diameters, near-field terms and thermal conductivity effects, so far neglected.

7.2 Ablation source

The compression directivity from a point ablation source is calculated to be the same as for a small compression transducer, ie:

3.   THERMAL EFFECTS

3.1 Metals

In order to determine the resulting temperature distribution, I first assume that the optical pulse is so short that thermal conductivity into the bulk of the sample can be neglected, that the laser uniformly irradiates an area A and that absorption takes place uniformly throughout a depth equal to the skin depth, δ. Then the temperature rise in the irradiated material is uniform, and given by $\delta E/(C\rho A\delta)$, where δE, C and ρ are the absorbed energy, specific thermal capacity and density of the specimen. Substituting typical experimental values yields surface temperatures in excess of 10^4 K, i.e. well above the threshold for damage and vaporisation. These effects are not observed in practice. The discrepancy is due to thermal conductivity, which has the effect of making the source extend deeper and the average temperature rise much smaller.

The true temperature distribution is calculated by solving the differential equation for heat flow in a semi-infinite slab. Suppose the laser is switched on at $t = 0$ and switched off at $t = t_0$ to generate a square pulse of intensity I_0 which irradiates the specimen uniformly. Then the temperature distribution $T(z,t)$ is given by [1]:

$$T(z,t) = \frac{2I_0\sqrt{\kappa t}}{K}\text{ierfc}\left[\frac{z}{2\sqrt{\kappa t}}\right] - \frac{2I_0\sqrt{\kappa(t-t_0)}}{K}\text{ierfc}\left[\frac{z}{2\sqrt{\kappa(t-t_0)}}\right] \qquad (1)$$

where K and κ are the thermal conductivity and diffusivity. The temperature distribution for other pulse shapes and non-uniform irradiation can be obtained by convolution and integration.

Substitution of typical experimental values yields maximum surface temperatures in the range 10 - 100 K for unfocused Q-switched (< 50 ns) laser pulses where ~ 10 mJ of energy is absorbed. The effective depth (source "thickness") to which the heat penetrates for the duration of each pulse is a few •m in typical metals.

3.2 Non-metals

For an insulator in which γ is small, the radiation penetrates well into the bulk of the material, in contrast to a metal where γ is very large. I assume that the flow of heat away from the irradiated region where energy is being absorbed is negligible during the timescale of the source, which is reasonable for a short laser pulse irradiating a poor conductor. If the intensity of the absorbed laser light at the surface is $I(0,t)$ then the temperature distribution as a function of depth is given by:

$$T(z,t) = \frac{\gamma e^{-\gamma z}}{C\rho}\int_0^t I(0,t')dt' \qquad (2)$$

Thermal diffusion lengths in ceramics and plastics are of the order of 0.1 •m, much less than optical penetration depths $(1/\gamma)$. This contrasts with metals where the optical penetration depth is much less than the thermal diffusion length. The temperature decays exponentially with depth in an insulator while its time dependence is the same as that of the pulse energy. Note that the surface temperature rise is large if the absorption is high and the thermal capacity and density are low. Care must be taken not to damage the surface of materials such as plastics and composites.

4.   THERMOELASTIC STRESSES

The modest rises in temperature calcuated above must be accompanied by thermal expansion, which in turn generates stresses and strains within the sample. Since for most practical NDE applications, the laser pulse diameter is likely to be much smaller than the ultrasonic propagation path, a 3-D treatment is required. It is convenient to consider the source first as a centre of expansion within a very small volume, V. If the centre of expansion is buried within the bulk of an isotropic material then thermal expansion is equivalent to the insertion of a small extra volume of material δV at that point. This generates 3 equal strains given by $\alpha \delta T$, where α is the linear expansion coefficient. However, in reality the source is situated at the surface, where boundary conditions imply no net perpendicular stress, leaving only stresses parallel to the surface. In practice the source has a small, non-zero thickness, the material is able to expand upwards perpendicular to the surface, there is a measurable displacement and hence a small perpendicular stress component also.

If energy δE from the pulse is absorbed within a volume V to give a temperature rise δT, then the bulk strain is given by [2]:

$$\frac{\delta V}{V} = 3\alpha \ \delta T = \frac{3\alpha \ \delta T}{\rho C V} \tag{3}$$

where C is the thermal capacity. Substituting typical experimental data for aluminium yields thermoelastic strains $\sim 2 \times 10^{-3}$, and hence (multiplying by the bulk elastic modulus) stresses ~ 100 MPa per mJ absorbed energy. These values are high, implying that common metals can be damaged plastically in this regime if too high an incident pulse energy is applied, even though there is no surface melting or ablation.

The thermoelastic stresses from an extended source in a good conductor can be calculated by integration over a surface distribution of centres of expansion. In an insulator where appreciable bulk penetration occurs to give an extended source below the surface, integration with respect to the depth coordinate is essential.

High temperature NDE is one important application area for laser ultrasonics, and I therefore briefly address the effect of raising the specimen temperature on thermoelastic stress generation. Firstly as the temperature is raised in common metals such as aluminium and steel, the reflectivity decreases so that δE increases, thereby raising the thermoelastic stresses. At the Curie point and martensitic transformation steel undergoes additional changes in reflectivity. In addition there are changes in all 3 physical parameters which make up the thermoelastic coupling factor, $3\alpha/\delta C$. Preliminary estimates suggest that

Theory   Experiment

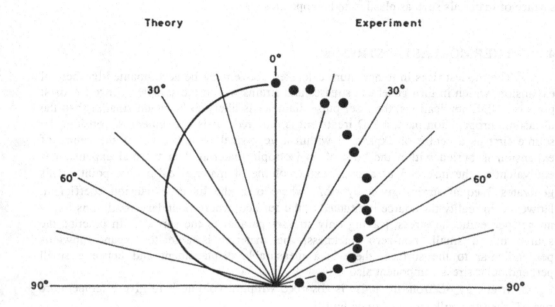

Figure 2
Comparison of experimental data [3] with calculation [2] for laser-induced ablation source
on 50 mm hemisphere.

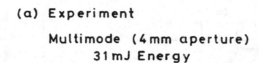

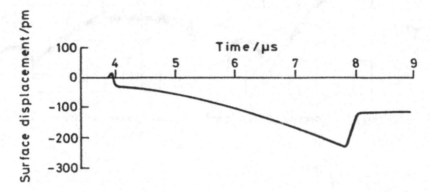

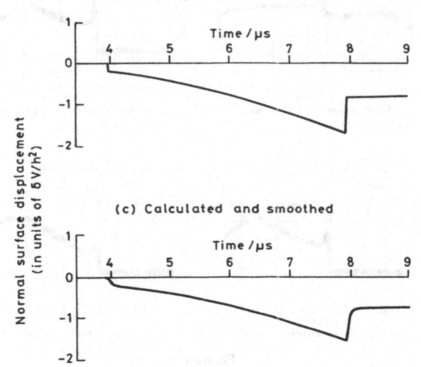

Figure 3

Comparison of epicentre waveform from 4 mm diameter 31 mJ laser pulse incident on 25 mm aluminium disc [4] with theory [2].

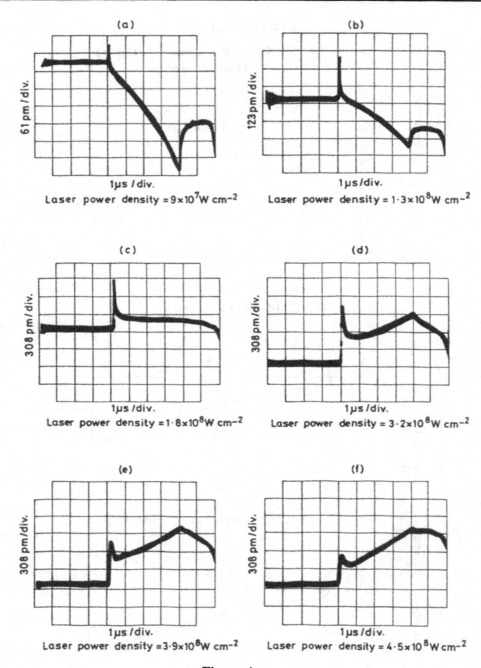

Figure 4

Epicentre waveforms recorded on 25 mm thick aluminium sample, as 33 mJ laser pulse is focused to generate higher power densities [5].

The epicentre is however not good position for investigating shear wavefields from these cylindrical sources. It also constitutes a special case for the thermoelastic wavefield. There have been fewer investigations at other orientations (eg Figure 5). The most significant change is to the P-wave from the thermoelastic source, which transforms for ?>0 into a short, strong positive pulse with excellent broadband characteristics. Otherwise the waveforms tend to be complicated by mode-converted arrivals. For instance, the shear is preceded by a head wave outside the critical angle. The changes to P-waves from ablation are minor in comparison. Interestingly, at the common NDE orientation of 45° the P-wave magnitude in the thermoelastic regime can approach ablation.

9. SURFACE AND GUIDED WAVES

It is possible to generate all types of surface and guided wave by laser. Generally Rayleigh waves have the largest amplitude on thick specimens, and are therefore of most interest. Lamb waves dominate on thin plates, and various extensional and torsional modes in rods. Surface acoustic wave pulses of similar magnitude and high bandwidth to laser-generated bulk waves can be generated. Figure 6 shows typical surface waves in the thermoelastic regime, detected by two types of non-contact receiver. Note the bipolar form of the dominant Rayleigh wave, and the relatively weak surface-skimming P-wave (lateral wave).

Equally well-defined surface waves can be generated in the ablation regime (Figure 7), generally with higher amplitude that thermoelastic waveforms. Note that there is again a bipolar Rayleigh wave, but of opposite polarity. Standard elastic wave theory can again be used [2] successfully to simulate these waveforms (Figures 6c, 7c). In the case of Lamb waves, individual symmetric and antisymmetric modes can be resolved. The waveforms tend to become dominated by a sheet mode [7] as the specimen becomes progressively thinner (Figure 8).

10. SOURCE SIZE AND BANDWIDTH

Three main effects of changing the dimensions of the laser-generated source have been observed experimentally. Theoretically, they can be taken into account by performing a spatial integration over the source area. Firstly the incident power density is raised or lowered and may thereby cause a transition from thermoelasticity to ablation or vice versa, secondly the bandwidth (and consequently magnitude) of the arrivals in the waveforms is altered, and thirdly the directivity of the source modified. For practical applications these effects may be in competition, depending upon which regime is in use, as I now briefly discuss.

In the thermoelastic regime a large source area is required to keep the power density below the ablation threshold. However, this causes problems for the bandwidth since it will inevitably broaden wave arrivals (hence also reducing their amplitude). These effects are minimised at epicentre, but for angled bulk waves and surface waves they are significant. Various approaches have been adopted to optimise the bandwidth. The

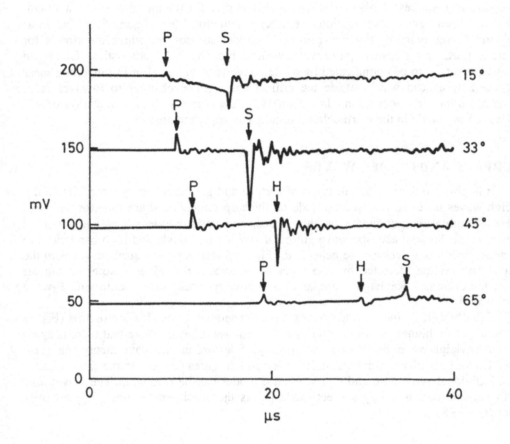

Figure 5
Thermoelastic waveforms as function of orientation on 48 mm steel plate. P, S, H denote
compression, shear and head waves [2].

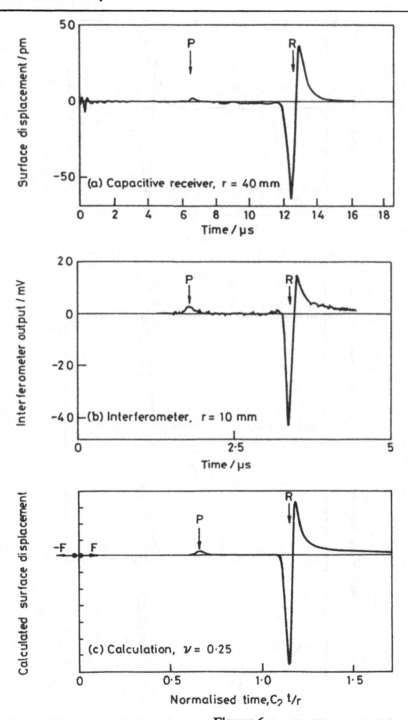

Figure 6
Comparison of experimental (a,b) surface waveforms with calculation (c) for
laser-ultrasonic source in thermoelastic regime [2].

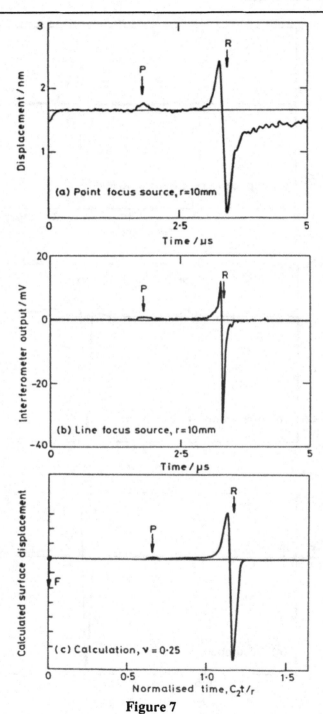

Figure 7

Comparison of experimental (a,b) surface waveforms with calculation (c) for
laser-ultrasonic source in ablation regime [2].

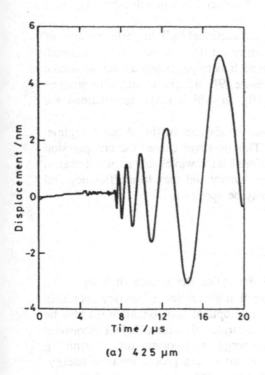

(a)   425 µm

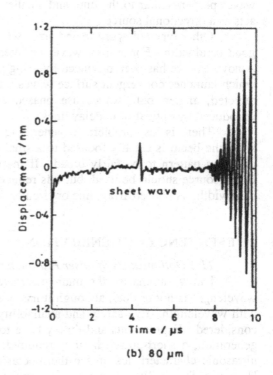

sheet wave

(b)   80 µm

Figure 8
Lamb waves generated by pulsed laser (thermoelastic regime) and detected by
interferometer 24 mm from acoustic source in aluminium samples of different thickness
[7].

earliest, and still one of the most effective, was to focus (or aperture) the incident radiation into a line [8]. This generates thermoelastically excellent broadband bulk and surface waves perpendicular to the line, and smaller signals of reduced bandwidth otherwise, ie it acts as a directional source.

Other types of aperture and array will produce directional bulk or surface waves of good bandwidth. For surface waves an interdigital array can for instance been simulated. A novel source has been produced by using a an axicon lens to generated a circular source which launches convergent surface waves into the centre [9]. Special attempts to produce directed, angled bulk waves use phased arrays [10], in which small retardations are introduced by optical fibre delay lines.

There is less problem in generating sufficient bandwidth in the ablation regime, since the beam is usually focused to a small area. This however causes the compression radiation pattern to be fairly broad. If more directional bulk waves are required then a larger source should be used, but this reduces power density and may lose efficiency and bandwidth. A phased array, line or circular source may be optimum.

11. EFFICIENCY OF GENERATION

11.1 Optimisation of Laser Parameters

I now summarise the main parameters that determine the choice of laser. The wavelength is not critical, although it has been shown that generation efficiency decreases with wavelength. Eye-safety and availability of suitable optical components must also be considered. Experiments and theory have together confirmed that for efficient ultrasonic generation, a short-pulsed laser is required. Pulse energy is important for determining ultrasonic characteristics: in the thermoelastic regime signals are proportional to energy. However, the likelihood of surface damage also increases and a compromise has to be reached. Experience has shown that 100-500 mJ is a useful nondestructive range. Closely linked to this are irradiated area and beam profile, whose effects have been discussed above.

The second important parameter is pulse duration. It influences the generation mechanism: too long a pulse generates mostly heat and little ultrasound, whereas a very short pulse is prone to ablation. Experience has shown that the range 5-50 ns is convenient. The pulse repetition rate is important if generation is to be combined with laser reception. The insensitivity of the latter indicates the use of signal averaging techniques, which are most efficiently carried out at high repetition rate (typically 1 kHz), with due regard to average powers.

11.2 Comparison with Conventional Transducers

There are important differences between lasers and piezoelectric probes in terms of the proximity and coupling to the sample and the nature of the wavefield (intensity, generation efficiency, bandwidth, directivity), which are summarised in the Table. Because of the lower sensitivity of laser reception, one particularly important aspect of generation has proved to be efficiency. By way of example thermoelastic bulk wave amplitudes in the range 1-10 nm can generated at a range of 10 mm using a modest 100 mJ laser pulse. Similar surface wave amplitudes are also achievable. In the ablation regime, the amplitudes may approach an order of magnitude larger. These values compare favourably with the amplitudes measured from a short-pulse piezoelectric probe with comparable or slightly smaller bandwidth.

Physical parameter	Comparison of laser and piezoelectric probe
Proximity to sample	Laser non-contact and totally remote: probe must remain in contact via solid or liquid medium.
Couplant requirement	Laser requires no couplant: problems of couplant variability with piezoelectric probe.
Ultrasonic amplitudes	Laser generates amplitudes comparable with short pulse piezoelectric probe.
Beam directivities	Laser variable by optical means; probe not variable, but better for high directivity.
Signal bandwidths	Laser generates broader band signals than probe.
Size and weight	Laser larger and heavier but usable remotely. Actual source much smaller than probe.
Effect of environment	Laser isolatable from harmful environment; contact probe rapidly deteriorates.
Economics	Laser technique more expensive.
Safety considerations	Extra precautions needed with high power lasers

Higher amplitude signals can certainly be generated in the ablation regime using laser pulses in the range 1-10 J, although accompanied by a fair level of damage (use of an optical fibre would impose a more stringent limit, to prevent fibre damage). Thus the replacement of a piezoelectric probe by the laser source should give very little loss of sensitivity. The laser competes more closely with the EMAT since this is also strictly speaking non-contact and capable of use at elevated temperatures. However, EMATS are relatively inefficient generators of ultrasound over any reasonable bandwidth, generating signals at least an order of magnitude smaller than the thermoelastic laser source.

For many reasons the laser will not replace piezoelectric materials as ultrasonic transmitters. However its non-contact nature, bandwidth and resolution capabilities assure its place in quantitative NDE .

REFERENCES

1.   Carslaw H S and Jaeger J C: *Conduction of Heat in Solids*, Clarendon Press, Oxford 1959.
2.   Scruby C B and Drain L E: *Laser Ultrasonics*, Adam Hilger, Bristol 1990.
3.   Hutchins D A, Dewhurst R J and Palmer S B. J Acoust Soc Am **70** (1981) 1362.

4. Scruby C B, Dewhurst R J, Hutchins D A and Palmer S B. J Appl Phys **51** (1980) 6210.
5. Dewhurst R J, Hutchins D A, Palmer S B and Scruby C B. J Appl Phys **53** (1982) 4064.
6. Edwards C, Taylor G S and Palmer S B. J Phys D **22** (1989) 1266.
7. Dewhurst R J, Edwards C, McKie A D W and Palmer S B. Appl Phys Lett **51** (1987) 1225.
8. Aindow A M, Dewhurst R J, Hutchins D A and Palmer S B. Proc Acoustics 80 Conf (Edinburgh: Institute of Acoustics) 1989 277.
9. Cielo P, Nadeau F and Lamontagne M. Ultrasonics **23** (1985) 55.
10. Vogel J A and Bruinsma A J A: *Nondestructive Testing*; Euro NDT '87 (eds Farley J M and Nichols R W), Pergamon Press, Oxford 1988, 2267.

RECEPTION OF ULTRASOUND BY LASER

C.B. Scruby
National NDT Centre, Oxfordshire, UK

1. INTRODUCTION

Most of the techniques to be considered in this lecture employ the principle of optical interferometry, the light being reflected from or scattered by the surface subject to ultrasonic displacement. Such interferometers may be divided into two types. In the first, light from a surface is made to interfere with a reference beam to give a measure of optical phase and hence instantaneous surface displacement. The second type is really a high resolution optical spectrometer which detects changes in the frequency of the scattered or reflected light, which in turn depend upon the velocity of the surface. First however, I shall briefly introduce a non-interferometric method, known best as the knife edge technique, which although limited in application can nevertheless be a simple and sensitive ultrasonic receiver. Although the techniques are all admittedly rather insensitive compared with piezoelectric devices, they do offer a number of advantages:

- They are non-contacting and thus do not disturb the ultrasonic field.
- Being remote, the point of measurement may be quickly moved and there are no restrictions on surface temperature.
- High spatial resolution may be obtained without reducing sensitivity. The measurements may be localized over a few μm if necessary.
- Interferometric measurements may be directly related to the wavelength of the light, so that no other calibration is required.
- They have a truly broadband frequency response, difficult to achieve with piezoelectric probes especially at high frequencies.

2. KNIFE-EDGE TECHNIQUE

A laser beam directed at a surface will in general be diffracted by any periodic disturbance tilting the surface. In the knife-edge technique the beam is focused to a small spot on the surface whose diameter is comparable with or less than the acoustic wavelength, so that the diffracted beams diverge from the surface. Only the zero- and two first-order beams are significant. A knife-edge is inserted to break the symmetry and cut out one beam; interference fringes are now formed between one first-order and half the zero-order diffracted beam. The sensitivity is determined by the rms background noise, which is given by [1]:

$$0.133 \ \Xi = 0.133 \left[\frac{hc\lambda\Delta f}{\eta W_0} \right]^{\frac{1}{2}} \tag{1}$$

where h is Planck's constant, c the velocity of light, λ the optical wavelength, Δf the ultrasonic bandwidth, η the quantum efficiency of the detector, W_0 the beam power and Ξ the expression in square brackets. It is assumed that the signal is maximised by adjusting the spot radius r so that $r = 0.279\Lambda$, Λ being the ultrasonic wavelength. Note that because the technique depends upon surface tilt it is best applied to surface and related waves. It also needs good reflectivity.

3.   REFERENCE BEAM (MICHELSON) INTERFEROMETRY

Interferometers of this type have various designs but can usually be described as variations of the classic Michelson reference beam interferometer (RBI). In Figure 1 light from a monochromatic source is collimated to form a parallel beam which is divided by a beam splitter, part going to a static reference mirror or retroreflector, and part to the surface whose displacement is required. The beams are reflected back to the beam splitter where they are recombined, a fraction of each going to the detectors. The relative phase of the signal and reference beams at the detector depends on the optical path difference. For an integral number of wavelengths difference, the beams are in phase and constructively interfere. If the difference increases by $\frac{1}{2}\lambda$, interference is destructive and the light intensity is a minimum. Thus as the surface moves, the detector output varies sinusoidally. Since the optical path difference must take account of both the outgoing and return beams, one cycle corresponds to a movement of $\frac{1}{2}\lambda$. The detector output is related to the displacement, x, of the surface thus:

$$V = V_1 + V_0 \sin(4\pi x/\lambda) \tag{2}$$

where V_0 is the amplitude of the interference or beat signal, and V_1 a dc bias which is zero for the difference signal in balanced detector arrangements, eg Figure 1.

Coherence is very important in practical interferometry. When a low coherence source is used in interferometry, the contributions to the interference from the various wavelengths present lose register after a number of cycles resulting in a decrease in the interference signal. The optical path difference that may be accepted without appreciable loss of interference quality, the "coherence length", may be too small for practical application since the path lengths of the signal and reference beams must be equalized to this accuracy. Lasers used for ultrasonic reception usually have a coherence length of at least several centimetres so that the matching of optical paths is not so critical and the distance from interferometer to specimen can be varied.

Many variations of the basic Michelson design are possible [1] to maximize the use of the available light, simplify the alignment and reduce sensitivity to noise, some of which I shall now describe. Firstly, the signal beam is usually focused onto the surface.

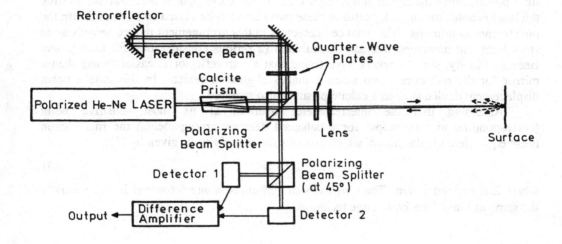

Figure 1
Design of RBI using retroreflector and calcite prism to restore coincidence if incident
and reflected beams from surface [1].

This enables good performance to be achieved with surfaces that lack a good optical finish and also makes the signal much less sensitive to surface tilt. If the surface has a diffusing or matt finish, focussing is essential to obtain reasonable sensitivity.

The use of a polarizing beam splitter in conjunction with a polarized laser and quarter-wave plates as shown in Figure 1 enables best use to be made of the available light power, Since the signal and reference beams have orthogonal polarization, rotating the laser enables the intensity ratio in these two beams to be varied so as to optimise the interference conditions. The balanced detectors in this arrangement reduce sensitivity to stray light and unwanted intensity modulation (a serious consideration for high power lasers). Finally, some workers have found that a retroreflector is easier to use than a mirror for the reference beam since it makes alignment easier. In this case a beam displacement device such as a calcite prism is also required.

Assuming that the interferometer operates at its most sensitive point (corresponding to zero output for a balanced detector arrangements) the rms photon noise equivalent displacement, which controls the sensitivity, is given by [1]:

$$0.113 \, \Xi \hspace{5cm} (3)$$

where Ξ is defined above. The sensitivity of a Michelson interferometer is thus virtually the same as that of the knife-edge technique.

4.   STABILISATION AGAINST VIBRATION

It is particularly difficult to use laser interferometry to measure very small ultrasonic displacements in the presence of low frequency background vibrations of many micrometres in laboratory and industrial environments. At these displacements the response of the interferometer is no longer linear and simply filtering out the low frequency signal distorts the ultrasonic signal. There are also large variations in sensitivity including changes of sign which depend on the timing of the ultrasonic signal with respect to the low frequency displacement.

The most direct way is to change the optical path length of the reference beam to compensate for vibration by moving the mirror or retroreflector by an electromagnetic [2] or piezoelectric [3] device. Although instability problems may be experienced with mechanical devices due to phase changes connected with mechanical resonances, this type of stabilisation technique has been successfully implemented. Quite large low frequency displacements may be accommodated, although misalignment may occur if the mirror becomes tilted.

Stabilisation may be accomplished without mechanical movement by means of electro-optic cells. These are devices containing a material in which birefringence may be induced by the application of an electric field. This means an increased velocity for light polarised parallel to an axis of the cell and a decreased velocity for the orthogonal polarisation. This may be used directly to change the optical path length of a suitably polarised beam passing through the electro-optic cell, and hence introduce compensation for vibration. Most cells now make use of the linear Pockels effect in crystals such as

KDP. Because of the limitations of the electro-optic effect, the range of control available is one or two cycles, barely sufficient to compensate for vibration levels usually encountered.

An alternative way [4] of using the electro-optic effect does not have this limitation (Figure 2). Two or more cells are combined to form a phase shifting device. This is fed by a phase-locked low frequency feed-back from the interferometer output. At high frequencies, the response is proportional to displacement whilst at low frequencies, the output is proportional to velocity. The advantages of this stabilization method include its unlimited range of compensation, stability and speed, which are very useful in applications requiring rapid scanning.

5.   HETERODYNE INTERFEROMETRY

The basic optical principles are still that of the RBI. However the light beam from the specimen is made to interfere with a reference beam that has been frequency-shifted (typically 40 MHz) by an electro-optic or (more commonly) an acousto-optic cell such as a Bragg cell [5]. This produces an output signal centred in the shift frequency, but phase-modulated in proportional to the displacement of the surface.

This interferometer (Figure 3) offers good stabilization against background vibration, and is therefore a popular design. It is limited in bandwidth by the frequency-shift, but can be made very narrow-band, and hence extremely sensitive. However, direction and phase information cannot be deduced from the more common single side-band versions. The sensitivity is typically a factor of $\sqrt{2}$ lower than the homodyne system.

6.   OTHER TYPES OF REFERENCE BEAM INTERFEROMETRY

An alternative method of running a Michelson interferometer is to use two interference signals 90° out of phase (in quadrature) [1]. This approach has sometimes been adopted as a relatively cheap and robust method for stabilizing against large extraneous vibrations. The quadrature signals are usually obtained by means of polarizing beam splitters and retardations plates. There are several different processing techniques for directional discrimination or the linearization of the output when using quadrature signals.

Various other types of interferometer have been applied to ultrasonic measurement, including Mach-Zehnder and Fizeau configurations, and dual-beam differential systems for surface waves.

7.   LONG-PATH-DIFFERENCE INTERFEROMETRY

Time-delay or long-path-difference interferometry (LPDI) determines the small frequency changes in the light scattered from moving surfaces. The output is

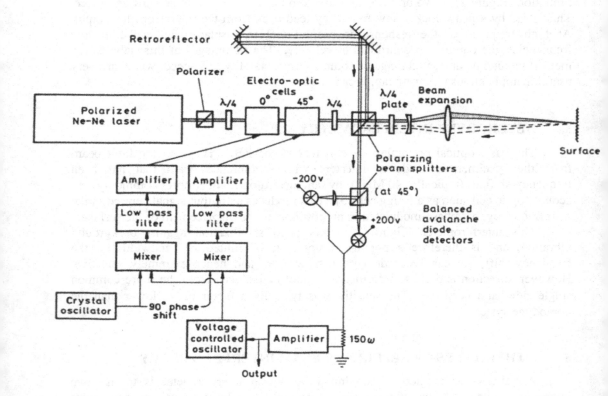

Figure 2

RBI for the detection of ultrasonic displacements with electro-optic compensation for low frequency vibration [4].

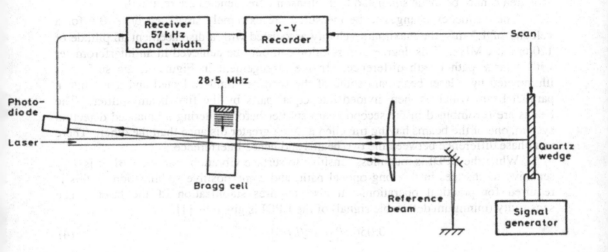

Figure 3
Heterodyne RBI using water-filled Bragg cell for beam-splitting and frequency-shifting
[5].

proportional to surface velocity rather than displacement. For highly reflecting surfaces, velocity interferometers have no particular advantage over those using a reference beam previously described. Indeed, because of the long optical path required, they may be much less practical, particularly at low frequencies. Velocity interferometers are however not affected by low frequency vibrational motion of the surface, nore are they limited by the speckle effect. They thus offer a potential advantage in sensitivity when dealing with poorly scattering or diffusing surfaces, particularly when the illuminated spot area cannot be made small and high ultrasonic frequencies are required.

The frequency changes to be measured are extremely small, eg $\delta f \sim 10^{-8}$ for a velocity of 1.5 m/s, the maximum velocity associated with a displacement amplitude of 120 nm at 2 MHz. This degree of resolution can only be achieved in an interferometer with a large path length difference. In the arrangement in Figure 4, the surface is illuminated by a laser beam and some of the scattered light collected and made into a parallel beam which is then divided into equal parts by the first beam splitter. The beams are recombined in the second beam splitter before entering a balanced detection system, one of the beams having travelled a much greater distance than the other. There is a phase difference between the two beams and hence interference.

While the LPDI is much less sensitive to surface vibration than the RBI, it is very sensitive to changes in the long optical path, and some positive stabilization of this is required for practical operation. It also requires stabilization of the laser. The sensitivity (minimum detectable signal) of the LPDI is given by [1]:

$$0.056\Xi/\left[\sin\left(\pi f L/c\right)\right] \tag{4}$$

The maximum sensitivity occurs when $f = \frac{1}{2}c/L$.

8.   FABRY-PEROT INTERFEROMETRY

A alternative way of increasing the effective path length difference in a velocity interferometer is to reflect the beam across the interference volume or "cavity" a large number of times using two highly reflecting surfaces, separated by a distance h. This is the principle of the Fabry-Perot interferometer (FPI) or "étalon", as used in high resolution spectroscopy. The interference between a large number of reflected beams produces a response curve consisting of a sequence of narrow peaks which are separated by a frequency spacing $\Delta \upsilon = c/2h$ (Figure 5). Fully constructive interference is obtained when all the transmitted beams are in phase.

When the reflectivity of the étalon surfaces R is close to unity, the width to half-height is given by [1]:

$$\upsilon_s = \frac{c(1-R)}{2\pi h\sqrt{r}} = \frac{c}{2h\Im} \tag{5}$$

where $\Im$ is "finesse", the ratio of order spacing to line width ($\Delta \upsilon/\upsilon_s$). To obtain good resolving power, it is necessary to make R as high as possible, employing a

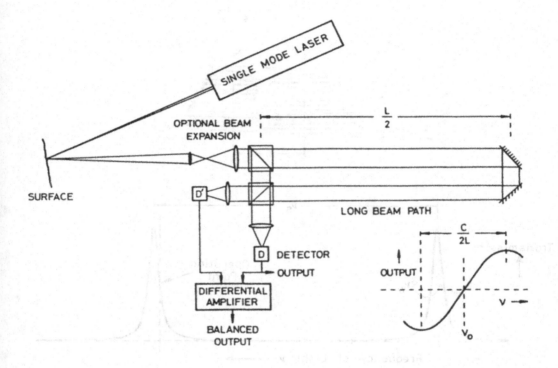

Figure 4

Design of LPDI to measure velocity of moving surface by means of phase shift between shorter and longer optical paths [1].

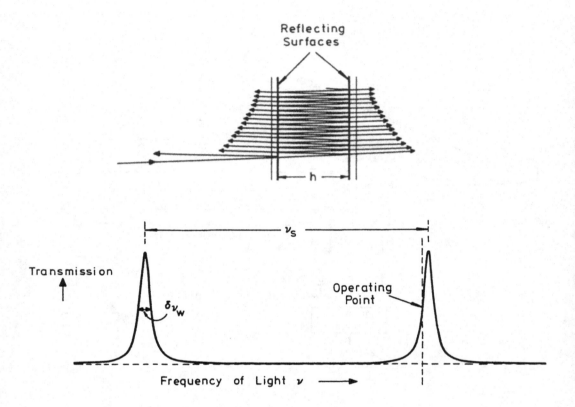

Figure 5
Multiple interference of light beams in a FPI, and example of spectral response curve
[1].

low-absorption multiple dielectric coating. When measuring small changes in the frequency of light reflected or scattered from a moving surface, h should be adjusted so that the laser frequency is near a point of maximum slope of the response curve in Figure 5. Even if a single mode laser is used, the Fabry-Perot tends to drift unless it is temperature-stabilized. Thus an active compensation technique to maintain the correct operating point on the response curve is very desirable.

The minimum detectable ultrasonic signal of frequency f for a plane FPI is [1]:

$$0.113 \; \Xi \frac{\sqrt{4+\beta^4}}{\beta} \tag{6}$$

where $\beta = 2f/\delta\upsilon_s$ and it is assumed that $f \ll \Delta\upsilon$. It can be seen that the sensitivity rises with f to a maximum response when $\beta = \sqrt{2}$, giving a minimum detectable signal of $0.226 \; \Xi$, before falling again at higher f. If, instead of conventional transmission, the beam reflected back towards the source is used, better high frequency response is obtained, using a form of "side-band stripping"[6].

9.   OPTICAL FIBRES AND INTERFEROMETRY

Optical fibres have begun to make an impact on the types of interferometer that can be used for ultrasonic reception. They can be used in several different ways. Firstly, a single-mode fibre can be used to transmit light from a relatively bulky laser to a very small probe head that contains the optical components that constitute the interferometer, enabling it to be used in confined spaces.

Secondly, an optical fibre can be used to transmit light from the interferometer to the surface of specimen and back again in all type of system, or just to collect the light from the specimen surface (in the LPDI or CFPI). The interferometer can then be remote from the specimen and just the end of the fibre scanned, a great advantage for some applications. For the RBI a single-mode fibre must be used since it constitutes one arm of the interferometer. Care must be taken to compensate for phase changes caused by vibration of the fibre. In the case of a CFPI or LPDI and a rough surface, a multimode fibre is preferable because of its greater light-gathering capacity (étendue).

Thirdly, an interferometer can be constructed with the optical fibre as an integral part. In the Sagnac design of time delay interferometer, a relatively large delay can be accommodated in a small volume, although compensation against vibration will be necessary [7]. Optical fibres can also be used to form a heterodyne Mach-Zehnder interferometer [5].

10.   ULTRASONIC RECEPTION ON ROUGH SURFACES

So far I have only considered the optical detection of ultrasound on reflecting specimen surfaces. Although this can be achieved for many laboratory measurements, it is a distinct luxury for field applications. I now consider in turn the performance of each system in turn on surfaces that scatter rather than reflect light. The knife-edge technique requires a good reflecting surface and is therefore not included.

10.1 Reference Beam Interferometry

The problem for all types of RBI (whether homodyne or heterodyne) on rough surfaces is one of speckle. As large a speckle as possible is required to maximise coherent light return, and the size of each speckle is inversely proportional to the dimensions (radius r_0) of the illuminated spot. Assuming a perfectly diffusing, non-absorbent surface, the sensitivity is thus multiplied by a further factor [1]:

$$0.25 \; \Xi(r_0/\lambda) \tag{7}$$

Note that for absorbent surfaces, W_0 in the expression for Ξ must be reduced by some factor to take into account losses at the surface.

10.2 Long-Path-Difference Interferometry

Provided the spot size is not a limitation, the sensitivity of the LPDI is governed by its étendue, ie the optical intensity entering the interferometer. This in turn depends on the distance of the collecting lens from the surface and its aperture, providing of course that the interferometer is able to accommodate a beam of this aperture and intensity. The expression for sensitivity becomes [1]:

$$0.113 \; \Xi \left[\frac{(F/D)}{\sin(\pi f L/c)} \right] \tag{8}$$

where F and D are focal length and diameter of the receiving optics.

10.3 Confocal Fabry-Perot Interferometry

The étendue of the plane FPI is limited. As the aperture is increased, the spread of inclinations of the light rays soon reduces its resolution. An alternative arrangement [8] which uses two partially transmitting coaxial confocal mirrors of equal radii, has a very much less severe restriction on light gathering power. It offers greater efficiency when light from an extended source is being analyzed, permitting the use of a larger illuminated spot on the surface without reduction of sensitivity. For an exactly confocal separation of the mirrors, the optical path length around the cavity is insensitive to the angle of inclination and distance from the axis, and thus light may be accepted over a wide input angle.

The typical CFPI (Figure 6 [9]) is very sensitive to variations in mirror separation. To ensure that it is maintained at the most sensitive part of the response curve, a stabilization system is incorporated. A sampling beam with polarization orthogonal to the signal beam is also passed through the interferometer. Its transmission is maintained at half its peak value by feed-back to a piezoelectric displacer.

The resolution, frequency response and sensitivity of a CFPI may be calculated in the same way as for a plane FPI, with some modifications as follows. First, the path of a typical ray makes two alternate transits across the confocal cavity, so that the effective

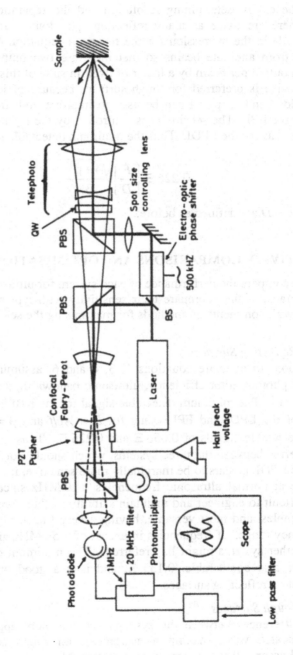

Figure 6
Design of CFPI for ultrasonic measurement of surface motion [9].

length of the cavity determining resolution and the separation of orders is doubled. Secondly there are twice as many reflections per double transit so that R must be replaced by R• in the expression for the finesse, f (equation 4). Thirdly, interference only arises from alternate beams so that there are two output beams, resulting in a reduction in output per beam by a factor of two. In spite of this slight loss of sensitivity, the CFPI device is preferred for rough surfaces because of its much greater étendue. Note that side-band stripping can be used to improve high frequency response of the CFPI as for the FPI. The sensitivity is controlled by the light intensity gathered by the collecting lens as for the LPDI. Thus the minimum detectable signal is given by [1]:

$$0.226 \; \Xi \left[\frac{F}{D} \right] \frac{\sqrt{4 + \beta^4}}{\beta} \qquad (9)$$

where ß, F and D are defined as before.

11. SENSITIVITY COMPARISONS AND OPTIMISATION

I now compare the performance of each system for ultrasonic reception on various types of surface. I then compare their sensitivity with typical contact probes, before concluding with comments on methods for maximising the sensitivity of interferometric detection.

11.1 Reflecting Surfaces

I propose to compare equations 1, 3, 4 and 6, assuming that the laser power, wavelength, photodetector efficiency, ultrasonic bandwidth, and hence Ξ are the same for each system. The minimum detectable signal for the RBI is 0.113 Ξ. The optimum sensitivity of the LPDI and FPI occurs for $L = \frac{1}{2} c/f$ and $\beta = \sqrt{2}$ respectively, giving minimum detectable signals of 0.056 Ξ and 0.226 Ξ. Thus there is little to choose in sensitivity terms between the three systems, which should not be surprising. Note that although the LPDI appears to be marginally more sensitive, it is difficult to achieve this performance at normal ultrasonic frequencies < 10 MHz since this implies $L > 15$ m, which is difficult to engineer and maintain stability. I therefore recommend the RBI as being the simplest and most versatile, having a very flat displacement response over a wide frequency range. If frequencies in excess of ~ 50 MHz are being considered then one of the other systems might be preferred. The minimum detectable signal for the knife-edge is also comparable, 0.133 Ξ, and it is a good option for surface waves measurement on reflecting surfaces.

11.2 Rough Surfaces

The differences between the systems become more apparent and the situation more complicated when making measurements on rough surfaces. For a useful comparison I assume that the laser light is scattered by a perfectly diffusing surface (no absorption), that a small sinusoidal movement is being detected, and no light losses in the optics.

LPDIs and CFPIs generally have adequate intrinsic étendue, and the limiting factors are the aperture available for the reception of the scattered light and the practicability of long interferometer paths or mirror separations. The étendue depends upon D/F, which varies with the particular application: I will assume a somewhat generous value of 0.1. Thus the minimum detectable signals for LPDI and CFPI (both under optimum conditions) are respectively 1.13 Ξ and 4.52 Ξ.

The maximum sensitivity of the RBI when the spot size is diffraction limited after expanding the illuminating beam to fill the available aperture assumed to be $D/F = 0.1$. The illuminated spot diameter ($2r_0$) at the $1/e^2$ intensity points is then 18 λ. Then the minimum detectable signal is 2.5 Ξ, comparable with the LPDI and CFPI. However, various factors may prevent a perfectly focused spot being obtained, and a more realistic minimum signal of 25 Ξ corresponds to a spot diameter of 180 λ, somewhat worse than the other systems. Note that reference beam signals are always subject to random variations due to speckle, a problem that can largely be avoided with the other types of interferometer.

For general application to ultrasonic measurements on rough surfaces at a typical 5 MHz, the CFPI appears to have the best all round performance, the LPDI being difficult to build in an optimised form for this frequency range. The RBI is a good option if there is close enough access to use a short focal length lens and a near-diffraction limited spot, and time to search for a suitable speckle. Above ~ 20 MHz the LPDI may be a better option, whereas below ~ 2 MHz the RBI has increasingly better sensitivity than either the CFPI or LPDI, both of whose displacement responses are approximately proportional to frequency.

11.3 Comparison with Conventional Transducers

There is relatively little information in the literature on the absolute sensitivity of other types of ultrasonic receiver. Unpublished work has shown that a typical damped, short-pulse (ie "broadband") piezoelectric probe can detect an ultrasonic signal as small as 10^{-13} m (0.1 pm) for a bandwidth of ~ 10 MHz. Resonant PZT transducers with bandwidth of ~0.1 MHz can probably detect displacements of 10^{-14} m, irrespective of surface condition. Data for EMATs are even more difficult to obtain, but it is estimated that an EMAT of bandwidth ~ 1 MHz can detect signals greater than ~ 5 x 10^{-12} m, several orders of magnitude less sensitive. For a reflecting surface I only consider the RBI, since all four methods have similar sensitivity. Assuming 2 mW He-Ne laser power, a bandwidth of 10 MHz and 75 % efficient photodiodes, yields Ξ = 2.9 x 10^{-11} m. Thus from equation 3 this RBI will detect displacements of ~ 3 x 10^{-12} m, which is between 1 and 2 orders of magnitude less sensitive than a piezoelectric probe of similar bandwidth, but at least as sensitive as an EMAT of comparable bandwidth. In principle the sensitivity of the RBI could be raised by using a higher power laser, but this brings saturation problems for the photodiode detectors.

For non-absorbent rough surfaces consider a CFPI using a more powerful 1 W argon laser. This system is insensitive to speckle effects and is more practicable than the LPDI for ultrasonic frequencies in the range 1 - 10 MHz. For the same ultrasonic bandwidth, Ξ is now 1.4 x 10^{-12} m. Thus for $D/F = 0.1$ at peak sensitivity this system could detect ~ 6×10^{-12} m. In practice, absorption at the surface and other factors may reduce this by as much as an order of magnitude. Even so, it is still competitive in sensitivity terms with the EMAT, its main rival for non-contact reception, especially at elevated temperatures.

11.4 Methods for Improving Sensitivity

It is clear from the previous section that optical reception methods are appreciably less sensitive than conventional piezoelectric probes, and important therefore to try and maximise their sensitivity. Having ensured that the system is well designed with minimal losses so that all the available light reaches efficient photodetectors and its sensitivity photon noise limited, there are two main avenues to explore. The first is higher power lasers, since the signal/noise is proportional to the square root of the power. There are two caveats: high power lasers can be noisy, and detectors have a limited capacity for high light levels. On rough and absorbent surfaces, the latter is not a problem, especially if a CFPI or LPDI is used. Practical limits on useful cw systems restrict the power to tens of watts. However, normal-pulsed lasers have been successfully used instead of cw systems to give a high instantaneous power for reception without a high average power.

The second general method is to develop faster, higher capacity signal averaging. The disadvantage here is the time taken to collect the data during which everything must remain stable. Even with extremely fast digital and hybrid averaging techniques, the reverberation time places a lower limit on the interval between consecutive pulses.

To summarise, optical methods offer outstandingly good performance as entirely non-contact ultrasonic receivers in terms of bandwidth and spatial resolution, but with a considerable sensitivity penalty.

REFERENCES

1.   Scruby C B and Drain L E: *Laser Ultrasonics*, Adam Hilger, Bristol 1990.
2.   White R G and Emmony D C. J Phys E; Sci Instrum **18** (1985) 658.
3.   Palmer C H and Green R E Jr. Appl Optics **16** (1977) 2333.
4.   Drain LE and Moss B C. Opto-electronics **4** (1972) 429.
5.   Whitman R L, Laub I J and Bates W J. IEEE Trans Sonics and Ultrasonics **SU-15** (1968) 186.
6.   Monchalin J-P, Héon R, Bouchard P and Padioleau C Appl Phys Lett **55** (1989) 1612.
7.   Jungerman R L, Khuri-Yakub B T and Kino G S J Acoust Soc Am **73** (1983) 1838.
8.   Hercher M Appl Optics **7** (1968) 951.
9.   Monchalin J-P. Appl Phys Lett **47** (1985) 14.

APPLICATIONS OF LASER ULTRASOUND

C.B. Scruby
National NDT Centre, Oxfordshire, UK

1.   INTRODUCTION

In this lecture I shall attempt to summarise the wide range of potential applications for laser ultrasound. The combination of laser generation with reception offers a completely remote ultrasonic inspection system, thereby opening up a wide range of exciting applications to flaw detection and materials characterisation. Although I propose to concentrate on the combination, I also discuss applications of laser-received or laser-generated ultrasound alone as appropriate, especially in the first two sections.

2.   MEASUREMENTS OF ULTRASONIC FIELDS

2.1 Measurements of Vector Displacement on Solid Surfaces

While the main interest in using laser interferometry is likely to be in the measurement of normal surface displacements due to the incidence of ultrasonic waves, in-plane (transverse) displacements may sometimes be important. One example is the detection of plane shear waves at normal incidence to the surface. A second is the oblique incidence of shear waves with their plane of polarisation perpendicular to the plane of incidence (SH waves). By combining normal and transverse components, the magnitude and direction of any vector displacement can hence be deduced [1]. Care must be taken to solve the boundary conditions equations correctly, taking into account phase changes due to critical angle effects, etc. Nevertheless, this is a powerful approach since such information is difficult to derive from conventional contact probes. Ultrasonic generation by laser is less flexible since both compression and shear waves are usually produced.

2.2 Measurements in Liquids

Although the main focus of attention is likely to be solid surfaces, it may be required to make measurements of ultrasonic fields in a liquid, eg during immersion testing. There are two main methods: the first is to use the surface of the liquid (or a membrane at the surface) to reflect light back into the interferometer. The second involves immersing a reflecting membrane (pellicle) into the liquid which is light enough to be displaced by the ultrasonic waves.

2.3 Scanning Methods

One practical benefit of using laser techniques is ease of scanning. Thus it is possible to generate 2-D ultrasonic images in a relatively short space of time. For ultrasonic reception by laser, the method of scanning is important. Angular or sector scanning requires a simple arrangement involving a rotating mirror or prism. Because the angle of incidence of the laser beam on the specimen surface varies during the scan, angular scanning is only suitable for diffusing surfaces. For reflecting surfaces parallel scanning is required to ensure the laser beam is always at perpendicular incidence. This is more difficult to engineer: one method involves placing the scanning mirror at the focus of a large diameter objective lens [1]. Parallel scanning is not necessary for laser-ultrasonic generation, since the source is insensitive to optical angle of incidence.

2.4 Bulk Wave Studies

Although great advances have been made with theoretical calculations of elastic wave propagation, it is still not possible to predict the ultrasonic response of complicated geometries, especially if the media are inhomogeneous and/or anisotropic. An alternative approach is to carry out an experimental simulation, using lasers to generate and receive the ultrasound. In the ablation regime, the ultrasonic source closely mimics the δ-function source used in theoretical studies, while the interferometer measure broadband displacement at a point, again as required by theory. This simulation approach has mainly been used to study the diffraction of ultrasound by defects [2]. It hs also been used to study ultrasonic propagation in fibre- and particulate-reinforced composite materials prior to acoustic emission measurements [3].

2.5 Surface Wave Studies

Laser techniques are particularly suitable for studying surface acoustic waves since the focused spot from a laser can be made much smaller than most ultrasonic wavelengths. If an interferometer receiver is used, care must be taken to ensure that surface tilt does not distort the signals. Directional surface waves are also ideal candidates for study with a two-beam or differential interferometer system. Very high sensitivities have been recorded, eg 6×10^{-14} m over 1 Hz bandwidth.

One of the simplest applications is the measurement of surface wave velocity, an important parameter for the characterisation of certain optical and electronic materials. Accuracies of ~ 1 part in 10^5 should be achievable [4]. If narrow bandwidths are used, then high sensitivities can also be obtained. Various studies have also been carried out of surface wave attenuation and scattering, with special reference to surface-breaking cracks [1]. There is more detail in section 4.2.

2.6 Full Field Visualisation

As an alternative to scanning the laser beams to produce an image, there are a number of methods for measuring small displacements over the whole surface simultaneously. Although the only NDE applications where they are likely to be relevant are certain surface wave measurements, I will list the main methods for completeness: holographic interferometry, double pulsed holography and speckle pattern interferometry [1]. Weaknesses compared with single point interferometry are lower sensitivity and inability to deliver continuous spatial and temporal coverage simultaneously.

3.   TRANSDUCER CALIBRATION

Laser techniques can make a very important contribution to the calibration of ultrasonic transducers. Although these are now widely used, they are not always fully calibrated, because it is both difficult and costly, particularly in the case of NDE contact transducers.   The non-laser techniques available (eg reciprocity methods, standard test-blocks, water tank procedures) all have draw-backs of one kind or another. The laser interferometer uniquely offers absolute measurement over a wide frequency band, high spatial resolution for beam plotting, rapid scanning, freedom from couplant (eliminating thereby a critical source of error), and non-contact sensing (so that the transducer and its wave-field are undisturbed).

3.1 Employing Laser Interferometry

A common method of calibrating a contact probe is to plot the ultrasonic beam from the transducer in pulse-echo mode using a small spherical reflector in a water tank. The beam profile in a solid such as steel or aluminium is deduced using a refraction correction to take into account the different wave speeds in water and the solid. It is clearly important to know whether this indirect method correctly predicts the beam in the solid in view of the change of loading of the front face of the transducer when switching from water to solid.

Interferometry can be used to measure the ultrasonic field generated by the transducer in situ on a specimen of the required material. Because it is impossible to scan a receiver inside a solid sample, a surface measurement has to suffice. Thus in a typical experiment (Figure 1), a rectangular test-block is used and the laser beam is scanned across the face opposite to the transducer [5]. Calibration of a compression wave probe in transmission mode is relatively straightforward, although a correction is needed for the boundary conditions at the receiver surface, especially if the transducer generates a broad beam, or if an angled shoe is being used. Shear wave probes can be calibrated using the same method, although in this case extra care must be taken with the boundary correction, especially outside the ultrasonic critical angle [5].

Transducers can be calibrated in reception mode, either by a substitution method (which slightly perturbs the ultrasonic field), or by employing a specially designed transparent test-block that enables the laser beam to be focused on the front face of the device under test. It may be preferable to use the laser-generated source (section 3.2).

3.2 Employing Laser-Generated Ultrasound

The laser source is ideal for calibrating ultrasonic transducers in receiver mode, or acoustic emission sensors. For the former interest usually centres on beam profile and pulse shape, whereas for the latter bandwidth and sensitivity tend to predominate. However, a broadly similar arrangement can be used in both instances, with test-blocks and geometries chosen to suit the eventual application. The laser is fired at a point on one surface of the test-block [1]. If a small amount of damage is tolerable, then the ablation regime (possibly assisted by fluid constraint to enhance generation at a reduced damage level) is better for calibrating ultrasonic probes, especially for their sensitivity to compression waves. Otherwise the thermoelastic regime must be used, remembering its

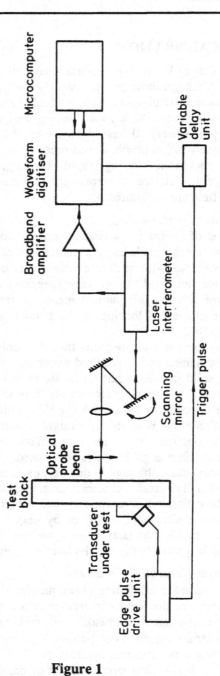

Figure 1

Experimental arrangement for calibrating ultrasonic fields generated by shear-wave transducers. The interferometer beam executes a raster scan across the specimen by means of a parallel scanner [5].

complex lobe structure, and its efficiency in shear-wave generation. The source is then scanned across the test-block to give a 2-D image of the response of the transducer. For acoustic emission calibration, the thermoelastic regime is ideal since it simulates a surface-breaking crack event particularly well.

4.   DEFECT DETECTION

The largest application area for ultrasonic NDE is the detection and characterization of defects that might affect the integrity of engineering structures and components. The most important defects are cracks, whether at the surface or in the bulk, since stress concentrations at crack tips seriously reduce structural strength. If laser ultrasonic techniques are to have a major impact on NDE then they must demonstrate ability to detect and characterise major types of defect.

Piezoelectric transducers are cost-effective for most NDE applications, so that laser techniques are unlikely to displace them. Nevertheless there are a few cases where these probes are extremely difficult to use or ineffective. Furthermore these will probably increase in number with time as industry seeks higher standards of quality in a wider range of materials and components under greater environmental extremes. Although temperature is currently the hostile environment of most interest, there are also problems for piezoelectric transducers in corrosive and radioactive environments, which cause the materials used in transducer construction to degrade slowly. It is also difficult to scan contact probes rapidly over large areas and obtain sufficient spatial and temporal resolution for the inspection of some advanced materials.

4.1 Bulk Defects

The majority of conventional ultrasonic measurements of bulk defects consist of pulse-echo or transmission data, and laser techniques can be used in the same manner. In the pulse-echo case the laser source and interferometric receiver are the same side of the potential defect and any reflected ultrasound recorded and interpreted, using the time delay to estimate the depth of the defect and the echo amplitude to estimate its size. The main differences when using laser techniques lie in the simultaneous generation of compression and shear waves. Furthermore unless an array is used the beam profile is fixed by the physical processes occurring in the source. Thus it is difficult to use laser ultrasonic techniques for generating and receiving angled shear waves alone, one of the most commonly used conventional methods.

The first use of laser ultrasound to detect bulk defects in metals appeared in the scientific literature in the 1970s [eg 6]. Various experiments were carried out to demonstrate that typical real and simulated defects could be detected, either in transmission or pulse-echo. These early systems tended to use rather powerful lasers (eg 1 J) for generation so that considerable damage ensued. By directing the interferometers (mostly RBIs) at reflecting surfaces, reasonable signal/noise ratios were obtained.

A more recent method for detecting and characterising defects is the Time-of-Flight Technique, where attention is focused upon the ultrasound diffracted from any defects. The Technique is more accurate and reliable than pulse-echo for sizing defects because it measures the arrival times of the diffracted ultrasound rather than the reflected (scattered) amplitudes. It is less likely to miss significant defects.

For optimum performance, the Technique requires broadband transducers which generate wide beams in the specimen. From theoretical studies the maximum diffracted wave amplitude for a typical geometry occurs for beam angles in the range 50 - 70° to the surface normal. The pulsed laser source and interferometer fulfil these criteria better than many piezoelectric transducers. The interferometer has a very wide reception beam as required, and ablation generates a very broad compression wave beam, both normal to the surface. In the thermoelastic regime, the compression wave beam is a broad lobe centred on 60°, which is ideal for the Technique. Figure 2 shows data obtained during the laser-ultrasonic inspection of a welded joint that was suspected of having a crack in the heat-affected zone region [7]. In the central A-scan (b) a signal is observed ahead of the back-wall echo. From its arrival time it is identified as diffraction from a point in the HAZ of the weld. When a B-scan is generated (a) the diffracted signal is seen to emanate from the tip of a crack whose depth varies across the weld.

4.2 Surface Defects

Detection is important in many materials because stresses are often maximised at the specimen surface making them particularly prone to growth. This is especially true in high strength, brittle materials such as glasses and ceramics, where crack growth could be catastrophic. The critical defect size for a brittle material such as a ceramic could be ~ 100 μm or even less, providing a considerable challenge to NDE. In the case of ceramics, the strong preference is for a technique which is non-contact to avoid degradation by couplants such as water. Laser ultrasonic techniques have good potential for the detection of surface defects in these materials, being capable of generating high amplitude Rayleigh waves that probe the surface region. Published data show that defects as small as 100 μm in depth can be detected in metals, using entirely non-contact laser techniques. While this resolution is inadequate for the highest strength ceramics, there is no fundamental reason why defects as small as 10 μm cannot be detected by using mode-locked lasers for generation and exceptionally broadband detection.

In a comprehensive study [8] a series of variable-depth slots were machined in the surface of an aluminium block to simulate cracks. In pulse-echo mode two signals are detected from each slot. The first corresponds to an echo from the corner at the top of the defect, while the second appears to be reflection/diffraction from the bottom of the defect, the time difference between them being an indication of defect size. The slots also transmit Rayleigh waves (Figure 3). Furthermore, because the depth penetration of the Rayleigh wave into the surface is a function of frequency, the transmission spectrum loses the higher frequencies, which are scattered into the reflected wave (Figure 3c).

Many of the above data have been obtained using a small circular ablative source of ultrasound. The laser light can alternatively be focused by a cylindrical lens into a line on the surface [1], ideal for generating directional Rayleigh surface waves and thus for detecting linear surface defects that are are parallel to the source line. A further alternative is to focus the incident laser light into a circle on the surface of the specimen by means of an axicon lens [9]. Thermoelastic stresses in this ring launch circular

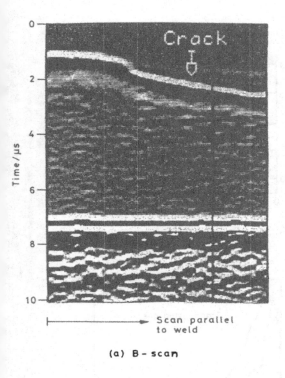

(a) B - scan

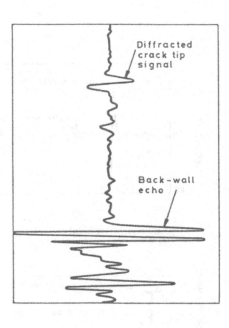

(b) Central A-scan from (a)

Figure 2
(a) Change in arrival time of wave diffracted from crack tip in Time-of- Flight B-scan
shows variation of depth of crack in heat-affected zone of weld; (b) single A-scan from
centre of (a) showing crack tip diffraction and back-wall signals in detail [7].

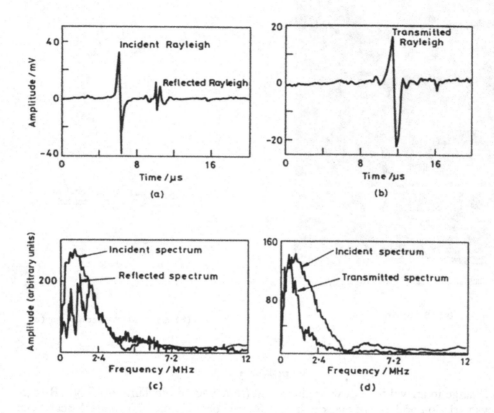

Figure 3
(a) Interaction of laser-generated Rayleigh pulse with 0.75 mm deep slot in aluminium block; source and receiver on same side of slot; (b) receiver on opposite side of slot showing transmission of Rayleigh wave; (c) comparison of frequency spectra of incident and reflected Rayleigh waves (a); (d) ditto for Rayleigh waves in (b) [8].

Rayleigh waves that converge onto the centre of the circle, where they produce a very high surface displacement, which can be detected by means of a Michelson laser interferometer directed at the centre. The amplification factor for the Rayleigh waves is of the order of 20, obtained without crossing into the damage-inducing ablation regime. Thus small defects (depth ~ 0.1 mm) can be detected with much better signal-to-noise than otherwise. The alignment of the axicon lens with respect to the surface is very critical, which may restrict its broader application.

4.3 Subsurface and Bonding Defects

Defects immediately below the surface of the sample can sometimes prove a problem for contact transducers, because they are located within a "dead zone", and yet do not break the surface for detection by surface waves. In the first study of this kind [10], a series of simulated sub-surface defects was machined in an aluminium plate at a range of depths from approximately 1 - 5 mm. The source laser and interferometer were directed at the same point on the surface to make a pulse-echo measurement. When the laser beams were directed at an unflawed region of the sample, the waveform was a simple step-function. However when the beams were directed at a point over one of the flaws the shape of the waveform changed to comprise a small amplitude high frequency oscillation superimposed on a larger amplitude, lower frequency oscillation. The depth of the flaw can be determined directly from the higher frequency oscillation, while the diameter can be deduced from the lower frequency, provided the depth has first been calculated. A good method for sizing the defect is to scan the laser probe across the surface [10], the size being estimated from the points where oscillations first begin to appear (Figure 4). This method has potential for near-surface lamellar flaws where the oscillation is a very obvious feature. In contrast the sensitivity of the technique falls rapidly with increasing depth for a given defect diameter.

One of the most important types of sub-surface defect is a disbond or delamination of a surface layer. The layer may be some form of coating (eg a ceramic plasma-sprayed onto a metal), or one ply of a composite. There is a growing interest in protective coatings and in many different types of composite, and techniques for detecting unbonded areas are currently being sought. As for some monolithic ceramics, non-contact inspection is often preferred to avoid contamination. Thermography and laser-ultrasonics are possible alternative techniques. The principle upon which laser ultrasonic inspection would operate is very similar to that described above, in which characteristic oscillations are measured.

5. MATERIALS CHARACTERISATION

The nondestructive measurement of mechanical and materials parameters is a growing field because of the increasing industrial requirement for high quality materials, whether traditional engineering materials such as aluminium and steel, or new materials that have been developed for specialised applications in for instance the aerospace industry. Ultrasonics is an obvious candidate for internal measurements since most

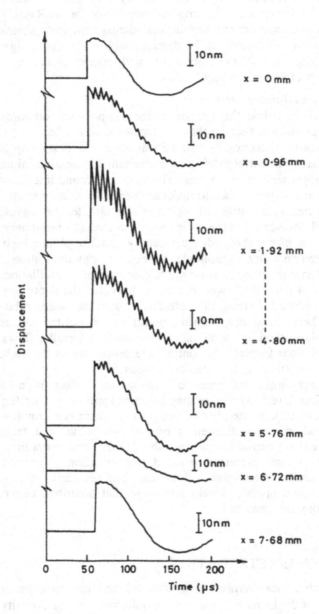

Figure 4
Waveforms obtained from linear scan of coincident source and receiver across specimen
containing 6.35 mm diameter sub-surface defect [10].

engineering materials are opaque to light and highly attenuating to other radiations. Techniques have been under laboratory development for some years which are based on the measurement of either ultrasonic velocity or attenuation. Already some are under trial for production control.

The limitations of conventional transducers, especially regarding couplant, are likely to be more serious for on-line measurements than for off-line flaw detection. This is because many engineering materials are fabricated at higher temperatures than can be accommodated by conventional probes, and also because the product may be travelling at too high a speed to maintain consistent coupling between probe and product.

There is an extremely wide range of measurements of potential interest to industry, which include dimensional, compositional, microstructural, mechanical and surface properties, together with characterisation of all defects, discontinuities, interfaces and joints. Finally, there is an interest in monitoring and controlling various thermo-mechanical processes such as solidification, extrusion and forging, including internal termperature measurement. I shall briefly discuss some examples where laser techniques have been applied.

5.1 Thickness Measurement

Ultrasonic thickness gauging has been in use for some years. The simplest approach is to measure the time-of-flight of a compression wave pulse using a pulse-echo transducer in contact with the sample. In addition to being limited to modest temperatures, this method is restricted to thicknesses > 0.5 mm, because of transducer dead-times and problems in resolving overlapping echoes. Laser ultrasonics has the potential for surmounting these problems. Thus a nitrogen laser has been successfully used to generate ultrasound and measure the thickness of stainless steel films in the range 10 - 300 μm with a typical accuracy of 1 %. An alternative approach to thin sheet measurement, which does not demand such high frequencies, is to employ Lamb waves.

Interestingly, one industrial trial of laser ultrasonics has been on the dimensioning of hot (~ 1200 °C) steel tubes [12]. The time delay interferometer (LPDI) is directed at the same point on the tube as the generator to simulate the conventional pulse-echo probe configuration. This enabled wall thicknesses in the range 15 - 25 mm to be measured while the tube was travelling past the system at a speed of 2 m/s.

5.2 Internal Temperature Measurement

The steel industry in particular has a pressing need to measure temperatures inside their products, both to optimise the process, and to save re-heat energy. Ultrasonics is a candidate method for achieving this because velocity is temperature dependent. Furthermore, the hostile environment of a steel mill demands a non-contact and preferably remote technique such as offered by laser ultrasonics. Various laboratory and preliminary industrial trials have been carried out, using combinations of laser and EMAT methods [eg 13]. Using tomographic reconstruction errors as small as ~ 20 °C can be achieved under ideal conditions. The biggest problem for industrialisation is a consistently high level of returned light for interferometric reception.

5.3 Elastic Constants and Phase Change Measurement

The determination of elastic constants is of great importance to the characterisation of a wide range of materials, and it can most readily be achieved nondestructively by making a series of ultrasonic measurements. Laser techniques are valuable when non-contact methods are required, and/or when elevated temperature measurements are desired. In the first study to demonstrate this capability, the elastic constants of a plutonium-gallium alloy were successfully measured over a temperature range entending up to 500 °C [14]. In a careful study of materials such as single crystal germanium, velocities have been measured to an accuracy of 0.1 %, which is comparable with classical methods [15].

There have been several studies to illustrate the ability of laser ultrasonics not only to measure elastic properties, but also to monitor changes through a phase transformation. Thus in one study [16] of iron from ambient up to 1000 °C, it was possible to detect both the Curie point and the martensitic phase transition. The aluminium and steel industries are particularly interested in being able to monitor the transformation boundary between liquid and solid metal, in order to control better processes such as continuous casting. Transmission measurements would permit the detection of the amount of liquid because of the lower ultrasonic velocity in this phase, or pulse-echo measurements could detect the solid-liquid interface. The biggest problem is again likely to be the harsh environment of a typical steel-mill. In addition to surface scale, there are likely to be strong convention currents and steam present.

5.4 Texture and Residual Stress Determination

Some preliminary studies have been carried out to demonstrate the potential of laser ultrasonic techniques for textural determination in materials such as extruded aluminium and rolled steel sheet. In principle the same method could be used to determine residual stresses, although no results have been published. However, this is a far more difficult problem, even with state-of-the-art contact probes, chiefly because the effects of stress are much smaller than those of texture.

5.5 Microstructural Monitoring

The second general method of characterising an ultrasonic signal by its attenuation has proved valuable for making nondestructive measurements of microstructure. The most useful parameter to measure is the frequency dependence of the attenuation, since it is this which characterises the source of the losses. Interfaces such as grain boundaries scatter ultrasonic energy in directions different from the main beam; they also convert compression waves into shear and vice versa. This leads to attenuation which increases as the fourth power of the ultrasonic frequency in the Rayleigh regime (scattering length parameter < ultrasonic wavelength). However the dislocations present in a plastically deformed metal remove energy from the ultrasonic beam by an absorption process, i.e. the ultrasonic energy is dissipated as heat, rather than scattered. The dependence is different for absorption, the attenuation increasing with the square of the frequency.

Laser ultrasonics should be well-suited to attenuation measurement for three main reasons. Firstly the bandwidth of both laser source and interferometer cover the entire frequency range of current attenuation studies, thus obviating the need to change transducer frequencies. Secondly, being non-contact, no correction is necessary to take into account couplant losses and the loading of the transducer on the surface. Finally, provided the areas of the laser excitation and the receiver spot are both small, it is unnecessary to make the type of diffraction correction required for conventional transducers. Thus the amplitude of the wave arrivals decays uniformly with propagation distance (inverse square law) in the absence of scattering or absorption. This is in contrast to conventional measurements with approximately plane waves from much larger diameter piezoelectric probes.

Various authors have reported such measurements mainly, but not exclusively, on steel. In a typical experiment (Figure 5) the data is dominated by a series of compression wave pulses that decay in the manner anticipated [17]. If however the time axis is expanded, it is found that these pulses broaden as they propagate, the higher frequencies being increasingly attenuated by processes such as grain scattering. The second steel has a much coarser and hence more scattering microstructure, and this produces more marked attenuation.

As an alternative, the microstructure can also be characterised from the ultrasound scattered out of the transmitted beam, ie by analysing portions of the "noise" that follow the main arrivals [17]. Thus the finer microstructure is responsible for smaller amplitude, higher frequency "noise" than the coarser microstructure. Since this method is non-contact, it should have applications to the on-line monitoring of microstructure in metals at elevated temperatures.

5.6 Characterisation of Coatings and Composites

The final materials where the potential of laser ultrasonic techniques has been demonstrated are coated materials and composites. In both cases the main benefit is absence of surface contamination. If the coating is thin the approach consists of measuring perturbations to Rayleigh waves launched along the surface; if thick, then guided waves may also be generated which can be used to characterise coating thickness and quality [18]. Various laser ultrasonic studies of composites have been carried out. To summarise, the technique can reproduce measurements made with contact probes, but in a manner that is non-contact and amenable to rapid, large-area scanning.

6.   INDUSTRIALISATION

Although there are still some gaps in our understanding, most essential fundamental work on laser generation and reception of ultrasound has been carried out. Various potential applications have been demonstrated in the laboratory; however few have undergone industrial trials. The technology is therefore ready for realistic development trials prior to full industrial exploitation. In order for industry to invest in laser ultrasonic technology it must be convinced of the benefits and value for money.

Laser ultrasonics will only make an industrial impact where there is an economic or safety imperative that cannot be met by existing technologies such as conventional ultrasonics. One likely area for early application is therefore likely to be in the area of hot (> 650 °C) inspection and process control in for instance the steel industry. Indeed

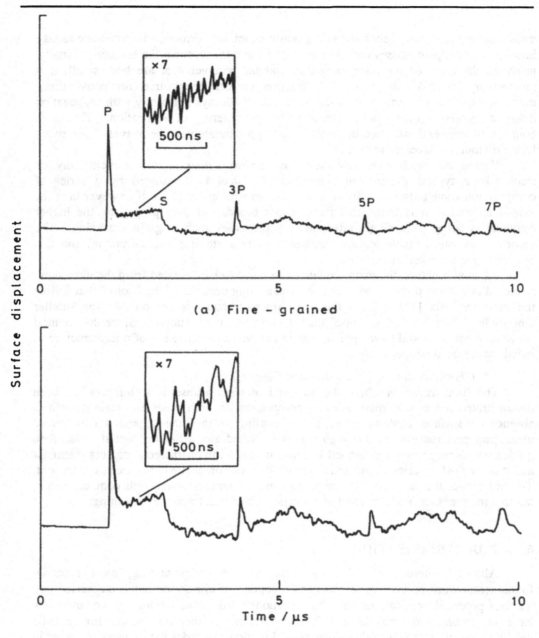

Figure 5
Data obtained from (a) finer-grained, (b) coarser-grained steel. Note successive
reflections of P-wave which progressively broaden and from which
frequency-dependent attenuation can be calculated, and ultrasonic forward scattering
which is of larger amplitude in (b) [17].

steel companies around the world are already looking seriously at laser ultrasonic technology. Indeed a number of trials have been carried out, some with laser systems, some with EMATS and some hybrid. There are a number of potential inspection and measurement problems on the steel line, the answers to which would improve product quality and process efficiency, such as locating the solid-liquid interface during continuous casting, determining the internal temperature prior to and during rolling and forging, dimensioning hot products and detecting internal defects. The aluminium industry has related problems, but at lower temperatures.

A second field where there is currently active industrial interest in laser ultrasonic NDE is the inspection of aerospace components, with special emphasis on composite structures. Trials have also been carried out to assess the technology. Here the key benefits are somewhat different: the absence of couplant and ability to be scanned readily over large areas. The third main area for the industrial acceptance of laser ultrasonics is in the inspection of advanced engineering or electronic materials. Many of these materials suffer some degradation if contacted with couplant; high temporal and spatial resolution are also required, so that some form of non-contact acoustic microscope is probably envisaged. Although I cannot predict where the technology will first make its true mark, it is likely to be in one of these three areas.

REFERENCES

1. Scruby C B and Drain L E: *Laser Ultrasonics*, Adam Hilger, Bristol 1990.
2. Ravenscroft F A, Newton K and Scruby C B. Harwell Report R13334 (1989).
3. Buttle D A and Scruby C B. J Acoust Emission 7 (1988) 211.
4. Murray D and Ash E A. Proc IEEE Ultrasonics Symp (1977) 823.
5. Moss B C and Scruby C B. Ultrasonics 26 (1988) 179.
6. Wilcox W W and Calder C A. Instrum Tech 25 (1978) 63.
7. Scruby C B. Ultrasonics 27 (1989) 195.
8. Cooper J A Dewhurst R J and Palmer S B. Phil Trans R Soc A320 (1986) 319
9. Cielo P, Nadeau F and Lamontagne M. Ultrasonics 23 (1985) 55.
10. Aindow A M, Dewhurst R J, Palmer S B and Scruby C B. NDT International 17 (1985) 328.
11. Tam A C. Appl Phys Lett 45 (1984) 510.
12. Keck R, Krüger B, Coen G and Häsing W. Stahl und Eisen 107 (1987) 1057.
13. Wadley H N G, Norton S J, Mauer F and Droney B. Phil Trans Roy Soc Lond A320 (1986) 341.
14. Calder C A Draney E C and Wilcox W W. Lawrence Livermore Laboratory Report UCRL-84139 (1980).
15. Aussel J D and Monchalin J-P. Ultrasonics 27 (1989) 165.
16. Dewhurst R J Edwards C McKie A and Palmer S B. J Appl Phys 63 (1988) 1225.
17. Scruby C B, Smith R L and Moss B C. NDT International 19 (1986) 307.
18. Scruby C B, Brocklehurst F K, Moss B C and Buttle D J. Nondestr Testing Eval 5 (1990) 97.

APPLICATIONS OF QUANTITATIVE ULTRASONICS

R.B. Thompson
Iowa State University, Ames, IA, USA

ABSTRACT

Practical applications of quantitative ultrasonic techniques are discussed. After a review of the role of nondestructive evaluation in the life cycle of structural components, attention is focussed on several specific examples. Included are discussions of uses of models in designing for inspectability and validating proposed procedures for flaw detection, techniques for measurements of deformation and fracture related material properties, and techniques for measurement of stress. References to more detailed review articles are provided.

1. INTRODUCTION

The ultrasonic measurement techniques which have been discussed in other lectures in this set of notes are motivated by a variety of technological problems. Figure 1 shows a historical interpretation of the evolution of problems and techniques, adapted from an article by R. S. Sharpe [1]. The essential idea is that the field of NDE has evolved from an emphasis on defect detection, location and sizing to the characterization of the materials in which those flaws reside. In particular, one seeks to measure the resistance of the material to deformation and fracture and to determine the loads which drive the growth of flaws. Future directions involve integration of the wide varieties of techniques which have evolved in response to these problems into a set of engineering tools for use throughout the entire life cycle of structural components.

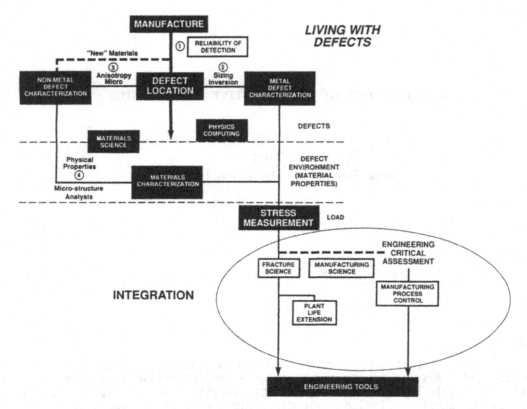

Fig. 1. Evolution of quantitative NDE capability (after Ref. [1]).

In this lecture, examples are given of the use of models during the design of a component to ensure inspectability, and of the use of quantitative measurements to control material manufacturing, to check the integrity of components after manufacture, and to monitor property degradation during use.

2.   CONSIDERATION OF INSPECTABILITY IN DESIGN

The first step in the life of a structural component is its design. In modern strategies such as concurrent engineering, simultaneous engineering or unified life cycle engineering, it is recognized that all aspects of a part's life cycle should be considered in the design process. For example, it is important to consider whether a part will be inspectable and what effect various inspection scenarios will have in extending the life or enhancing the reliability of that component. Figure 2 schematically shows how a variety of such factors must be simultaneously considered in the design process [2].

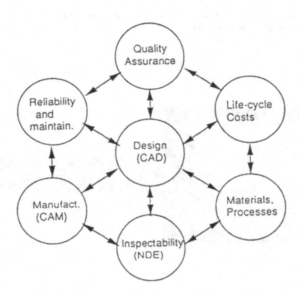

Fig. 2.   A general unified life cycle engineering environment (after Ref.
[2]).

     To achieve such a goal, it is necessary to develop realistic models
of the inspection process. The fundamentals underlying such models have
been described in the lectures presented by Achenbach [3]. As an example
of the output of such models, Fig. 3 shows a contour map of the
probability of detecting critical flaws as a function of position in a
simulated turbine disk [4]. The ultrasonic model, which included the
effects of transducer radiation characteristics, measurement noise, etc.,
predicted the probability of detecting flaws as a function of location
and as influenced by the part geometry. From such information, it is
possible to predict such quantities as the probability that particular
flaws would go undetected and hence the likelihood that a failure would
occur under various inspection scenarios.

3.   MEASUREMENT OF MATERIAL PROPERTIES TO CONTROL PROCESSING

     After a component is designed, it is essential to select a
manufacturing process which will produce material with the required
mechanical properties. An example may be found in the preparation of
sheet metal for deep drawing into automobile door panels and beverage
cans. For such applications, an important property is the anisotropy of
the plastic deformation. One would like plastic deformation to occur
more readily in the plane of the sheet than through its thickness, so
that excess thinning or tearing does not occur during deep drawing. The
anisotropic plastic deformation properties of a sheet are controlled, in
turn, by the preferred crystallographic orientation (texture) of the
grains which comprise it. As discussed in the lecture by Sayers [5],
information concerning the degree of texture can be inferred from

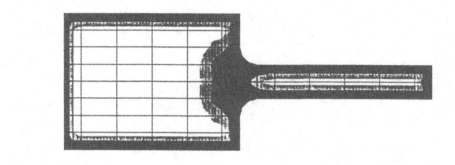

POD in Disk, 0.5cm Fillet, 0.25cm Index
ID NUMBER:1
UNKNOWN - MAG MIN: 0.00E+00 MAX: 1.00E+00

0.00E+00   2.50E-01   5.00E-01   7.50E-01   1.00E+00

Fig. 3.   Example of a CAD-generated display of inspectability (POD) as a
          function of position within the cross-section of a simulated
          turbine disk.   The POD scale ranges from black for the lowest to
          white for the highest POD values.   The low POD in the fillet
          region is due to the combination of a coarse scan mesh and a
          tight fillet radius. (after Ref. [4]).

measurements of the anisotropy of the ultrasonic wave speeds.   Hence
there is the potential to predict formability parameters
nondestructively.

      Considerable effort has been placed on the development of practical
instrumentation based on these ideas.   In one approach, the speeds of
guided modes propagating at 0°, 45° and 90° with respect to the rolling
direction are measured with electromagnetic-acoustic transducers (EMAT's)
[6].   These measurements allow one to obtain the three orientation
distribution coefficients (ODC's) W_{400}, W_{420}, W_{440}, as discussed by
Sayers.   It is then possible to compute the anisotropic elastic moduli of
the sheet [5].   This enables one to utilize existent correlations between
anisotropic values of Young's modulus and formability parameters [7].
However, since the use of EMATs allows the information to be obtained
nondestructively and without couplant [8], the possibility for on-line,
process control applications emerges.   Figure 4 presents results of a
static test of this technique.   The formability parameter $\bar{r}$, known as the

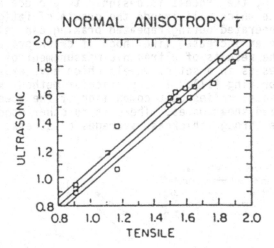

Fig. 4.   Comparison of average plastic strain ratio, r̄, inferred from
ultrasonic velocity, to that measured destructively on tensile
coupons (after Ref. [8]).

average plastic strain ratio determines the propensity for plastic
deformation to occur in the plane of the sheet as opposed to through the
thickness, with high values being desirable for deep drawing. An
excellent correlation is seen between the values inferred from the
ultrasonic measurements and those obtained in destructive tensile tests.

4.   MEASUREMENT OF STRESS

As implied by Fig. 1, a prediction of the fitness of a structural
component for service requires a knowledge of defects, the material's
resistance to fracture (material properties such as fracture toughness)
and the stresses which tend to cause the defect to grow. Under ideal
conditions, the stresses can be calculated from geometry and applied
loads. However, in some cases, those loads are not fully known and in
others, residual stresses are superimposed on the applied stresses.
Hence it is desirable to be able to directly measure the stresses.

X-ray diffraction provides one such tool. By directly measuring
stress induced changes in interatomic spacings, one obtains a fundamental
measurement from which stress can be readily computed. However, because
of the limited penetration of X-rays, only near surface stresses can be
determined, and these may not be characteristic of the bulk of the
material. Moreover, there are a variety of safety considerations
associated with the use of x-ray defraction.

Ultrasonic techniques provide an alternative [9] which can sense the
stress state in regions extending further into the material. Techniques
using Rayleigh waves to measure near surface stresses (but penetrating
much deeper than those sensed by X-rays) are discussed in the lecture by
Del Santo [10]). Another example is discussed here. When railroad

wheels are fabricated, the process is designed to produce a compressive
stress at the surface which will impede the growth of fatigue cracks.
However, the heat generated during repeated braking can relieve these
stresses, leading to an increased likelihood of crack propagation.
Figure 5 presents the results of ultrasonic measurement of the radial
distribution of stresses in a set of wheels which have experienced
various degrees of braking [11]. This particular method, known as the
birefringence technique, relies on a comparison of the speeds of
orthogonally-polarized shear waves. There is no other nondestructive way
to obtain comparable through-thickness averages of stress.

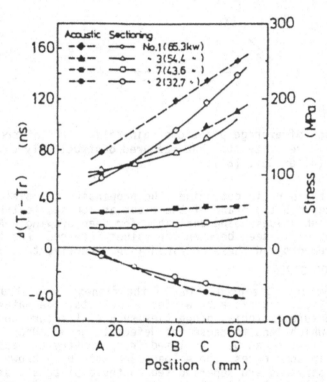

Fig. 5. Comparison of residual stresses obtained by acoustoelastic
         birefringent technique to those obtained with the sectioning
         method. (after Ref. [11]).

5.   USE OF MODELS TO VALIDATE IN-SERVICE INSPECTION PROCEDURES

    The failure of structural components are often governed by
statistical laws. Because of the wide distribution of failure times,
design lives must be selected quite conservatively, with most individual
components having considerable unused life. An example of great current
interest is found in the nuclear power industry, in which many power
plants are reaching their design lives. Economic realities makes it
desirable to extend their service period beyond the initial design life,

but this can only be done if inspection ensures that failure has not
initiated, i.e. that a crack of a particular size has not yet formed.
However, because of the serious consequences of failure, validation of
the ability of inspection procedures to find the required flaws is
absolutely essential. Such validations have been traditionally performed
experimentally, using full-scale mock-ups. However, the size and
geometrical diversity of nuclear power plant components, e.g. pressure
vessels and piping, renders construction of the required suite of samples
uneconomical. Models of the inspection process are thus finding
increasing use in validation [12]. As an example, Fig. 6 presents the
results of an experimental validation of three models for the pulse-echo
angle beam response of an internal crack as a function of probe position
[13]. This comparison was obtained as a part of an international
round-robin comparison to validate models. Included in the comparison is
a model developed at Kassel University, as described by Langenberg in
another lecture in this set of notes [14].

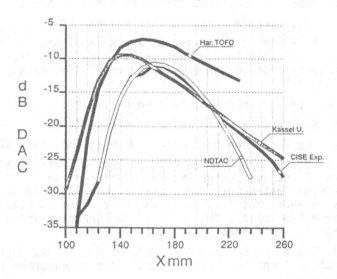

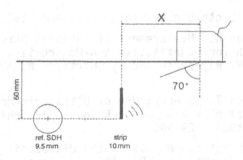

Fig. 6.   Comparison of ultrasonic model predictions to measured
          scattering from a buried crack (after Ref. [13]).

6. OTHER APPLICATIONS

The above are but a few examples, selected from the author's personal experience. Numerous others could be given. These would include techniques to detect and characterize flaws and to measure failure related material properties and stress throughout the entire life cycle of a structional component; design, processing and manufacturing, and in-service usage. More detailed discussions and numerous other examples may be found in References [15-17].

REFERENCES

1. Sharpe, R. S.: Closing Comments, in: Nondestructive Characterization of Materials (P. Höller, V. Hauk, G. Dobmann, C. O. Ruud and R. E. Green, Jr., Eds.) Springer-Verlag, Berlin, 1989, 886-892.

2. Schmerr, L. S. and D. O. Thompson: Incorporating Inspectability into Design: The New Role of NDE in Concurrent Engineering, in: NDE's Role in Concurrent Engineering, ASME, New York, in press.

3. Achenbach, J. D., Lecture notes in this volume, 1992.

4. Gray, J. N., T. A. Gray, N. Nakagawa, and R. B. Thompson: Models for Predicting NDE Reliability, in: Metals Handbook, Vol. 17, Nondestructive Evaluation and Quality Control, ASM, Metals Park, Oh., 1989, 702-715.

5. Sayers, C. M., Lecture notes in this volume, 1992.

6. Thompson, R. B., Lecture notes in this volume, 1992.

7. Mould, P. R. and T. R. Johnson: in: Sheet Met. Ind. 50 (1973), 328

8. Papadakis, E. P., R. B. Thompson, S. Wormley, K. Forouraghi, D. D. Bluhm, and H. D. Skank: Design and Fabrication of an Industrial-Grade Instrument to Measure Texture and Predict Drawability in Sheet Metal, in: Review of Progress in Quantitative Nondestructive Evaluation, Vol. 10B, (D. O. Thompson and D. E. Chimenti, Eds.) Plenum Press, NY, 1991, 2053-2060.

9. Thompson, R. B., W.-Y. Lu, and A. V. Clark: Ultrasonic Method, in: SEM Monograph on Techniques for Residual Stress Measurements (J. Lu, Ed.) SEM, in press.

10. Del Santo, P. P., Lecture notes in this volume, 1992.

11. Fukuoka, H.: Ultrasonic Measurement of Residual Stress, in: Solid Mechanics Research for Quantitative Non-Destructive Evaluation (J. P. Achenbach and Y. Rayapakse, Eds.), Martinus Nijhoff, Dordrecht, 1987, 275-299.

12. Thompson, R. B. and T. A. Gray: Use of Ultrasonic Models in the Design and Validation of New NDE Techniques, in: Phil. Trans. Roy. Soc. Lond. A320, 1986, 329-340.

13. Validation of Mathematical Models of the Ultrasonic Inspection of Steel Components, Document PISCDOC (90) 12, Programme for Inspection of Steel Components, Commission of the European Communities Joint Research Centre, Ispra, Italy, 1991.

14. Langenberg, K. J., Lecture notes in this volume, 1992.

15. Thompson, R. B.: Quantitative Ultrasonic Nondestructive Evaluation Methods, J. Appl. Mech. 50 (1983) 1191-1201.

16. Thompson, R. B. and D. O. Thompson: Ultrasonics in Nondestructive Evaluation, Proc. IEEE 73 (1985), 1716-1755.

17. Thompson, R. B. and H. N. G. Wadley: The Use of Elastic Wave-Material Interaction Theories in NDE Modelling, Critical Reviews in Solid State and Materials Sciences 16 (1989) 37-89.

ULTRASOUND IN SOLIDS WITH POROSITY, MICROCRACKING AND POLYCRYSTALLINE STRUCTURING

C.M. Sayers
Schlumberger Cambridge Research, Cambridge, UK

1 INHOMOGENEOUS MATERIALS

The scattering of an ultrasonic wave in an elastically inhomogeneous medium results in a frequency dependent velocity and attenuation of the wave. The ultrasonic attenuation and dispersion are therefore sensitive to the microstructure of the material. Since the microstructure also has an important effect on material properties there is considerable interest in the development of ultrasonic techniques for the determination of fracture toughness, hardness, impact strength, yield strength and tensile strength for example [1]. Variations in the microstructure within a sample and from sample to sample may arise from composition fluctuations, inclusions, grain growth due to faulty heat treatment, incorrect fibre fraction in composites, porosity and microcracking. In contrast to the elastic constants of the material, which can be obtained from ultrasonic measurements at a single frequency, the determination of the above mentioned properties requires the measurement of the frequency dependence of the velocity or attenuation. An example is the use of ultrasonics to predict the yield strength of plain carbon steel [2]. This prediction is based on the Hall-Petch relations, which relate the yield strength and impact transition temperature to the mean grain size, an important parameter determining the frequency dependence of the ultrasonic attenuation.

In general, the interpretation of frequency dependent velocity and attenuation data requires the use of multiple scattering theory. This is introduced below, and the simplifications which occur at low volume fractions are discussed and the magnitude of the frequency dependent effects assessed.

1.1 Single scattering theory

The wave equation for a homogeneous isotropic elastic solid may be written in vector form for the displacement u:

$$(\lambda + \mu)\nabla(\nabla.\mathbf{u}) + \mu\nabla^2\mathbf{u} = \rho\frac{\partial^2\mathbf{u}}{\partial t^2} \tag{1}$$

where λ and μ are Lamé's elastic constants and ρ is the density of the medium. Writing $\mathbf{u} = -\nabla v + \nabla \times \mathbf{A}$, where the vector potential $\mathbf{A}$ may be further written as $\mathbf{A} = \nabla \times (\mathbf{r}\pi)$, equation (1) may be written as two scalar Helmholtz equations for the compressional and shear components v and π [3]. Assuming a time harmonic displacement $u_i(\mathbf{r}, t) = u_i(\mathbf{r})e^{i\omega t}$ these equations reduce to

$$(\nabla^2 + k^2)v = 0. \qquad (\nabla^2 + K^2)\pi = 0$$

where

$$k = \omega(\frac{\rho}{\lambda + 2\mu})^{1/2}, \qquad K = \omega(\frac{\rho}{\mu})^{1/2}.$$

Consider a medium with density ρ_1 and elastic wave velocities $v_{l1} = (M_1/\rho)^{1/2}$, $v_{t1} = (\mu_1/\rho)^{1/2}$ where $M_1 = \lambda_1 + 2\mu_1$, λ_1 and μ_1 being Lamé's elastic constants of the medium, containing n_0 elastic inhomogeneities per unit volume. The scattered field generated by the j'th inhomogeneity may be calculated from the total wave field incident at j. This consists of the incident wave together with waves scattered from all other scatterers in the medium.

If the volume fraction of scatterers is sufficiently low, multiple scattering can be ignored. It will be assumed further that the distribution of scatterers is statistically independent, ie

$$p(\mathbf{r}_1, \mathbf{r}_2, \cdots, \mathbf{r}_N) = p(\mathbf{r}_1)p(\mathbf{r}_2) \cdots p(\mathbf{r}_N). \tag{2}$$

Here $p(\mathbf{r}_1, \mathbf{r}_2, \cdots, \mathbf{r}_N)$ is the probability of a configuration with scatterer 1 at $\mathbf{r}_1$, 2 at $\mathbf{r}_2$, etc and $p(\mathbf{r}_j)$ is the probability of j being at $\mathbf{r}_j$ independently of the positions of the other scatterers. Consider an incident longitudinal wave with wave number $k_1 = \omega/v_{l1}$ in the medium surrounding the scatterers, where ω is the angular frequency of the wave. The scattered longitudinal wave is

$$U_{scatt} = f(\theta, \varphi)\frac{e^{ik_1 r}}{r}$$

at large distances from the scatterer, $f(\theta, \varphi)$ being called the scattering amplitude. Spherical polar coordinates have been used with polar axis $0x_3$ along the direction of propagation of the incident wave and θ and φ being the polar and azimuthal angles respectively. For spherical scatterers $f(\theta, \varphi) = f(\theta)$ is a function of θ only.

The velocity v_l' and attenuation α of the medium may be obtained from the wave number $\beta = \omega/v' + i\alpha$ describing the propagation of the average wave in the inhomogeneous medium. Foldy [4] treated the scattering of scalar waves by isotropic elastic scatterers distributed at random. It was found that

$$(\frac{\beta}{k_1})^2 = 1 + \frac{4\pi n_0}{k_1^2}f \tag{3}$$

where f is the scattering amplitude and is independent of direction. Lax [5] generalized Foldy's treatment to include anisotropic scattering and a non-random distribution of scatterers. This treatment gave

$$(\frac{\beta}{k_1})^2 = 1 + \frac{4\pi n_0}{k_1^2}cf(0) \tag{4}$$

with $f(0)$ the scattered amplitude in the forward direction. The correction factor c is defined by c = effective field/coherent field, the effective field at a scatterer depending on the correlation in scatterer positions. In the limit of scatterer size being small compared to their mean spacing, $c \to 1$. Other discussions include those of Urick and Ament [6], Waterman and Truell [7], Twersky [8] and Lloyd and Berry [9]. In the low concentration limit these give equation (3) with f replaced by $f(0)$ for anisotropic scatterers.

For spherical scatterers in the low volume fraction limit, therefore, we take

$$(\frac{\beta}{k_1})^2 = 1 + \frac{4\pi n_0}{k_1^2} f(0) \tag{5}$$

Ying and Truell [3] solve the scattering problem by expanding the incident longitudinal wave v_{inc}, the scattered longitudinal and shear waves ψ_s and π_s and the longitudinal and shear waves v_2 and π_2 excited in the inclusion as

$$v_{inc} = \frac{1}{k_1} \sum_{l=0}^{\infty} (-i)^{l+1} (2l + 1) j_l(k_1 r) P_l(\cos\theta)$$

$$v_s = \sum_{l=0}^{\infty} A_l h_l(k_1 r) P_l(\cos\theta)$$

$$\pi_s = \sum_{l=0}^{\infty} B_l h_l(K_1 r) P_l(\cos\theta).$$

$$v_2 = \sum_{l=0}^{\infty} C_l j_l(k_2 r) P_l(\cos\theta)$$

$$\pi_2 = \sum_{l=0}^{\infty} D_l j_l(K_2 r) P_l(\cos\theta)$$

where $k_1 = \omega/v_{l1}$ and $K_1 = \omega/v_{t1}$ are the longitudinal and shear wavenumbers in the matrix, $k_2 = \omega/v_{l2}$ and $K_2 = \omega/v_{t2}$ are the longitudinal and shear wavenumbers in the inclusion phase and $P_l(\cos\theta)$ is the Legendre function of degree l. For a cavity $v_2 = \pi_2 = 0$. $j_l(\xi)$ is the spherical Bessel function of order l, $h_l(\xi) = j_l(\xi) - i n_l(\xi)$ is the l'th order spherical Hankel function and $n_l(\xi)$ is the spherical Neumann function of order l. The expansion coefficients A_l, B_l, C_l and D_l may be found from the boundary conditions at the surface of the scatterer.

The scattering amplitude $f(\theta)$ is related to the coefficients A_l via the expression

$$f(\theta) = -\sum_{l=0}^{\infty} (2l + 1) T_l P_l(\cos\theta) \tag{6}$$

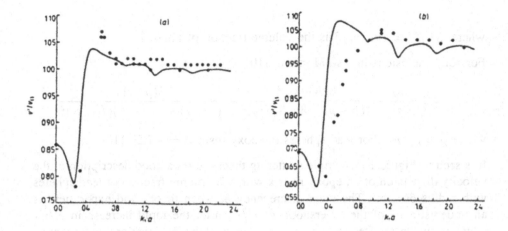

Figure 1: Normalized compressional wave velocity as a function of $k_1 a$ for (a) 5% and (b) 15% volume fraction of lead spheres of radius $660 \mu m$ in epoxy resin. The full circles give the experimental measurements of Kinra et al [11], the open circles the zero frequency result of equation (7) and the curve the results obtained from equation (5) [10].

where $T_l = (-i)^l A_l^+ / (2l + 1)$.

1.2 Comparison with experiment for lead in epoxy

Figure 1 shows a comparison between single scattering theory [10] and the experimental measurements of Kinra et al [11] for samples with 5% and 15% volume fraction of lead spheres of radius $660 \mu m$ in an epoxy resin. The open circles represent the zero frequency longitudinal velocity, which may be used experimentally for determining the volume fraction of lead present.

At low frequencies ($k_1 a \ll 1$) equation (5) takes the form [12]:

$$\left(\frac{k}{k_1}\right)^2 = 1 - \frac{4\pi n_0 a^3}{3}(A + Bk_1^2 a^2 - iCk_1^3 a^3). \qquad (7)$$

Here a is the radius of the scatterers and A, B and C are functions of ρ_1, ρ_2, k_1, K_1, k_2 and K_2; ρ_i, k_i and K_i being the density, longitudinal wave number and shear wave number for medium i with $i = 1,2$. At low frequencies, equation (7) gives

$$\frac{v_l'}{v_{l1}} = 1 + \frac{1}{2}\frac{\Delta V}{V}(A + Bk_1^2 a^2) \qquad (8)$$

$$\alpha = \frac{2\pi n_0}{3}Ck_1^4 a^6 \qquad (9)$$

where $\Delta V/V = 4\pi a^3 n_0/3$ is the volume fraction of phase 2.

For solid inclusions in a solid matrix [10]

$$A = 2 - \frac{\rho_2}{\rho_1} - \frac{3(K_1/k_1)^2}{4(1-p) + 3p(K_2/k_2)^2} + \frac{5(p-1)}{(p-1) + \frac{3}{2}(p + \frac{3}{2})(K_1/k_1)^2}$$

where $p = \mu_2/\mu_1$. For lead spheres in epoxy resin $A = -7.20$ [10].

It is seen in Figure 1 that single scattering theory gives a good description of the velocity dispersion of an epoxy matrix with 5% volume fraction of lead spheres with radius 660μm. The main discrepency between theory and experiment is an underestimate of the 'overshoot' of v'_l/v_{l1} after the rapid increase in v'_l/v_{l1} which occurs as the frequency is increased beyond the Rayleigh or low frequency region. The position of the rapid rise in velocity may be used to determine the size of the particles in the medium. A strong peak occurs in the ultrasonic attenuation in this frequency range and this may also be used for particle sizing (Figure 2). In view of the agreement seen in Figure 1a it seems reasonable to interpret the disagreement between theory and experiment seen in Figure 1b for $\Delta V/V = 15\%$ as arising from multiple scattering effects neglected in the single scattering approximation of equation (5). If this interpretation is correct, the effect of multiple scattering is seen to shift the rapid rise in v'_l/v_{l1} from $k_1 a \approx 0.3$ to $k_1 a \approx 0.5$ and to supress the magnitude of the 'overshoot' which occurs after this rise in velocity. The deviation from single scattering theory will be sensitive to the scatterer pair distribution function and therefore information on this important function should be obtainable from the velocity dispersion curve.

1.3   Multiple scattering correction

The multiple scattering theory of Waterman and Truell [7] has been widely used to treat ultrasonic wave propagation in a two phase medium consisting of spherical scatterers with density ρ_2, longitudinal wave velocity v_{l2} and shear wave velocity v_{t2}, in a matrix with density ρ_1, longitudinal wave velocity v_{l1} and shear wave velocity v_{t1}. This theory is based on the approximation that the wave field incident at a scatterer may be represented by the total wave field that would exist at the scatterer if the scatterer was not there. It is further assumed that the positions of the scatterers are statistically independent in accordance with equation (2). With these assumptions, Waterman and Truell [7] find that

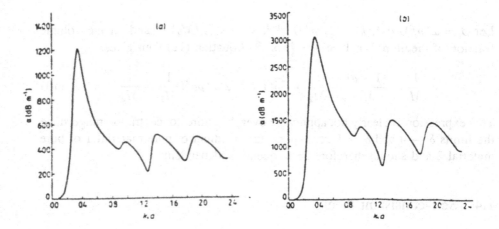

Figure 2: Compressional wave attenuation as a function of $k_1 a$ for (a) 5% and (b) 15% volume fraction of lead spheres of radius 660μm in epoxy resin obtained from equation (5) [10].

equation (5) is replaced by

$$\left(\frac{\beta}{k_1}\right)^2 = 1 + \frac{4\pi n_0}{k_1^2} f(0) + \frac{4\pi^2 n_0^2}{k_1^4}[f^2(0) - f^2(\pi)] \qquad (10)$$

for spherical scatterers. This same result was also obtained by Urick and Ament [6] by treating the scattering by many thin slabs joined together.

Lloyd and Berry [9] have applied the multiple scattering treatment of Lloyd [13,14], developed for the treatment of the electronic structure of liquids and disordered alloys, to the propagation of waves through a random array of spherical obstacles. To second order in the scattering amplitude the method gives, for the distribution of equation (2), the result

$$\left(\frac{\beta}{k_1}\right)^2 = 1 + \frac{4\pi n_0}{k_1^2} f(0) + \frac{4\pi^2 n_0^2}{k_1^4}[-f^2(0) + f^2(\pi) - \int_0^\pi \frac{1}{\sin\frac{1}{2}\theta} \frac{d}{d\theta} f^2(\theta) d\theta] \quad (11)$$

which differs from equation (10) in the high concentration limit. Lloyd and Berry [9] show further that there is an implicit assumption of a superposition of thin slabs in the treatment of Waterman and Truell [7].

For the simple case of scatterer and matrix having the same density and shear modulus but different longitudinal wave velocities the theories of Waterman and Truell [7] and Lloyd and Berry [9] both give [15]:

$$\left(\frac{\beta}{k_1}\right)^2 = 1 - \frac{4}{3}\pi a^3 n_0[1 - \left(\frac{k_2}{k_1}\right)^2][1 - \frac{2}{5}(k_1 a)^2[1 - \left(\frac{k_2}{k_1}\right)^2]]$$

$$+ \left(\frac{4}{3}\pi a^3 n_0\right)^2 \frac{(k_1 a)^2}{5}[1 - \left(\frac{k_2}{k_1}\right)^2]^2. \qquad (12)$$

Let $\beta = \omega(\rho/M)^{1/2}$, $k_1 = \omega(\rho/M_1)^{1/2}$, $k_2 = \omega(\rho/M_2)^{1/2}$ and let the volume fraction of medium 2 in 1 be $\delta = \frac{4}{3}\pi n_0 a^3$. Equation (12) then gives:

$$\frac{1}{M} = \frac{(1-\delta)}{M_1} + \frac{\delta}{M_2} + \frac{2}{5}\delta(1+\delta/2)\omega^2\rho a^2(\frac{1}{M_1} - \frac{1}{M_2})^2 \qquad (13)$$

This expression is clearly not applicable over the entire concentration range since the limits $\delta \to 0$ and $\delta \to 1$ correspond to a medium of pure material 1 or pure material 2 and should therefore be frequency independent.

1.4   Self-consistent theory

The incorrect behaviour displayed by equation (13) in the limit $\delta \to 1$ is due to the assumption in equation (12) that scatterers of type 2 are only surrounded by a medium of type 1. This is not true in the high concentration limit in which the wave incident on regions of both type 1 and 2 should have the wavenumber β of the effective medium. One approximate solution to this problem is to define the medium surrounding the scatterers in a self-consistent manner. In a two-phase medium the self-consistent scattering theory of Sayers and Smith [15-17] considers both types of material as deviations from the effective properties of the medium, and therefore considers two types of scatterers to be present. If the scatterers are surrounded by an effective medium with longitudinal wavenumber k, the wavenumber β of the coherent longitudinal wave passing through the composite is given by:

$$(\frac{\beta}{k})^2 = 1 + \frac{4\pi n_0}{k^2}\langle f(0)\rangle \qquad (14)$$

where $\langle f(0)\rangle$ is the forward scattering amplitude averaged over the two types of scatterers. Self-consistency requires $\beta = k$ allowing this equation to be solved for k. For the case of scatterer and matrix having the same density and shear modulus but different longitudinal wave velocities as considered above [15] the solution of equation (14) gives:

$$\frac{1}{M} = \frac{(1-\delta)}{M_1} + \frac{\delta}{M_2} + \frac{2}{5}\delta(1-\delta)\omega^2\rho a^2(\frac{1}{M_1} - \frac{1}{M_2})^2 \qquad (15)$$

which has the correct behaviour in the limits $\delta \to 0$ and $\delta \to 1$. It should be noted that the multiple-scattering correction in equation (15) is of opposite sign to that in equation (13) and arises from the need to define the effective properties self-consistently in the high concentration limit.

2   POROUS MATERIALS

In this lecture the effect of porosity on the propagation of ultrasound is treated and the possibility of using the frequency dependence of the velocity to determine the pore size distribution is investigated. An alternative method for determining the average pore size is to use the ultrasonic attenuation, and Ying and Truell [3] have given expressions for the scattering cross-section γ in the long wavelength limit. Measurements of γ for machined voids with diameters of about 400μm in titanium are in agreement with these predictions [18]. Franzblau and Kraft [19] have measured the attenuation of longitudinal waves in tungsten filaments. The porosity in these specimens develops after exposure to temperatures above $3000°$C, the dominant scattering species being a distribution of nearly spherical pores. Their number density $n_0 \approx 1.3 \times 10^7$ cm^{-3} and mean radius $\langle a \rangle \approx 8.2\mu$m were determined by metallographic counting proceedures. Measurements of the ultrasonic attenuation between 30 and 90 MHz gave $\langle a \rangle = 8.2\mu$m, $n_0 = 3.3 \times 10^6$ cm^{-3}. This agreement is rather good in view of the difficulty of the measurement and the analysis in terms of single scatterer theory.

Since it is a measure of particle displacement, a measurement of attenuation is more difficult than a measurement of transit time, and must also be corrected for increasing beam divergence for decreasing frequency. It is also more sensitive to the transducer coupling. In the long wavelength limit the cross-section varies as $\omega^4 a^6$, ω being the frequency of measurement and a being the radius of the scatterer. Attenuation is therefore insensitive to the presence of small pores.

Although a velocity measurement at a single frequency can only give the total volume of voids and not their size, this amount of information has proven useful in a number of applications. Papadakis and Petersen [20] investigated the possibility of using the ultrasonic velocity as a predictor of powder metal density in briquetted and sintered metal parts. A relation between density and velocity was expected since voids tend to add compliance and so decrease the elastic moduli. The data for briquetted specimens showed no universal relation between velocity and density because of variations in the size and shape of the iron particles in the different samples. For sintered specimens, however, the ultrasonic velocity was found to be a useful measure of pore volume. This is probably because the sintering process homogenises the size of the particles and joins them together at all places except at the voids.

The relationship between ultrasonic velocity and porosity has also been investigated by Reynolds and Wilkinson [21] for concrete, by Nagarajan [22] for

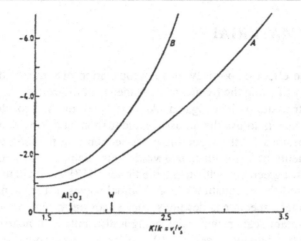

Figure 3: Plot of A and B in equation (8) for a porous medium as a function of $K_1/k_1 = v_{l1}/v_{s1}$ [12]. $K_1/k_1 \geq \sqrt{2}$ for an isotropic matrix.

sintered alumina disks with porosities up to about 40%, and by Ranachowski [23] for electrical porcelain with porosities up to about 15%. Equations for the ultrasonic velocity at finite volume fraction are given in section 2.2 below.

2.1   Single scattering theory

For a porous medium, the quantities A, B and C occurring in equations (7-9) are functions only of K_1/k_1 given by

$$\frac{K_1}{k_1} = \frac{v_{l1}}{v_{t1}} = \left(\frac{\lambda_1 + 2\mu_1}{\mu_1}\right)^{1/2} = \left(\frac{2(1-\nu_1)}{(1-2\nu_1)}\right)^{1/2}$$

where ν_1 is the Poisson's ratio of the matrix and $K_1/k_1 \geq \sqrt{2}$ for an isotropic medium. Explicit expressions for A, B and C are given in reference [12]. The attenuation obtained by substituting the expression for C in equation (9) agrees exactly with that given by equation (35) of Ying and Truell [3], who obtained the attenuation from the scattering cross-section of a single scatterer. This was calculated from the rate of energy transport by the scattered longitudinal and shear wave across a closed surface enclosing the cavity. In the calculation presented here, however, only the longitudinal wave scattered in the forward direction was used. For scalar wave scattering the equivalence of these two proceedures is expressed by the optical theorem [5]. When ultrasound is scattered by an obstacle, mode conversion occurs and the agreement follows only for incident longitudinal waves. For incident shear waves the interference with the scattered longitudinal waves must also be included since $v_l > v_t$ in an isotropic medium.

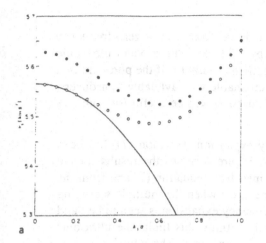
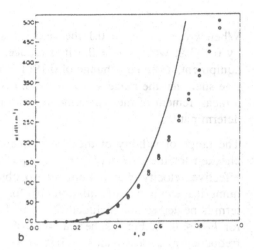

Figure 4: (a) Compressional velocity and (b) attenuation of a porous medium as a function of k_1a for $v_{l1} = 6$ Km/s, $\delta = 0.1$ and $K_1/k_1 = 2$. Shown is the single scatterer result of equation (7) valid for small k_1a (—) and the values obtained from equation (10) by neglecting (o) and including (•) multiple scattering [12].

Ying and Truell [3] have plotted the scattering cross-section against K_1/k_1. Figure 3 plots the terms A and B in the expression (8) for the velocity as a function of K_1/k_1.

In the limit $K_1/k_1 >> 1$ equations (8) and (9) give

$$\frac{v_l'}{v_{l1}} = 1 - \frac{3}{8}(\frac{K_1}{k_1})^2\frac{\Delta V}{V}[1 + (\frac{K_1}{k_1})^2\frac{k_1^2a^2}{4}] \tag{16}$$

$$\alpha = \frac{\pi n_0}{8}K_1^4a^6. \tag{17}$$

For rubber, for example, $K_1/k_1 = 40$. The velocity in this case therefore drops off very rapidly with increasing porosity and has a very large frequency dependent component.

The value of K_1/k_1 considered above is not typical. Most isotropic solids have a value of K_1/k_1 of about 2. For $K_1/k_1 = 2$, $A = -13/8$, $B = -883/360$. Equation (8) then gives

$$\frac{v_l'}{v_{l1}} = 1 - \frac{13}{16}(1 + \frac{883}{585}k_1^2a^2)\frac{\Delta V}{V}.$$

For a 10% volume fraction of pores

$$\frac{v_l' - v_{l1}}{v_{l1}} = -0.081 - 0.123(k_1a)^2.$$

When $k_1 a = \omega a / v_{l1} = 0.1$ the velocity is reduced from that at zero frequency by 0.133%; when $k_1 a = 0.5$ it is reduced by 3.337%. With modern ultrasonic equipment, a velocity change of 0.1% is readily measurable. If the pores are of a size such that the range $k_1 a = 0.1 - 0.5$ is attainable with available transducers, a measurement of the ultrasonic velocity dispersion will be useful for pore size determination.

The range of validity of the low frequency expansion (equation (7)) has been checked for the case $K_1 / k_1 = v_{l1} / v_{t1} = 2$. Figure 4 shows the results for the effective velocity v'_l and attenuation α obtained from equation (7) and from the numerical evaluation of equation (10) for the cases when the multiple scattering term is neglected and included. The low frequency expansion is seen to be valid for $k_1 a \leq 0.25$. For a spherical void of radius 10μm, this limits the ultrasonic frequency to be less than 40 MHz in Al_2O_3 for example. The porosity is most easily evaluated from the zero frequency limit of the velocity, whilst the size of the pores can be evaluated from the frequency dependence of either the velocity or the attenuation.

Winkler [24] has measured the frequency dependence of the longitudinal phase velocity and attenuation in several high-porosity (20-26%) sandstones from 0.4 to 2 MHz. The dry samples all show negative velocity dispersion (velocity decreasing with increasing frequency) and an attenuation varying as the third to fourth power of the frequency. These observations are in qualitative agreement with equations (8) and (9) and suggest a scattering mechanism as being important in these samples [24]. Brine-saturated rocks show positive velocity dispersion and an attenuation which is greater than that in the dry samples and which increases with a first to second power dependence [24]. Since the acoustic mismatch between the pores and the grains would be expected to be reduced upon saturation, these results suggest that a local fluid-flow loss mechanism is present in the saturated samples [24].

Figure 5 shows the variation of ultrasonic velocity as a function of porosity for the cases when the multiple scattering term in equation (10) is neglected and included. It is seen that the multiple scattering treatment of Waterman and Truell [7] predicts an unphysical rise in velocity for $\delta = \Delta V / V = (4/3)\pi n_0 a^3$ greater than 0.3. This is of similar origin to the difference between equations (13) and (15), the treatment of Waterman and Truell again giving non-physical results when $\delta \rightarrow 1$. The single scattering theory is seen to predict a decrease in velocity with increasing porosity as expected, but the velocity does not fall to zero even when $\delta \rightarrow 1$.

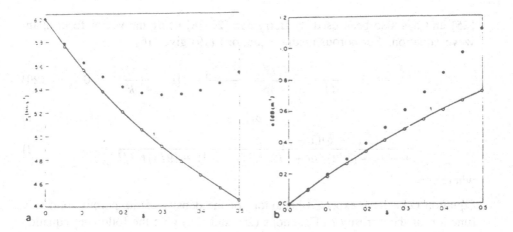

Figure 5: (a) Variation of ultrasonic velocity of a porous medium with porosity δ for $K_1/k_1 = 2$. The compressional wave velocity of the matrix is assumed to be 6 Km/s. (b) Variation of attenuation for the same case. The curves and symbols are as in Figure 4. A value of $k_1 a = 0.1$ is assumed [16].

2.2 Self-consistent theory

In a two-phase medium the self-consistent scattering theory of Sayers and Smith [15-17] considers both types of material as deviations from the effective properties of the medium and therefore considers two types of scatterers to be present. Consider the scattering from a representative elementary volume containing a representative volume of material and yet having dimensions small compared to the wavelength. If the properties of the surrounding effective medium is acoustically matched to that of the representative elementary volume, no net scattering will occur [16]. In the long wavelength limit, for which the particle size is small compared to the ultrasonic wavelength, Rayleigh has shown that only the first three terms of equation (6) are important. Thus

$$f(\theta) \approx -T_0 - 3T_1 \cos\theta - 5T_2(3\cos^2\theta - 1)/2. \tag{18}$$

Summing over all scatterers in the representative elementary volume, the requirement that the scattering vanish at all angles gives:

$$\langle T_0 \rangle = \langle T_1 \rangle = \langle T_2 \rangle = 0. \tag{19}$$

where $\langle T_i \rangle = (1 - \delta)T^{(1)} + \delta T^{(2)}$, δ being the volume fraction of pores. This is analogous to the coherent potential approximation (CPA) in quantum mechanics

[25] and has also been used by Berryman [26-28] using the vector form of the wave equation. For porous media, equations (19) give [16]:

$$(1 - \delta)[1 - \frac{3(K/k)^2}{4(1 - p) + 3p(K_1/k_1)^2}] + \delta[1 - \frac{3}{4}(\frac{K}{k})^2] = 0 \qquad (20)$$

$$\rho = \rho_1(1 - \delta) \qquad (21)$$

$$\frac{(1 - \delta)(1 - p)}{(p - 1) + (1/2)(3p + 9/2)K^2/k^2} - \frac{\delta}{[1 - (9/4)K^2/k^2]} = 0 \qquad (22)$$

where $p = \mu_1/\mu$.

Equation (21) is the expected behaviour of the density of the porous solid as a function of the porosity δ. Equations (20) and (22) give the following equation for p [16]:

$$3(1 - 2\delta)[1 - \frac{3}{4}(\frac{K_1}{k_1})^2]p^2 + [(6\delta - 5) + \frac{3}{4}(3 - \delta)(\frac{K_1}{k_1})^2]p + 2 = 0. \qquad (23)$$

$(K/k)^2$ is then given from the solution of equation (23) by

$$(K/k)^2 = \frac{4/3[(1 - p) + (3/4)p(K_1/k_1)^2]}{[1 - p\delta + (3p/4)\delta(K_1/k_1)^2]}.$$

Figure 6 shows the variation of $(K/k)^2 = (v_l/v_t)^2$ and $p^{-1} = \mu/\mu_1$ as a function of porosity δ. This variation is characterised by the following behaviour:

- As $\delta \to 0$, $p \to 1$ and $(K/k)^2 \to (K_1/k_1)^2$.

- The shear modulus of the porous medium is $\mu = \mu_1 p^{-1}$ and is seen to vanish at $\delta = 1/2$ independently of the value of K_1/k_1.

- As $\delta \to 1/2$, $(K/k)^2 \to 8/3$, independently of K_1/k_1. This value corresponds to the Poisson's ratio of the porous medium $\nu = 0.2$.

2.3   Comparison with experiment

Thompson et al. [29] have measured the longitudinal and shear wave velocties on a set of sintered iron compacts which were prepared at various temperatures and pressures to produce porosities ranging from 0% to 11%. Extended holding times at elevated temperatures were used to produce as nearly spherical pores

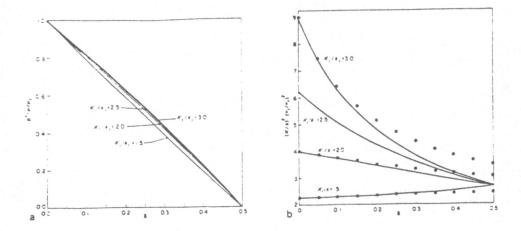

Figure 6: Variation of (a) $p^{-1} = \mu/\mu_1$, the ratio of the shear modulus of the porous and matrix material and (b) $(K/k)^2 = (v_l/v_t)^2$ as a function of porosity δ for several values of K_1/k_1. The circles are results obtained from the differential scheme [16].

as possible. However, due to the interconnection of individual pores, considerable non-sphericity of pores developed at the higher concentrations. Figure 7 compares the longitudinal and shear wave velocities with the self-consistent scattering theory of Sayers and Smith [16] discussed above. Significant disagreement exists at the highest porosity level of 11%. Thompson et al. [29] suggest that this may result from the development of non-spherical and interconnected pore shapes as the porosity increases.

Figure 8 compares the ratio of the measured velocities v_t/v_l to the ratio predicted by the self-consistent scattering theory of Sayers and Smith [16]. Thompson et al. [29] note from this plot that the spherical pore theory may predict the velocity ratio better than it predicts either of the velocities individually. This suggests the possibility of using the velocity ratio to determine porosity and an individual velocity to gain information on pore shape [29].

Boocock et al. [30] have reported measurements of the longitudinal and shear wave velocity in UO_2 as a function of porosity. Figure 9 compares the longitudinal and shear wave velocities with the self-consistent scattering theory of Sayers and Smith [16] whilst figure 10 compares the ratio of the measured velocities v_t/v_l to the ratio predicted by theory. Significant disagreement is again found at the highest porosity level and may result from the development of interconnected porosity as the volume fraction of pores increases. The three samples

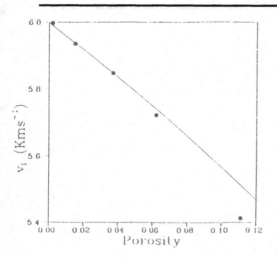
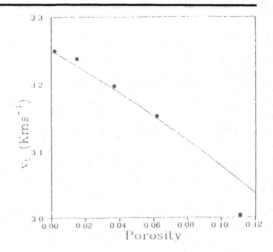

Figure 7: (a) Longitudinal and (b) shear wave velocities for iron compacts measured by Thompson et al. [29]. The curves show the prediction of reference [16].

with the highest porosity were found to be too friable to produce a micrograph.

2.4 Effect of a distribution in pore sizes

Thus far, it has been assumed that the pores all have the same radius a. In general a distribution of pore sizes will be present, and it is assumed in this section that the probability of a pore having a radius between a and $a + da$ is $n(a)da$ [12]. A rectangular distribution of a, centered on a value a_0 and with width r, will be assumed:

$$n(a) = 1/r \qquad a_0 - r/2 < a < a_0 + r/2$$
$$= 0 \qquad \text{outside this range.} \tag{24}$$

When the restriction of identical scatterers is relaxed, the propagation constant β is given by equation (14) where $\langle f(0)\rangle$ denotes an average over the distribution function $n(a)$. From equation (6) it is seen that this average involves $\langle a^3\rangle$, $\langle a^5\rangle$ and $\langle a^6\rangle$ to lowest order [12]. For the distribution of equation (24)

$$\langle a^n\rangle = \int_0^\infty a^n n(a)da$$
$$= \frac{a_0^{n+1}}{(n+1)r}[(1 + r/2a_0)^{n+1} - (1 - r/2a_0)^{n+1}].$$

If $\delta = \Delta V/V$, then the number of pores per unit volume is $n_0 = 3\delta/4\pi\langle a^3\rangle$. The velocity dispersion is therefore proportional to $\delta\langle a^5\rangle/\langle a^3\rangle$, the attenuation

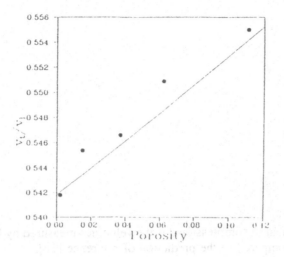

Figure 8: Ratio of shear to compressional wave velocities measured by Thompson et al. [29] for iron compacts compared to the prediction of reference [16].

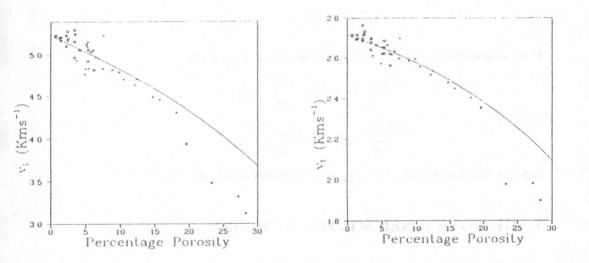

Figure 9: (a) Longitudinal and (b) shear wave velocities measured by Boocock et al. [30] for UO_2. The curves show the predictions of reference [16].

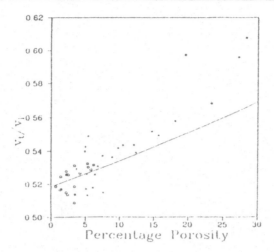

Figure 10: Ratio of shear to compressional wave velocities measured by Boocock et al. [30] for UO_2 compared to the prediction of reference [16].

to $\delta\langle a^6\rangle/\langle a^3\rangle$ [12]. A measurement of both, therefore, will not only give the average pore radius, but also the second moment of the distribution $n(a)$ [12]. For the rectangular distribution

$$(\frac{r}{a_0})^2 = 12(\frac{\langle a^2\rangle}{a_0^2} - 1).$$

If the distribution is not too wide, an expansion in r/a_0 gives

$$\frac{\langle a^5\rangle}{\langle a^3\rangle} = a_0^2[1 + 7(\frac{\langle a^2\rangle}{a_0^2} - 1)]$$

$$\frac{\langle a^6\rangle}{\langle a^3\rangle} = a_0^3[1 + 12(\frac{\langle a^2\rangle}{a_0^2} - 1)]$$

showing the dependence of the velocity dispersion on a_0 and $\langle a^2\rangle$.

3 MICROCRACKED SOLIDS - I: THEORY

In this lecture the ultrasonic velocity in microcracked media will be obtained from the static elastic moduli, the expressions given being restricted to the long-wavelength limit. For low concentrations of microcracks it may be assumed that the cracks are non-interacting and therefore the effective elastic properties may be calculated by assuming that each crack is subjected only to the externally applied stress field $\sigma^\times$. The contributions to the overall strain from individual cracks may then be summed to give the total strain. For higher crack densities when crack interactions cannot be neglected, O'Connell and Budiansky [31] and Budiansky and O'Connell [32] proposed a self-consistent scheme for the calculation of the elastic stiffness tensor for random crack orientation statistics. In this method the effect of crack interactions is included by assuming that each crack is embedded in a medium with the effective stiffness of the cracked body. This scheme was extended by Hoenig [33] to crack distributions for which the overall elastic stiffness tensor is transversely isotropic. Bruner [34] and Henyey and Pomphrey [35] have pointed out that the self-consistent scheme may overestimate the crack interactions and have proposed an alternative, differential scheme in which the crack density is increased in small steps and the elastic properties are recalculated incrementally. Finally, Hudson [36-38] has given results for both randomly orientated and parallel cracks that are correct to second order in the crack density; his results, however, are restricted to moderately small crack densities as will be shown below.

3.1   The self-consistent and differential schemes

The elastic constants for an isotropic homogeneous material containing elliptical cracks can be derived from the solution for elliptical cavities as described below. The average strain tensor in the solid for a homogeneous material of volume V containing arbitrary cavities may be written [39] as:

$$\frac{1}{V_s} \int_{V_s} \epsilon_{ij} dV = \epsilon_{ij} + \frac{1}{2V_s} \sum_r \int_{S_r} (u_i n_j + u_j n_i) dS \qquad (25)$$

where V_s is the volume of the solid, S_r is the surface of the r'th cavity lying within V and ϵ_{ij} is the macroscopic strain defined by

$$\epsilon_{ij} = \frac{1}{2V_s} \int_{S_e} (u_i n_j + u_j n_i) dS.$$

Here S_e is the solid portion of the exterior boundary. The macroscopic strains ϵ_{ij} are related to the macroscopic stress components $\bar{\sigma}_{ij}$ (defined by $\bar{\sigma}_{ij} = (1/V) \int_V \sigma_{ij} dV$) by the effective compliance tensor M_{ijkl}:

$$\bar{\epsilon}_{ij} = M_{ijkl} \bar{\sigma}_{kl}$$

Since ϵ_{ij} within the solid is given by $\epsilon_{ij} = M^0_{ijkl} \sigma_{kl}$, where M^0_{ijkl} is the compliance tensor of the solid, equation (25) gives [40]

$$M_{ijkl} \bar{\sigma}_{kl} = M^0_{ijkl} \bar{\sigma}_{kl} V/V_s + N_{ijkl} \bar{\sigma}_{kl}.$$

where

$$N_{ijkl}\bar{\sigma}_{kl} = -\frac{1}{2V_s}\sum_r \int_{S_r} (u_i n_j + u_j n_i)dS.$$

For an ellipsoidal cavity with principal axes $2a \geq 2b \geq 2c$ it is convenient to introduce a set of axes $Ox'_1 x'_2 x'_3$ with origin at the centre of the ellipsoid and Ox'_1, Ox'_2 and Ox'_3 along the a, b and c axes, respectively. In the limit $c/a \ll 1$, $c/b \ll 1$ (flat cracks) the above integral may be evaluated by taking the integration over the surface S_r of the crack and replacing displacements by displacement jumps across S_r while putting $V_s = V$. Hence in the crack reference frame the contribution of the r'th crack to $\bar{\epsilon}_{ij}$ is:

$$\frac{1}{2V}\int_{S_r} ([u'_i]n'_j + [u'_j]n'_i)dS = N'_{ijkl}\bar{\sigma}'_{kl} \tag{26}$$

where the bracket [ ] denotes jump discontinuities in the displacement. In this limit, $N_{ijkl} = \Delta M_{ijkl}$, the change in M_{ijkl} due to the presence of the cracks.

Under the action of uniform far stresses, an arbitrary ellipsoidal cavity characterised by semi-axes a, b, c in an anisotropic material deforms into another ellipsoid [41]. Thus the crack face displacements will also be ellipsoidal, so that the cavity displacements u_i are [42]

$$u'_i = \beta_i a(1 - x'^2_1/a^2 - x'^2_2/b^2)^{1/2}$$

where the β_i are dimensionless parameters. For a penny-shaped crack of radius a in an isotropic medium, substitution of this result in equation (26) gives:

$$N'_{3333} = 16(1 - \nu^2)a^3/3E$$

$$N'_{1313} = N'_{2323} = 8(1 - \nu^2)a^3/(3E(2 - \nu)). \tag{27}$$

all other components of N'_{ijkl} being zero. E and ν are the Young's modulus and Poisson's ratio of the matrix material, and the expressions for the β_i derived by Hoenig [42] have been used.

In the self-consistent method, each crack is assumed to be embedded in a medium with the effective stiffness of the cracked medium. E and ν in equation (27) are therefore replaced by $\bar{E}$ and $\bar{\nu}$, and the results are averaged over all crack orientations after transforming the N'_{ijkl} back to the unprimed coordinate system. For an isotropic random distribution of cracks this gives:

$$\bar{E}/E = 1 - 16na^3(1 - \bar{\nu}^2)(10 - 3\bar{\nu})/(45(2 - \bar{\nu}))$$

$$na^3 = 45(\nu - \bar{\nu})(2 - \bar{\nu})/(16(1 - \bar{\nu}^2)(10\nu - \bar{\nu}(1 + 3\bar{\nu})))$$

where $na^3 = Na^3/V$ (N is the number of cracks in V) is the crack density. This result was given by O'Connell and Budiansky [31] and Budiansky and O'Connell [32].

The differential scheme is obtained by adding the changes N_{ijkl} in an incremental fashion and recalculating the matrix constants M^0_{ijkl} at each increment; this results in a set of two differential equations for E and ν as functions of na^3 [34,35].

Figure 11 compares the predictions of the self-consistent and differential schemes for isotropic crack orientation statistics. Figure 11 also shows the results obtained by Hudson's scheme [36-38] which is correct to second order in the crack density. Note that the latter scheme clearly favours the differential scheme at small to moderate crack densities. A notable feature of the self-consistent scheme is the prediction of a vanishing elastic stiffness at a crack density $na^3 = 9/16$. O'Connell and Budiansky [31] and Budiansky and O'Connell [32] argue that the vanishing of the elastic stiffnesses corresponds to a loss of coherence of the solid produced by an intersecting crack network at a critical value of the crack density parameter of 9/16. Bruner [34] argues that since the self-consistent method treats the material as containing non-intersecting cracks in an elastic continuum, it cannot be expected to predict a loss of coherence. Charlaix [43] has shown that the percolation threshold of a 3-D assembly of widthless discs is given by $na^3 = 0.185$. Thus the crack density at which the elastic stiffnesses vanishes in the self-consistent scheme is much higher than the density at which a percolating crack network first forms. This can be expected since the formation of a percolating network does not necessarily mean that the solid loses elastic stiffness.

The extension of the self-consistent scheme to crack distributions for which the effective elastic stiffness is transversely isotropic has been carried out by Hoenig [33]. Figure 12 compares the self-consistent and differential schemes for the case when the crack normals are all parallel to Ox_3. In this case it follows from equation (27) that the only elastic compliances altered by the presence of cracks are $M_{3333} = 1/\bar{E}$, and $M_{3131} = M_{2323} = 1/4\bar{G}$ in the notation of Hoenig [33]. It is seen from Figure 12 that there is no finite crack density at which $\bar{E}$ or $\bar{G}$ as defined above vanish as would be expected from the vanishing probability of crack intersections for parallel cracks. The Hashin-Shtrikman upper bound for parallel cracks has been given by Laws and Dvorak [44] and coincides with the non-interacting result shown in Figure 12. Both the self-consistent and differential schemes give elastic stiffnesses that fall below the upper bound in

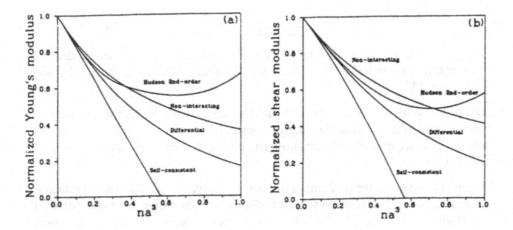

Figure 11: Comparison of the predictions of the self-consistent, differential and second-order Hudson schemes for an isotropic distribution of penny-shaped cracks in a medium with Poisson's ratio 0.25 [40]. Also shown is the result for non-interacting cracks.

agreement with the works of Milton [45] and Norris [46] who find that both the self-consistent and differential schemes are realizable in the sense that a microstructure can be specified having the elastic constants given by either of the schemes.

The elastic constants for cracks with normals all parallel to a given plane but otherwise randomly distributed have also been evaluated in the self-consistent scheme by Hoenig [33]. Despite the high probability for crack intersections for this orientation statistics, no finite crack density was found for which the effective elastic constants vanish, in contrast to the case of isotropic crack distributions.

3.2 Ultrasonic velocities for partially oriented microcracks

Microcracks in solids frequently display some degree of alignment reflecting the stress history of the material. In the presence of oriented cracks, the ultrasonic wave velocities will depend on the propagation direction and polarization of the wave. In this section the longitudinal and shear wave anisotropy is evaluated for materials containing an arbitrary orthotropic orientation distribution of cracks. This includes the important case of axial symmetry, for which simplified equations are given. It is shown that a quantitative relationship exists between the microcrack-induced longitudinal and shear wave anisotropy valid for an arbitrary

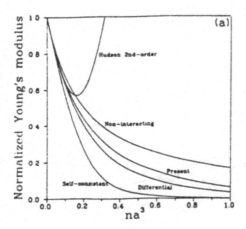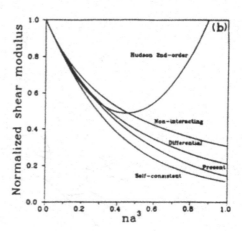

Figure 12: Comparison of the predictions of the self-consistent, differential and second-order Hudson schemes for the Young's modulus $\bar{E}$ and shear modulus $\bar{G}$ in the notation of Hoenig [33] for a distribution of parallel penny-shaped cracks in a medium with Poisson's ratio 0.25 [40]. Also shown is the result for non-interacting cracks.

orthotropic orientation distribution of microcracks.

It is assumed that the mechanical behaviour of brittle materials is determined by the formation, growth and coalescence of microcracks. As a result of the anisotropy of the stress field, these microcracks show some degree of preferred orientation. To model the effect of this on ultrasonic velocities these cracks are approximated by ellipsoids embedded in an isotropic background medium. For an ellipsoidal crack with principal axes $2a, 2b. 2c$ $(a \geq b \geq c)$ it is convenient to introduce a set of axes $OX_1X_2X_3$ with origin at the centre of the ellipsoid and OX_1, OX_2 and OX_3 along the $a. b$, and c axes, respectively (see Figure 13). The fourth-order elastic stiffness tensor of the microcracked material, C_{ijkl}, will be written in the form

$$C'_{ijkl} = C^0_{ijkl} + \gamma_{ijkl}. \tag{28}$$

where γ_{ijkl} is the difference between C'_{ijkl} and the elastic stiffness tensor C^0_{ijkl} of the uncracked material, which will be assumed to be isotropic. If the axes $OX_1X_2X_3$ for all cracks are aligned, the material will exhibit orthotropic symmetry with three orthogonal planes of mirror symmetry having plane normals in the OX_1, OX_2 and OX_3 directions. If the γ_{ijkl} in this case are denoted by Γ_{ijkl}, then in the crack reference frame $OX_1X_2X_3$ the non-zero Γ_{ijkl} are Γ_{11}, Γ_{22}, Γ_{33}, $\Gamma_{12} = \Gamma_{21}$, $\Gamma_{23} = \Gamma_{32}$, $\Gamma_{31} = \Gamma_{13}$, Γ_{44}, Γ_{55} and Γ_{66} in the Voigt (two-index)

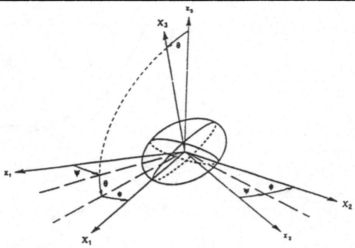

Figure 13: The orientation of coordinate system $OX_1X_2X_3$ with origin at the centre of an ellipsoidal crack with respect to the material coordinate system $Ox_1x_2x_3$ [61].

notation. For circular cracks with $a = b \gg c$, $\Gamma_{11} = \Gamma_{22}$, $\Gamma_{23} = \Gamma_{31}$, $\Gamma_{44} = \Gamma_{55}$ and $\Gamma_{66} = (\Gamma_{11} - \Gamma_{12})/2$. The anisotropy of Γ_{ijkl} in this case may be completely specified by three anisotropy parameters a_1, a_2 and a_3 [47,48], which are defined by:

$$a_1 = \Gamma_{11} + \Gamma_{33} - 2\Gamma_{13} - 4\Gamma_{44} \tag{29}$$

$$a_2 = \Gamma_{11} - 3\Gamma_{12} + 2\Gamma_{13} - 2\Gamma_{44} \tag{30}$$

$$a_3 = 4\Gamma_{11} - 3\Gamma_{33} - \Gamma_{13} - 2\Gamma_{44}. \tag{31}$$

In general, the cracks will not be perfectly aligned and a quantitative description of the elastic wave anisotropy requires a knowledge of the orientation distribution of cracks. The orientation of an ellipsoidal crack with principal axes $OX_1X_2X_3$ with respect to a set of axes $Ox_1x_2x_3$ fixed in the material may be specified by three Euler angles v, θ and o, as shown in Figure 13. The orientation distribution of cracks is then given by the crack orientation distribution function $W(\xi, v, o)$, where $\xi = \cos\theta$, θ being the angle between OX_3 and Ox_3 [49-51]. $W(\xi, v, o)d\xi\,dv\,do$ gives the fraction of cracks between ξ and $\xi + d\xi$, v and $v + dv$ and o and $o + do$. Clearly,

$$\int_0^{2\pi} \int_0^{2\pi} \int_{-1}^{1} W(\xi, v, o)d\xi\,dv\,do = 1.$$

It will be assumed that the orientation distribution of cracks is orthotropic with symmetry axes coincident with the reference axes Ox_1, Ox_2 and Ox_3. Expressions for the elastic stiffnesses C_{ijkl} of the cracked material can be derived from the Γ_{ijkl} and the crack orientation distribution function as follows [47,48,52]. If $T_{ijklmnpq}$ is given by the transformation rule for tensors of rank four, i.e.

$$T_{ijklmnpq} = \left(\frac{\partial x_i}{\partial X_m}\right)\left(\frac{\partial x_j}{\partial X_n}\right)\left(\frac{\partial x_k}{\partial X_p}\right)\left(\frac{\partial x_l}{\partial X_q}\right)$$

then, taking into account the orientation distribution of cracks, the first order correction γ_{ijkl} in equation (28) is given by

$$\gamma_{ijkl} = \overline{T}_{ijklmnpq}\Gamma_{mnpq}$$

where

$$\overline{T}_{ijklmnpq} = \int_0^{2\pi} \int_0^{2\pi} \int_{-1}^1 T_{ijklmnpq}(\xi, v, o)W(\xi, v, o)d\xi\,dv\,do.$$

These integrals may be evaluated by expanding the crack orientation distribution function $W(\xi, v, o)$ as a series of generalized spherical harmonics and using the orthogonality relations between these functions [52]. Since the elastic stiffness tensor is of fourth rank, it depends only on the coefficients W_{lmn} of the expansion of $W(\xi, v, o)$ for $l \le 4$. If the cracks are ellipsoidal and their orientation distribution is orthotropic with symmetry axes coincident with the reference axes $Ox_1x_2x_3$, the non-zero W_{lmn} are all real and are restricted to even values of l, m and n. For circular cracks with $a = b \gg c$, $W_{lmn} = 0$ unless $n = 0$. The elastic stiffnesses are therefore determined in this case by W_{200}, W_{220}, W_{400}, W_{420} and W_{440} and the three anisotropy factors a_1, a_2 and a_3 defined above [47,48]. The ultrasonic wave velocities in the material may then be obtained as solutions of the Christoffel equations [53]. For dry penny-shaped cracks the a_i, defined by equations (29-31), may be calculated, for example, from the work of Hudson [36-38]. As is seen in Figure 14, a_1 is found to be much smaller than a_2 and a_3 [47,48]. As a result, the coefficients W_{4m0} make only a small contribution to the measured ultrasonic wave velocities and may be ignored, the resultant equations for the velocities being

$$\rho v_{11}^2 = \rho \bar{v}_P^2 + 8\sqrt{10}\pi^2 a_3(W_{200} - \sqrt{6}W_{220})/105 \tag{32}$$

$$\rho v_{22}^2 = \rho \bar{v}_P^2 + 8\sqrt{10}\pi^2 a_3(W_{200} + \sqrt{6}W_{220})/105 \tag{33}$$

$$\rho v_{33}^2 = \rho \bar{v}_P^2 - 16\sqrt{10}\pi^2 a_3 W_{200}/105 \tag{34}$$

$$\rho v_{12}^2 = \rho v_{21}^2 = \rho \bar{v}_S^2 + 4\sqrt{10}\pi^2(7a_2 + 2a_3)W_{200}/315 \tag{35}$$

$$\rho v_{23}^2 = \rho v_{32}^2 = \rho \bar{v}_S^2 - 2\sqrt{10}\pi^2(7a_2 + 2a_3)(W_{200} - \sqrt{6}W_{220})/315 \tag{36}$$

$$\rho v_{31}^2 = \rho v_{13}^2 = \rho \bar{v}_S^2 - 2\sqrt{10}\pi^2(7a_2 + 2a_3)(W_{200} + \sqrt{6}W_{220})/315 \tag{37}$$

Here ρ is the density of the material and $\bar{v}_P$ and $\bar{v}_S$ are the longitudinal and shear wave velocities for a material with the same density of cracks but with random crack orientations.

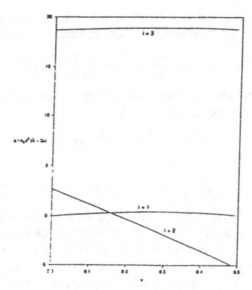

Figure 14: The variation of the anisotropy factors a_i defined by equations (29-31) as a function of Poisson's ratio [47].

4 MICROCRACKED SOLIDS - II: EXPERIMENT

Laboratory measurements of ultrasonic wave velocities in rock samples as a function of an increasing hydrostatic compression applied to the sample [54,55] have demonstrated that the presence of microcracks significantly decreases the ultrasonic wave velocities in the rock. As the compression is increased the measured velocities increase markedly. This behaviour has been attributed to the closure of microcracks with increasing pressure until they have no effect on the elastic properties of the rock. In general these microcracks are not randomly oriented and the rock displays an elastic anisotropy determined by the shape and content of the cracks and by the crack orientation distribution function introduced in lecture 3. In section 4.1, the problem of inverting the measured variation in ultrasonic velocity with direction of propagation and polarization to obtain the microcrack orientation distribution function is addressed. In section 4.2 the measurements of Thill et al. [56] on Salisbury granite are used to plot microcrack pole figures, and these are compared with the pole figures obtained by Thill et al. [56] using petrographical techniques.

Batzle et al. [57] have studied directly the closure of microcracks in rocks under increasing uniaxial compressive stress using a scanning electron microscope. Closure characteristics depend on crack orientation, shape, surface roughness and the nature of crack interactions and intersections. For an applied uniaxial stress, Nur and Simmons [58] found the change in longitudinal wave velocity for propagation parallel to the stress axis in Barre granite to be greater than that for propagation perpendicular to this axis. For any propagation direction there are two shear waves which travel at different velocities, the difference in velocities being called the shear wave birefringence. This anisotropy occurs because the effect of an applied anisotropic stress is to close cracks in some directions while leaving cracks open in other directions. An initially random orientation distribution of cracks will therefore become anisotropic in the presence of a uniaxial compressive stress and this could lead to anisotropic permeability properties of the rock. In section 4.3 the theoretical approach outlined in section 4.1 is used to obtain the orientation distribution of open cracks in rock under compressive stress from the experimental results of Nur and Simmons [58].

The failure of brittle rocks during compression is believed to be preceded by the formation, growth and coalescence of microcracks [59,60]. Tensile stresses necessary for microcrack growth include shear along pre-existing microcracks and stress concentrations around inhomogeneities. In the measurements of Nur and Simmons [58], however, the stresses applied were rather low and crack propagation did not occur. In section 4.4 ultrasonic longitudinal and shear wave velocities measured in three orthogonal directions through samples of Berea sandstone stressed to failure as a function of an applied stress are reported and used to test the theory presented in lecture 3.

4.1   Inversion for the microcrack orientation distribution

The ultrasonic wave velocities in the material may be obtained from the elastic stiffnesses C_{ijkl} given above by solving the Christoffel equations [53]. This then allows the W_{lmn} to be determined from the measured velocities. These coefficients can be used to calculate the microcrack normal orientation distribution $q(\zeta, \eta)$ defined by:

$$q(\zeta, \eta) = n(\zeta, \eta) / \int_0^{2\pi} \int_{-1}^{1} n(\zeta, \eta) d\zeta d\eta$$

where $n(\zeta, \eta) d\zeta d\eta$ is the number of cracks with normal between ζ and $\zeta + d\zeta$ and η and $\eta + d\eta$. Here $\zeta = \cos\chi$, χ and η being the polar and azimuthal angles in the reference frame $Ox_1x_2x_3$. $q(\zeta, \eta)$ may be expanded as a series of spherical harmonics

$$q(\zeta, \eta) = \sum_{l=0}^{\infty} \sum_{m=-l}^{l} Q_{lm} P_l^m(\zeta) \epsilon^{-im\eta} \tag{38}$$

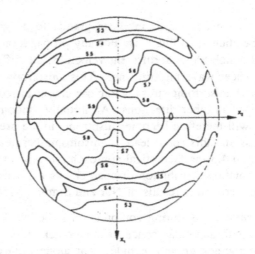

Figure 15: Contoured equal area projection of the longitudinal velocities (Km/s) measured by Thill et al. [56] on Salisbury granite [47].

where the $P_l^m(\zeta)$ are the normalised Legendre functions [49-51]. For circular cracks the Q_{lm} are given in terms of the W_{lmn} by $Q_{lm} = 2\pi W_{lm0}$. Only the coefficients W_{lmn} for $l \leq 4$ can be obtained from the angular dependence of the ultrasonic velocities. Expanding equation (38) to order $l = 4$ gives:

$$4\pi q(\zeta, \eta) = 1 + 4\pi[Q_{20}P_2^0(\zeta) + 2Q_{22}P_2^2(\zeta)\cos 2\eta$$
$$+ Q_{40}P_4^0(\zeta) + 2Q_{42}P_4^2(\zeta)\cos 2\eta + 2Q_{44}P_4^4(\zeta)\cos 4\eta]$$

4.2 Application to the measurements of Thill et al.

Thill et al. [56] have measured longitudinal wave velocities in a large number of directions on a spherical sample of Salisbury granite. Figure 15 shows a contoured equal area projection of the measured longitudinal velocities. The velocity distribution exhibits orthotropic symmetry, the coordinate planes being approximately parallel to the three mirror planes. The orientation distribution of microcrack normals was also measured using a petrofabric microscope and correlates closely with the longitudinal velocity symmetry pattern. Two sets of microcracks were found to be present. The principal set consisted of well developed microcracks frequently extending through many of the grains. For this set the crack normals showed a marked alignment along the direction Ox_1 in Figure 15. The second set had crack normals aligned in the direction Ox_3 in Figure 15 and consisted of relatively few, discontinuous, weakly defined microcracks. Thill et al. [56] concluded that the effect of this second set on the measured velocity is negligible, the principal set controlling the velocity anisotropy of the rock.

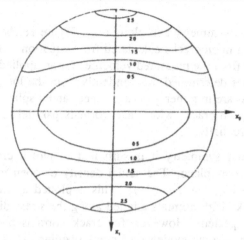

Figure 16: Contoured equal area projection of the orientation distribution of microcrack normals calculated from equation (38) using the values of W_{200} and W_{220} determined from the longitudinal velocities measured by Thill et al. [56] on Salisbury granite [47].

For the case of dry cracks appropriate to the measurements of Thill et al. [56] the values of the a_i are plotted in Figure 14. Because of the small value of a_1, the coefficients W_{4mn} only make a small contribution to the measured longitudinal velocity and cannot therefore be determined accurately from the data of Thill et al. [56] for which an error in the measured velocities of 1.4% is quoted. Therefore, only W_{200} and W_{220} were evaluated and the microcrack pole figure, plotted in Figure 16, was calculated using the values $W_{200} = -0.0105$, $W_{220} = 0.00646$ determined from the measurements. The Ox_3 direction lies at the centre of this figure. A random distribution of crack normals would correspond to a value of one everywhere in this figure. The peaks therefore correspond to a preferential orientation of crack normals. The ultrasonic pole figure therefore shows a preferred orientation of crack normals along Ox_1 in agreement with the results of Thill et al. [56].

4.3   Application to the measurements of Nur and Simmons

Nur and Simmons [58] subjected a cylinder of Barre granite to a uniaxial compressive stress normal to the axis of the cylinder. Compressional and shear wave velocities were measured for propagation normal to the cylinder axis at various angles to the stress axis. These measurements are shown in Figure 17. It is convenient to choose a set of reference axes $Ox_1x_2x_3$ with Ox_3 along the cylinder axis and Ox_1 in the direction of applied stress. Upon application of the stress, an initially isotropic distribution of open microcracks will develop axial symmetry with symmetry axis along Ox_1. Figure 18 shows the orientation distribution of open microcracks calculated from equation (38) using the values of Q_{lm} determined from the measured velocities for an applied stress of 30 MPa. The two pole figures determined independently from the longitudinal and shear wave measurements are in rather good agreement and display an axial symmetry about the x_1 axis with cracks with plane normals parallel to the applied stress being closed preferentially.

Because of the axial symmetry it is sufficient to plot a cross-section of the pole figure in the x_3x_1 plane. The actual density of open microcracks with a specific orientation is given by $n(\zeta.\eta)$. This is plotted at intervals of 10 MPa in Figure 19. Cracks with normals aligned along the stress direction are closed preferentially as expected. However, for crack normals perpendicular to the applied stress there is some evidence of crack opening. A possible mechanism for the opening of a crack with normal perpendicular to the applied stress has been presented elsewhere [48].

4.4   Stress-induced anisotropy of Berea sandstone

Ultrasonic velocities were reported by Sayers and van Munster [61] in three 50-mm cubes of Berea sandstone cut parallel to the bedding plane. Berea is a light-brown, fine-grained, well-sorted sandstone. Its mineralogy is mainly quartz with small amounts of feldspar and clay minerals. Ultrasonic velocities were measured with the samples in an air-dry condition using the true triaxial loading frame described by Sayers et al. [62]. The initial velocity was measured at a hydrostatic compressive stress of 4 MPa, after which the stress perpendicular to the bedding plane was raised in steps of 4 MPa to peak, keeping the other two stress components fixed. The velocities are plotted in Figure 20 as a function of the major principal stress component, using a reference set of axes chosen with Ox_3 perpendicular to the bedding plane and parallel to the direction of maximum stress and Ox_1 and Ox_2 parallel to the remaining cube edges. The velocities measured were v_{11}, v_{22} and v_{33} (longitudinal waves) and v_{12}, v_{23} and v_{31} (shear waves). Samples I and II contained a small and approximately equal amount of clay while sample III contained a relatively large amount [61].

It is seen in Figure 20 that v_{33}, v_{23} and v_{31} rise monotonically throughout most of the test, but with a small drop immediately prior to failure; v_{11}, v_{22} and v_{12} rise at first, but then drop after the first third of the test. The measurements for the two samples with low clay content are in agreement, but the sample

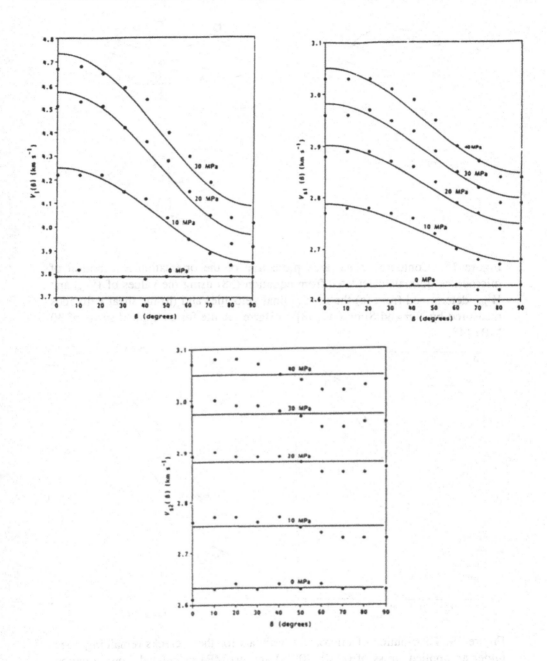

Figure 17: Compressional and shear wave velocities measured by Nur and Simmons [58] for propagation normal to the axis of a cylinder of Barre granite at various angles ϑ to a uniaxial stress applied perpendicular to the axis of the cylinder.

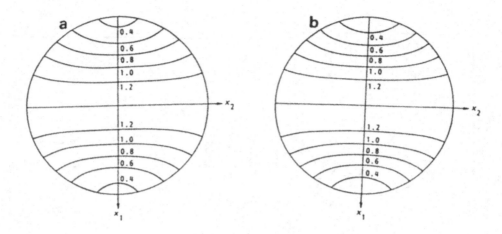

Figure 18: Contoured equal area projection of the orientation distribution of microcrack normals calculated from equation (38) using the values of W_{200} and W_{220} determined from (a) the longitudinal velocities and (b) the shear velocities measured by Nur and Simmons [58] on Barre granite for an applied stress of 30 MPa [48].

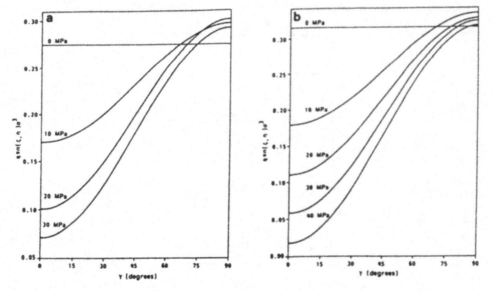

Figure 19: Distribution of microcrack normals for those cracks remaining open under an applied stress of 0, 10, 20, 30 and 40 MPa calculated from equation (38) using the values of W_{200} and W_{220} determined from (a) the longitudinal and (b) the shear velocities measured by Nur and Simmons [58] on Barre granite [48].

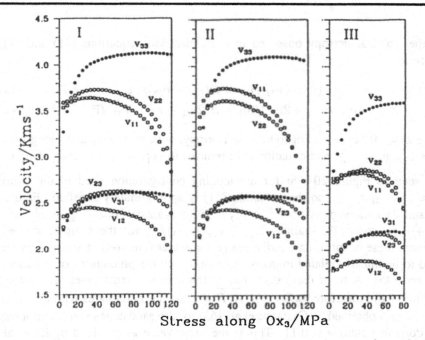

Figure 20: Ultrasonic velocities plotted as a function of the major principal stress component in a coordinate system with Ox_3 perpendicular to the bedding plane and parallel to the direction of maximum stress and Ox_1 and Ox_2 parallel to the remaining cube edges [61].

with higher clay content is seen to have substantially lower velocities and to fail at a significantly lower value of the maximum compressive stress. At low stress the observed increase in wave velocity can be explained by the closure of pre-existing microcracks and grain boundaries in the rock and, as the stress increases still further, by the formation of new microcracks and the growth and coalescence of pre-existing cracks. The measurements may therefore be analysed using the theory presented in section 3.2.

$\bar{v}_P$ and $\bar{v}_S$ may be obtained from equations (32-37):

$$\bar{v}_P^2 = (v_{11}^2 + v_{22}^2 + v_{33}^2)/3 \tag{39}$$

$$\bar{v}_S^2 = (v_{12}^2 + v_{23}^2 + v_{31}^2)/3. \tag{40}$$

It is seen from equations (32-37) that the P and S-wave anisotropies are not independent, but are related by the following equations:

$$(v_{11}^2 - v_{22}^2) = A(v_{31}^2 - v_{23}^2) \tag{41}$$

$$(v_{11}^2 + v_{22}^2 - 2v_{33}^2) = A(2v_{12}^2 - v_{23}^2 - v_{31}^2) \tag{42}$$

where

$$A = 12a_3/(7a_2 + 2a_3). \tag{43}$$

For the small anisotropy observed in most materials, equations (41) and (42) reduce to

$$(v_{11} - v_{22})/\overline{v}_P = B(v_{31} - v_{23})/\overline{v}_S \tag{44}$$

$$(v_{11} + v_{22} - 2v_{33})/\overline{v}_P = B(2v_{12} - v_{23} - v_{31})/\overline{v}_S \tag{45}$$

where $B = A(\overline{v}_S/\overline{v}_P)^2$. For transverse isotropy, with symmetry axis along Ox_3, $v_{11} = v_{22}$, $v_{23} = v_{31}$, and equation (42) reduces to $(v_{11}^2 - v_{33}^2) = A(v_{12}^2 - v_{31}^2)$.

It is seen in Figure 20 that for the loading configuration used in this work $v_{11} \approx v_{22}$, $v_{31} \approx v_{23}$, so that equation (41) is approximately satisfied. Figures 21a and 21b show plots of $(v_{11}^2 + v_{22}^2 - 2v_{33}^2)$ against $(2v_{12}^2 - v_{23}^2 - v_{31}^2)$ and $(v_{11} + v_{22} - 2v_{33})/\overline{v}_P$ against $(2v_{12} - v_{23} - v_{31})/\overline{v}_S$ for the samples studied. Although these quantities are individually non-linear functions of stress, they are found to be linearly related in good agreement with the predictions of equations (42) and (45). A fit of these equations to the measurements gives $A = 3.628$, $B = 1.583$ and therefore $\overline{v}_P/\overline{v}_S = \sqrt{A/B} = 1.51$. This compares with the value $\overline{v}_P/\overline{v}_S = 1.5$ obtained by Castagna et al. [63] for dry sandstones and corresponds to a Poisson's ratio $\nu = 0.11$. This is the same value as obtained by Lo et al. [64] for Berea sandstone. The value $A = 3.628$ should be compared with the value $A = 5.036$ obtained using the theory of Hudson [36-38] for dry, non-contacting cracks and a value of Poisson's ratio $\nu = 0.11$. This disagreement may be the result of grain boundary closure since, this is not included in the theory.

4.5   Conclusion

The ultrasonic pole figures presented in this lecture are approximations to those obtained by petrofabric analysis since only the first few coefficients W_{lmn} in an expansion of the crack orientation distribution function in generalized spherical harmonics are included. The W_{lmn} are given by

$$W_{lmn} = \frac{1}{4\pi^2} \int_0^{2\pi} \int_0^{2\pi} \int_{-1}^{1} W(\xi, \psi, o) Z_{lmn}(\xi) e^{im\psi} e^{ino} d\xi \, d\psi \, do.$$

where the $Z_{lmn}(\xi)$ are the generalized Legendre functions defined by Roe [49-51]. W_{lmn} therefore represents the value of a polynomial of trignometrical functions of θ, ψ and o averaged over all crack orientations. The use of a limited number of W_{lmn} therefore corresponds to the specification of the orientation distribution by its first few moments and represents the maximum amount of

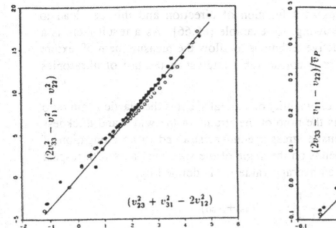
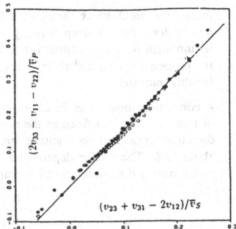

Figure 21: Plot of (a) $(2v_{33}^2 - v_{11}^2 - v_{22}^2)$ against $(v_{23}^2 + v_{31}^2 - 2v_{12}^2)$ and (b) $(2v_{33} - v_{11} - v_{22})/\overline{v}_P$ against $(v_{23} + v_{31} - 2v_{12})/\overline{v}_S$ for the measurements of reference [61]. The lines are fits of equations (42) and (45) to the data. This Figure corrects that in reference [61].

information that can be obtained using ultrasonic waves with wavelengths greater than the crack size. For the pole figures shown only W_{200} and W_{220} were used. More detail could be obtained by including the terms W_{400}, W_{420} and W_{440}. However, this would require more accurate measurements of the wave velocities.

5 TEXTURED POLYCRYSTALLINE AGGREGATES

In a polycrystalline aggregate the elastic constants in the specimen reference frame vary from grain to grain due to the random orientation of the grains. Poly-crystalline materials are therefore elastically inhomogeneous, the elastic constant mismatch at the grain boundaries leading to scattering of the ultrasonic wave. In the long wavelength limit, however, the material can be modelled as an elastic continuum with elastic constants determined by the elastic constants of the grains and by the crystallite orientation distribution function (CODF). This function gives the probability of a crystallite having a given orientation with respect to the specimen reference frame, and gives a quantitative description of the texture, or crystallographic alignment, of the material. In a strongly textured metal, the yield stress varies as a function of direction and this can lead to non-uniform flow in deep drawing, for example [65,66]. As a result there is a requirement for a non-destructive technique to allow the measurement of texture for process control, and there is considerable interest in the use of ultrasonics for this purpose.

A convenient measure of the drawability of a metal sheet is the plastic strain ratio or r-value. This is defined as the ratio of the strains in the width and thickness directions measured on a tensile stress specimen, strained to an elongation of about 15%. The r-value depends on the angle of the specimen axis with respect to the rolling direction, and an average value r_m is defined by;

$$r_m = \frac{1}{4}(r_0 + 2r_{45} + r_{90})$$

where r_α is the r-value measured for a tensile specimen cut at an angle α to the rolling direction. High values of r_m correlate with good drawability. This occurs for cubic metals if a high proportion of crystallographic (111) planes lie in the plane of the sheet. A convenient measure of the plastic anisotropy in the plane of the sheet is the quantity Δr defined by

$$\Delta r = \frac{1}{2}(r_0 + r_{90} - 2r_{45}).$$

This is found to correlate with the amount of earing that occurs in deep drawing. Ears tend to form along directions perpendicular to directions of high r-value. A positive value of Δr implies earing at 0° and 90° whereas a negative value corresponds to earing at 45° to the rolling direction. The possibility of using the angular variation of the ultrasonic velocity to characterize the plastic anisotropy is suggested by the results of Stickels and Mould [67] who showed that the angular variation of the Young's modulus in the rolling plane of low carbon steel sheet correlates with that of the plastic strain ratio. It was found that the angular variation of the elastic modulus could be used to characterize the formability of the sheet. Davies et al. [68] found, by using the crystallite orientation distribution defined by Roe [49-51], that both the elastic and plastic properties of an orthotropic aggregate of cubic crystallites are determined primarily by the fourth-order coefficients of an expansion of this function in spherical harmonics, and therefore demonstrated that the empirical correlation of Stickels and Mould [67] could be justified using established theories of elastic and plastic behaviour.

In metals with a texture or non-random distribution of crystallite orientations, the ultrasonic velocity depends on the orientation of the propagation and polarization directions with respect to the principal texture axes. It is shown in

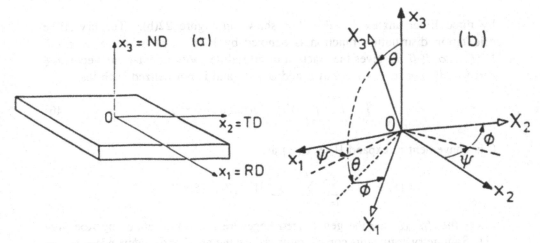

Figure 22: (a) Choice of specimen axes for a rolled plate with Ox_1 along the rolling direction (RD), Ox_2 along the transverse direction (TD) and Ox_3 along the normal direction (ND). (b) The orientation of the crystallite coordinate system $OX_1X_2X_3$ with respect to the sample coordinate system $Ox_1x_2x_3$ [69].

section 5.2 how information on the CODF can be obtained from ultrasonic velocity measurements [69]. This relies on the theory of ultrasonic propagation in textured polycrystalline solids presented in section 5.1.

5.1   Ultrasonic anisotropy of textured polycrystalline aggregates

Let $Ox_1x_2x_3$ be an orthogonal set of reference axes fixed in the sample. For a rolled plate these axes could be chosen as the rolling, transverse and normal directions as shown in Figure 22(a). It will be assumed that the sample has orthotropic symmetry, ie it is assumed to possess three orthogonal planes of mirror symmetry given by the planes x_1x_2, x_2x_3 and x_3x_1 in Figure 22(a). This is the symmetry of rolled plate. It will be assumed, in addition, that the sample is an aggregate of crystallites of cubic crystallographic symmetry, as is the case for steel and aluminium. The case of hexagonal and orthorhombic crystal symmetry may be treated similarly [70-73].

The crystallographic alignment, or texture, of the plate is most conveniently described by the crystallite orientation distribution function (CODF). This method of analysis has been developed extensively by Roe [49-51] and Bunge [74,75]. In this lecture the notation of Roe is used.

Let $OX_1X_2X_3$ be an orthogonal set of reference axes for a crystallite given by the (100), (010) and (001) crystallographic directions. The orientation of a given crystallite with respect to the sample reference axes $Ox_1x_2x_3$ can be specified by three Euler angles v , θ and o shown in Figure 22(b). The crystallite orientation distribution function is denoted by $W(\xi, v, o)$, where $\xi = \cos\theta$. $W(\xi, v, o)d\xi dv do$ gives the fraction of crystallites with orientations between ξ and $\xi + d\xi$, v and $v + dv$ and o and $\phi + d\phi$ and is normalized such that

$$\int_0^{2\pi} \int_0^{2\pi} \int_{-1}^1 W(\xi, \psi, \phi)d\xi dv d\phi = 1. \tag{46}$$

It is convenient to expand $W(\xi, v, o)$ as:

$$W(\xi, v, o) = \sum_{l=0}^{\infty} \sum_{m=-l}^{l} \sum_{n=-l}^{l} W_{lmn} Z_{lmn}(\xi) \epsilon^{-im v} \epsilon^{-imo} \tag{47}$$

where the $Z_{lmn}(\xi)$ are the generalized Legendre functions defined by Roe [49-51]. Symmetry arguments considerably reduce the number of independent W_{lmn}. For an orthotropic aggregate of cubic crystals the elastic stiffness tensor depends only on W_{400}, W_{420} and W_{440} in equation (47) [76]. Explicit expressions for the elastic stiffness tensor C_{ijkl} in terms of these coefficients have been given elsewhere [52,69]. The ultrasonic wave velocities in the material may be then obtained from the C_{ijkl} by solving the Christoffel equations [53].

The ultrasonic velocities in the three principal directions of a polycrystalline aggregate with orthotropic symmetry and cubic crystallographic symmetry can be written in terms of the single crystal anisotropy $c = c_{11} - c_{12} - 2c_{44}$ and the coefficients W_{400}, W_{420} and W_{440} [69]:

$$\rho v_{11}^2 = \lambda + 2\mu + \frac{12}{35}\sqrt{2}\pi^2 c(W_{400} - \frac{2}{3}\sqrt{10}W_{420} + \frac{1}{3}\sqrt{70}W_{440}) \tag{48}$$

$$\rho v_{22}^2 = \lambda + 2\mu + \frac{12}{35}\sqrt{2}\pi^2 c(W_{400} + \frac{2}{3}\sqrt{10}W_{420} + \frac{1}{3}\sqrt{70}W_{440}) \tag{49}$$

$$\rho v_{33}^2 = \lambda + 2\mu + \frac{32}{35}\sqrt{2}\pi^2 c W_{400} \tag{50}$$

$$\rho v_{12}^2 = \rho v_{21}^2 = \mu + \frac{4}{35}\sqrt{2}\pi^2 c(W_{400} - \sqrt{70}W_{440}) \tag{51}$$

$$\rho v_{23}^2 = \rho v_{32}^2 = \mu - \frac{16}{35}\sqrt{2}\pi^2 c(W_{400} + \sqrt{\frac{5}{2}}W_{420}) \tag{52}$$

$$\rho v_{31}^2 = \rho v_{13}^2 = \mu - \frac{16}{35}\sqrt{2}\pi^2 c(W_{400} - \sqrt{\frac{5}{2}}W_{420}). \tag{53}$$

Here ρ is the density of the matrix and λ and μ are the second-order (Lamé) constants of a polycrystalline aggregate with a random orientation distribution of crystallites.

It should be noted that these equations were obtained to first order in the texture using the Voigt approximation [69]. It follows that equations (48-53) are *exact* to first order in the texture since Hill [39,77] has shown that $C^V - C$ is always a positive semi-definite tensor, C^V being the Voigt approximation to the elastic constant tensor C of the aggregate. Hence the first order term in an expansion of $\epsilon(C^V - C)\epsilon$ as a power series in the single crystal anisotropy vanishes for all strains e, since c can take both positive and negative values. It then follows from the Voigt symmetry of C^V and C that $C_{ijkl} = C_{ijkl}^V + O(c^2)$ for each component separately [78].

It is seen that the compressional and shear wave anisotropies are not independent, but are related by the following equation:

$$(2v_{33}^2 - v_{11}^2 - v_{22}^2) = (2v_{12}^2 - v_{23}^2 - v_{31}^2). \tag{54}$$

Expressions for the velocity of waves travelling in any direction with respect to the specimen axes may also be derived. The velocities of waves travelling in the $x_1 x_2$ plane are given below.

Let θ be the angle between the propagation and rolling direction. The Rayleigh wave velocity for propagation in the direction θ is given by $v_R = v_R^0 + \delta v_R$, where v_R^0 is the Rayleigh wave velocity for an isotropic distribution of crystallites and δv_R is given by [79-81]:

$$\delta v_R(\theta) = \frac{1}{2v_R^0}(R_1 + R_2 \cos 2\theta + R_4 \cos 4\theta) \tag{55}$$

The amplitudes R_1, R_2 and R_4 are proportional to the coefficients W_{400}, W_{420} and W_{440} respectively, and are functions of Poisson's ratio ν. Similar expressions for the change in the velocity of bulk waves propagating in the plane of the plate may also be derived [78,82]. Thus

$$\delta v_{SH}(\theta) = v_{SH}(\theta) - v_s^0 = \frac{2\sqrt{2}\pi^2 c}{35\rho v_s^0}(W_{400} - \sqrt{70}W_{440} \cos 4\theta). \tag{56}$$

$$\delta v_{SV}(\theta) = v_{SV}(\theta) - v_s^0 = \frac{-8\sqrt{2}\pi^2 c}{35\rho v_s^0}(W_{400} - \sqrt{5/2}W_{420} \cos 2\theta). \tag{57}$$

$$\delta v_l(\theta) = v_l(\theta) - v_l^0 = \frac{6\sqrt{2}\pi^2 c}{35\rho v_l^0}(W_{400} - \frac{2}{3}\sqrt{10}W_{420} \cos 2\theta + \frac{1}{3}\sqrt{70}W_{440} \cos 4\theta).$$
$$\tag{58}$$

Here $v_{SH}(\theta)$, $v_{SV}(\theta)$ and $v_l(\theta)$ are the velocities of the shear horizontal, shear vertical and longitudinal waves propagating in the plane of the plate at an angle

θ to the rolling direction. v_l^0 and v_s^0 are the isotropic longitudinal and shear wave velocities.

The above equations are sufficient to determine the texture parameters W_{400}, W_{420} and W_{440}. These may then be used to plot an ultrasonic pole figure as described below [69].

5.2   Inversion of wave velocities to determine crystallite orientation

Consider the normal **t** to a crystallographic plane with orientation with respect to the specimen reference coordinate system $Ox_1x_2x_3$ specified by polar angle χ and azimuthal angle η [83]. In an X-ray or neutron diffraction measurement of the pole figure or distribution of plane normals, χ and η specify the orientation of the sample with respect to the incident and diffracted beams. If $I(\zeta, \eta)$, with $\zeta = \cos \chi$, is the intensity measured at this orientation from this plane, the plane normal orientation distribution $q(\zeta, \eta)$ is defined by:

$$q(\zeta, \eta) = \frac{I(\zeta, \eta)}{\int_0^{2\pi} \int_{-1}^1 I(\zeta, \eta) d\zeta d\eta}.$$

The neutron or X-ray pole figure is usually presented as a stereographic projection of $I(\zeta, \eta)$ on a convenient coordinate plane of the specimen. Following Roe [49-51], $q(\zeta, \eta)$ may be expanded as a series of spherical harmonics:

$$q(\zeta, \eta) = \sum_{l=0}^{\infty} \sum_{m=-l}^{l} Q_{lm} P_l^m(\zeta) \epsilon^{-im\eta} \tag{59}$$

where the $P_l^m(\zeta)$ are the normalised Legendre functions [49-51].

For a given crystallographic plane, the Q_{lm} may be determined from the W_{lmn}. If only terms up to fourth-order in the expansion of $q(\zeta, \eta)$ are included, it is found that [69]

$$4\pi q(\zeta, \eta) = 1 + 4\pi S[\frac{3}{8\sqrt{2}}(35\zeta^4 - 30\zeta^2 + 3)W_{400}$$

$$+ \frac{9\sqrt{5}}{2}(1 - \zeta^2)(1 - \frac{7}{6}(1 - \zeta^2))W_{420}\cos 2\eta$$

$$+ \frac{3\sqrt{35}}{8}(1 - \zeta^2)^2 W_{440}\cos 4\eta] \tag{60}$$

where, for example, $S = -4\pi/3$ for the [111] pole figure, $S = 2\pi$ for the [100] pole figure and $S = -1/2$ for the [110] pole figure.

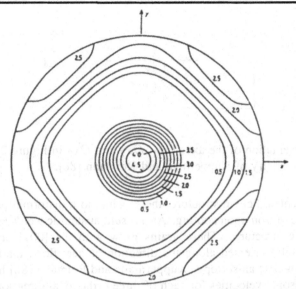

Figure 23: [100] pole figure for the values $W_{400} = 0.02259$, $W_{420} = 0.00106$ and $W_{440} = -0.00277$ determined from the ultrasonic measurements of Kupperman and Reimann [86]. After [69].

The theory presented in this section allows the coefficients W_{400}, W_{420} and W_{440} to be determined from the measured ultrasonic wave velocities. These coefficients can then be used to plot an ultrasonic pole figure using equation (60) which may be compared with pole figures obtained using X-ray or neutron diffraction.

5.3   Comparison with experiment

5.3.1   Use of waves travelling along the principal axes

The theory presented in section 5.1 has been checked by Allen and Langman [84,85] who determined the parameters W_{420} and W_{440} using ultrasonic velocity measurements on two plates, a 99.5% pure aluminium plate (BS 1470/1050A) and the second an aluminium 4.5% Mg alloy plate (BS 1470/5083/0). Both plates were 50mm thick. The values obtained from the (200) and (111) pole figures were found to be in reasonable agreement with each other and with the values obtained using ultrasonics. This confirms that the main effect of the texture on the ultrasonic velocities is due to crystallographic alignment of the grains and not due to factors such as grain shape anisotropy, etc.

The use of ultrasonic bulk waves to determine the texture of austenitic welds was presented in reference [69]. In austenitic stainless steel, the crystallites

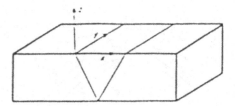

Figure 24: Orientation of the axes Ox_1, Ox_2 and Ox_3 in Figure 22 with respect to the weld geometry of Kupperman and Reimann [86].

have a face-centered cubic structure, there being no transformation to the body-centered phase of iron upon cooling. As a result, austenitic welds exhibit a coarse grain dendritic structure. These grains grow with the [001] crystallographic axis preferentially oriented along the direction of heat flow, the resultant weld exhibiting ultrasonic anisotropy. Kuppermann and Reimann [86] have measured the three ultrasonic velocities for each of three orthogonal directions in a sample cut from a 'V'-type multipass shielded-arc weld of type 308 stainless steel weld metal. The quantity $4\pi q(\zeta, \eta)$ for the [001] plane normal calculated using the values of W_{400}, W_{420} and W_{440} computed from the ultrasonic velocities is shown in Figure 23 projected on the xy plane. The strong alignment of the [001] plane normals along the z axis is seen together with some evidence of alignment at $45°$ to the axis of the weld. The relation of the axes x, y and z to the geometry of the weld is shown in Figure 24 [86].

5.3.2 Angular dependence of the SH wave velocity

It is seen from equation (56) that an ultrasonic SH wave propagating in the x_1x_2 plane has the same velocity for propagation parallel ($\theta = 0°$) and perpendicular ($\theta = 90°$) to the rolling direction. This corresponds to an interchange of the direction of propagation and polarization of the wave. For an orthotropic material in the presence of an anisotropic stress state, this symmetry is broken. If the principal stress directions are parallel to the principal texture axes then:

$$\rho(v_{ij}^2 - v_{ji}^2) = \sigma_{ii} - \sigma_{jj} \tag{61}$$

where v_{ij} is the phase velocity of a shear wave propagating in direction x_i and polarized in direction x_j [87-89]. This relationship is independent of the magnitude of the texture of the material. Its use for separating texture and stress was suggested by MacDonald [87] following earlier work of Biot [88] and Thurston [89].

Thompson et al. [90-92] have used this technique for thin plates by using electromagnetic acoustic transducers (EMATs) to excite the fundamental horizontally polarized shear (SH) wave of the plate travelling in the plane of the plate. The velocity of this mode is frequency independent, and is identical to the SH wave velocity in an unbounded medium with the same texture and stress distribution. Measurements on 6061-T6 aluminium, 304 stainless steel and commercially pure copper plate gave essentially equal velocities for propagation along the rolling and transverse directions, in agreement with theory. The stress predicted using equation (61) was found to be in agreement with the uniaxial tensile stress applied.

Allen and Langman [84,85] generated the zero-order SH plate wave using a conventional piezoelectric shear wave transducer aligned to produce horizontally polarized waves. Two EMATs mounted in a rigid base were used as receivers. The three transducers lie along the same axis and the phase delay associated with the propagation of the elastic wave between the two EMATs arising from a stimulus of the piezoelectric transducer is measured. Their results are in good agreement with theory at zero stress. The angular dependence of the SH wave velocity measured on four of the plates is shown in Figure 25. Deviations from the theoretical curves shown in Figure 25 occur because in anisotropic media the phase and group velocities are not parallel, this phenomenon being referred to as beam skewing. The effect of this on the angular variation of the SH wave velocity is discussed in reference [82].

5.3.3 Rayleigh waves

The accuracy of equation (55) has been checked by determining the parameters W_{420} and W_{440}, which determine the angular dependence of the Rayleigh wave velocity, using bulk waves [81] and comparing the predicted variation with that measured. For a stress-free plate W_{440} can be determined from the angular variation of the SH wave velocity (equation (56)) whilst the shear wave birefringence gives W_{420}. Two samples were used. The first was a 99.5% pure aluminium plate (BS 1470/1050A) and the second an aluminium 4.5% Mg alloy plate (BS 1470/5083/0). Both plates were 50mm thick. Figure 26 shows a fit of equation (56) to the data for the aluminium and aluminium alloy plates. Figure 27 shows a comparison of the prediction of equation (55), using the values of W_{420} and W_{440} determined from the SH and shear wave birefringence measurements,

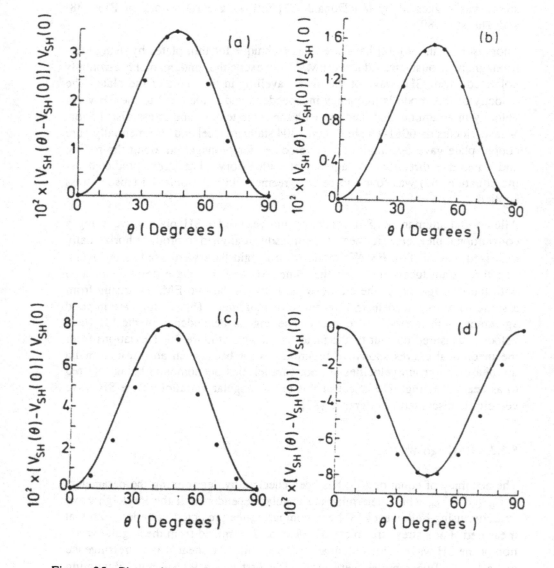

Figure 25: Plots of the measured fractional change in ultrasonic SH wave velocity with angle θ from the rolling direction for (a) aluminium sample number 3, (b) aluminium sample number 6, (c) stainless steel sample number 3 and (d) copper sample number 2. The sample numbers are those of references [84] and [85]. Equation (56) has been plotted for comparison [78].

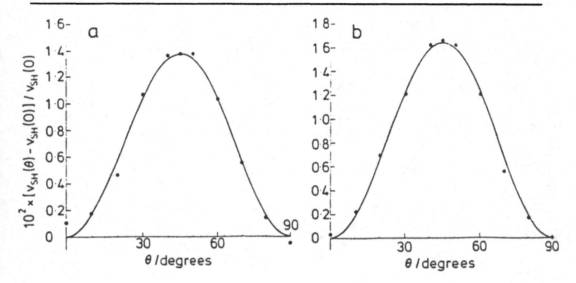

Figure 26: Plots of the measured fractional change in ultrasonic SH wave velocity with angle θ from the rolling direction for (a) a 99.5% pure aluminium plate and (b) an aluminium 4.5% magnesium alloy [81].

with the Rayleigh wave velocity measured at several different frequencies. The main angular dependent term is seen to be that arising from the term R_2 involving W'_{420}, which was determined from the shear wave birefringence. Since this represents a through thickness average of the texture of the plate, the low frequency results should be in better agreement with the prediction of equation (55), and this is verified by the results shown in Figure 27. Assuming the plates to be free of residual stress, the results show that both plates have a surface texture different to the bulk average. However, in contrast to the pure aluminium plate, which apparently contains a texture variation over a thin layer only, the aluminium alloy plate has a texture which varies over a longer length scale.

5.4 Conclusion

In this lecture the theoretical basis for treating the propagation of ultrasound in textured polycrystalline metals has been presented. It has been demonstrated how quantitative information concerning the crystallite orientation distribution function can be determined from the angular dependence of the ultrasonic velocities. Three methods for determining texture have been described. Of these the Rayleigh wave method is capable of determining the variation of texture with depth by using the frequency dependence of the Rayleigh wave velocity.

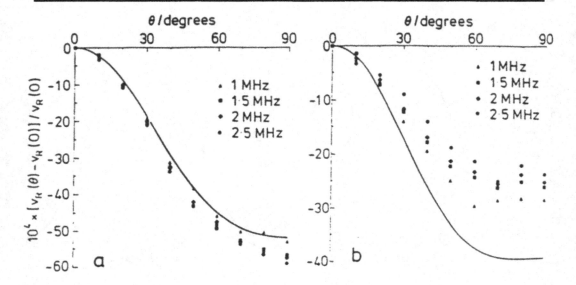

Figure 27: Plots of the measured fractional change in Rayleigh wave velocity with angle θ from the rolling direction for (a) a 99.5% pure aluminium plate and (b) an aluminium 4.5% magnesium alloy at several different frequencies. The curves show the prediction of equation (55) [81].

References

[1] SAYERS, C.M., Characterisation of microstructures using ultrasonics, in *Ultrasonic Methods in Evaluation of Inhomogeneous Materials* edited by A. Alippi and W.G. Mayer, pp 175-183, Martinus Nijhoff, 1987.

[2] KLINMAN, R., WEBSTER, G.R., MARSH, F.J. and STEPHENSON, E.T. (1980) Ultrasonic prediction of grain size, strength and toughness in plain carbon steel, *Mat. Eval.*, **38**, 26.

[3] YING, C.F. and TRUELL, R. (1956) Scattering of a plane longitudinal wave by a spherical obstacle in an isotropically elastic solid, *J. Appl. Phys.*, **27**, 1086-1097.

[4] FOLDY, L.L. (1945) The multiple scattering of waves, *Phys. Rev.*, **67**, 107-119.

[5] LAX, M. (1952) Multiple scattering of waves II. The effective field in dense systems, *Phys. Rev.*, **85**, 621-629.

[6] URICK, R.J. and AMENT, W.S. (1949) The propagation of sound in composite media, *J. Acoust. Soc. Am.*, **21**, 115-119.

[7] WATERMAN, P.C. and TRUELL, R. (1961) Multiple scattering of waves, *J. Math. Phys.*, **2**, 512-537.

[8] TWERSKY, V. (1962) *J. Opt. Soc. Am.*, **52**, 145.

[9] LLOYD, P. and BERRY, M.V. (1967) *Proc. Phys. Soc.*, **91**, 678.

[10] SAYERS, C.M. and SMITH, R.L. (1983) Ultrasonic velocity and attenuation in an epoxy matrix containing lead inclusions, *J. Phys. D*, **16**, 1189-1194.

[11] KINRA, V.K., KER, E. and DATTA, S.K. (1982) *Mech. Res. Comm.*, **9**, 109.

[12] SAYERS, C.M. (1981) Ultrasonic velocity dispersion in porous materials, *J. Phys. D*, **14**, 413-420.

[13] LLOYD, P. (1967) *Proc. Phys. Soc.*, **90**, 207.

[14] LLOYD, P. (1967) *Proc. Phys. Soc.*, **90**, 217.

[15] SAYERS, C.M. (1980) On the propagation of ultrasound in highly concentrated mixtures and suspensions, *J. Phys. D*, **13**, 179-184.

[16] SAYERS, C.M. and SMITH, R.L. (1982) The propagation of ultrasound in porous media, *Ultrasonics*, **20**, 201-205.

[17] SAYERS, C.M. and TAIT, C.E. (1984) Ultrasonic properties of transducer backings, *Ultrasonics*, **22**, 57-60.

[18] TITTMAN, B.R., MORRIS, W.L. and RICHARDSON, J.M. (1980) *Appl. Phys. Lett.*, **36**, 199.

[19] FRANZBLAU, M.C. and KRAFT, D.W. (1971) *J. Appl. Phys.*, **42**, 5261.

[20] PAPADAKIS, E.P. and PETERSEN, B.W. (1979) Ultrasonic velocity as a predictor of density in sintered powder meta parts, *Mat. Eval.*, **37**, 76.

[21] REYNOLDS, W.N. and WILKINSON, S.J. (1977) A.E.R.E. Report R8974.

[22] NAGARAJAN, A. (1971) *J. Appl. Phys.*, **42**, 3693.

[23] RANACHOWSKI, J. (1971) *Proc. Conf. on Acoustics of Solid Media, Warsaw*, ed L. Filipczynski et al. p203.

[24] WINKLER, K.W. (1983) Frequency dependent ultrasonic properties of high-porosity sandstones, *J. Geophys. Res. B*, **88**, 9493-9499.

[25] ELLIOT, R.J., KRUMHANSL, J.A. and LEATH, P.L. (1974) The theory and properties of randomly disordered crystals and related physical systems, *Rev. Mod. Phys.*, **46**, 465-543.

[26] BERRYMAN, J.G. (1979) Theory of elastic properties of composite materials, *Appl. Phys. Lett.*, **35**, 856-858.

[27] BERRYMAN, J.G. (1979) Long-wavelength propagation in composite elastic media I. Spherical inclusions, *J. Acoust. Soc. Am.*, **68**, 1809-1819.

[28] BERRYMAN, J.G. (1979) Long-wavelength propagation in composite elastic media II. Ellipsoidal inclusions, *J. Acoust. Soc. Am.*, **68**, 1820-1831.

[29] THOMPSON, R.B., SPITZIG, W.A. and GRAY, T.A. Relative effects of porosity and grain size on ultrasonic wave propagation in iron compacts.

[30] BOOCOCK, J., FURZER, A.S. and MATTHEWS, J.R. (1972) The effect of porosity on the elastic moduli of UO_2 as measured by an ultrasonic technique, A.E.R.E. Report M2565.

[31] O'CONNELL, R.J. and BUDIANSKY, B. (1974) Seismic velocities in dry and saturated cracked solids, *J. Geophys. Res.*, **79**, 5412-5426.

[32] BUDIANSKY, B. and O'CONNELL, R.J. (1976) Elastic moduli of a cracked solid, *Int. J. Solids Structures*, **12**, 81-97.

[33] HOENIG, A. (1979) Elastic moduli of a non-randomly cracked body, *Int. J. Solids Structures*, **15**, 137-154.

[34] BRUNER, W.M. (1976) Comment on 'Seismic velocities in dry and saturated cracked solids' by R.J. O'Connell and B. Budiansky, *J. Geophys. Res.*, **81**, 2573-2576.

[35] HENYEY, F.S. and POMPHREY, N. (1982) Self-consistent moduli of a cracked solid, *Geophys. Res. Lett.*, **9**, 903-906.

[36] HUDSON, J.A. (1980) Overall properties of a cracked solid, *Math. Proc. Camb. Phil. Soc.*, **88** 371-384.

[37] HUDSON, J.A. (1981) Wave speeds and attenuation of elastic waves in material containing cracks, *Geophys. J. Royal Astr. Soc.*, **64**, 133-150.

[38] HUDSON, J.A. (1986) A higher order approximation to the wave propagation constants for a cracked solid, *Geophys. J. Royal Astr. Soc.*, **87**, 265-274.

[39] HILL, R. (1963) Elastic properties of reinforced solids: some theoretical principles, *J. Mech. Phys. Solids*, **11**, 357-372.

[40] SAYERS, C.M. and KACHANOV, M. (1991) A simple technique for finding effective elastic constants of cracked solids for arbitrary crack orientation statistics, *Int. J. Solids Structures*, **12**, 81-97.

[41] ESHELBY, J.D. (1957), The determination of the elastic field of an Ellipsoidal inclusion and related problems. *Proc. Roy. Soc.*, **A241**, 376-396.

[42] HOENIG, A. (1978) The behaviour of a flat elliptical crack in an anisotropic body, *Int. J. Solids Structures*, **14**, 925-934.

[43] CHARLAIX, E. (1986) Percolation threshold of a random array of discs: a numerical simulation, *J. Phys. A*, **19**, L351-L354.

[44] LAWS, N. and DVORAK, G.J. (1987) The effect of fibre breaks and aligned penny-shaped cracks on the stiffness and energy release rates in unidirectional composites, *Int. J. Solids Structures*, **23**, 1269-1283.

[45] MILTON, G. (1984) The coherent potential approximation is a realizable effective medium scheme, preprint.

[46] NORRIS, A.N. (1985) A differential scheme for the effective moduli of composites, *Mechanics of Materials*, **4**, 1-16.

[47] SAYERS, C.M. (1988a) Inversion of ultrasonic wave velocity measurements to obtain the microcrack orientation distribution function in rocks, *Ultrasonics*, **26**, 73-77.

[48] SAYERS, C.M. (1988b) Stress-induced ultrasonic wave velocity anisotropy in fractured rock, *Ultrasonics*, **26**, 311-317.

[49] ROE, R.J. (1964) Description of crystallite orientation in polycrystalline materials having fibre texture, *J. Chem. Phys.*, **40**, 2608-2615.

[50] ROE, R.J. (1965) Description of crystallite orientation in polycrystalline materials - III: General solution to pole figure inversion, *J. Appl. Phys.*, **36**, 2024-2031.

[51] ROE, R.J. (1966) Inversion of pole figures for materials having cubic crystal symmetry, *J. Appl. Phys.*, **37**, 2069-2072.

[52] MORRIS, P.R. (1969) Averaging fourth-rank tensors with weight functions, *J. Appl. Phys.*, **40**, 447-448.

[53] MUSGRAVE, M.J.P., Crystal Acoustics, Holden-Day, San Fransisco, 1970.

[54] BIRCH, F. (1960) The velocity of compressional waves in rocks to 10 kilobars. Part 1, *J. Geophys. Res.*, **65**, 1083-1102.

[55] BIRCH, F. (1961) The velocity of compressional waves in rocks to 10 kilobars. Part 2, *J. Geophys. Res.*, **66**, 2199-2224.

[56] THILL, R.E., WILLARD, R.J. and BUR, T.R. (1969) Correlation of longitudinal velocity variation with rock fabric, *J. Geophys. Res.*, **74**, 4897-4909.

[57] BATZLE, M.L., SIMMONS, G. and SIEGFRIED, R.W. (1980) Microcrack closure in rocks under stress: direct observation, *J. Geophys. Res.*, **85**, 7072-7090.

[58] NUR, A. and G. SIMMONS, Stress-induced velocity anisotropy in rock: an experimental study, *J. Geophys. Res.*, **74**, 6667-6674, 1969.

[59] PATERSON, M.S., Experimental Rock Deformation - The Brittle Field, Springer, Berlin, 254 pp., 1978.

[60] KRANZ, R.L., Microcracks in rock: A review, *Tectonophysics* **100**, 449-480, 1983.

[61] SAYERS, C.M. and J.G. van MUNSTER, Microcrack-induced seismic anisotropy of sedimentary rocks, accepted for publication in *J. Geophys. Res.*

[62] SAYERS, C.M., J.G. van MUNSTER and M.S. KING, Stress-induced ultrasonic anisotropy in Berea sandstone, *Int. J. Rock Mech.*, **27**, 429-436, 1990.

[63] CASTAGNA, J.P., M.L. BATZLE and R.L. EASTWOOD, Relationships between compressional-wave and shear-wave velocities in clastic silicate rocks, *Geophysics*, **50**, 571-581, 1985.

[64] LO, T., K.B. COYNER and M.N. TOKSOZ, Experimental determination of elastic anisotropy of Berea sandstone, *Geophysics*, **51**, 164-171, 1986.

[65] HATHERLEY, M. and HUTCHINSON, W.B., An introduction to textures in metals, monograph no. 5, London, Institution of Metallurgists.

[66] HUTCHINSON, W.B., (1984) *Int. Metall. Rev.*, **29**, 25-42.

[67] STICKELS, C.A. and MOULD, P.R., (1970) *Met. Trans.*, **1**, 1303-1312.

[68] DAVIES, G.J., GOODWILL, D.J. and KALLEND, J.S. (1972) *Met. Trans.*, **3**, 1627-1631.

[69] SAYERS, C.M. (1982) Ultrasonic velocities in anisotropic polycrystalline aggregates, *J. Phys. D*, **15**, 2157-2167.

[70] SAYERS, C.M. (1986) Angular dependent ultrasonic wave velocities in aggregates of hexagonal crystals, *Ultrasonics* **24**, 289-291.

[71] SAYERS, C.M. (1987) The elastic anisotropy of polycrystalline aggregates of Zirconium and its alloys, *J. Nucl. Mat.* **144**, 211-213

[72] SAYERS, C.M. (1987) Elastic wave anisotropy in the upper mantle, *Geophys. J. Roy. Astr. Soc.* **88**, 417-424

[73] SAYERS, C.M. (1987) Orientation of olivine in dunite from elastc wave velocity measurements, *Geophys. Res. Lett.* **14**, 1050-1052.

[74] BUNGE, H.J. (1965) *Z. Metallk.*, **56**, 872.

[75] BUNGE, H.J. (1968) *Krist. Tech.*, **3**, 431.

[76] PURSEY, H. and COX, H.L. (1954) *Phil. Mag.*, **45**, 295.

[77] HILL, R. (1957) The elastic behaviour of a crystaline aggregate, *Proc. Phys. Soc.*, **A65**, 349-354.

[78] SAYERS, C.M. and PROUDFOOT, G.G. (1986) Angular dependence of the ultrasonic SH wave velocity in rolled metal sheets, *J. Mech. Phys. Solids* **34**, 579-592.

[79] SAYERS, C.M. Texture independent determination of residual stress in polycrystalline aggregates using Rayleigh waves. *J. Phys. D.*, L179-L184 (1984).

[80] SAYERS C.M. Angular dependence of the Rayleigh surface velocity in polycrystalline metals with small anisotropy. *Proc. Roy. Soc. A* **400**, 175-182 (1985).

[81] SAYERS, C.M., ALLEN, D.R., HAINES, G.E. and PROUDFOOT, G.G., Texture and stress determination in metals by using ultrasonic Rayleigh waves and neutron diffraction, *Phil. Trans. Royal Soc.* **320**, 187-200 (1986).

[82] ALLEN D.R., LANGMAN R. and SAYERS C.M. Ultrasonic SH wave velocity in textured aluminium plates. *Ultrasonics*, September 1985, 215-222.

[83] ALLEN, A.J., HUTCHINGS, M.T., SAYERS, C.M., ALLEN, D.R. and SMITH, R.L. (1983) Use of neutron diffraction texture measurements to establish a model for calculation of ultrasonic velocities in highly oriented austenitic weld material, *J. Appl. Phys.*, **54**, 555-560.

[84] ALLEN, D.R. and LANGMAN, R. (1985) AERE-R11573.

[85] LANGMAN, R. and ALLEN, D.R. (1985) AERE-R11598.

[86] KUPPERMAN, D.S. and REIMANN, K.J., (1980) *I.E.E.E. Trans Sonics Ultrasonics* **SU-27**, 7.

[87] MACDONALD, D.E., (1981) *I.E.E.E. Trans Sonics Ultrasonics* **SU-28**, 75.

[88] BIOT, M.A., (1940) *J. Appl. Phys.* **11**, 522.

[89] THURSTON, R.N., (1965) *J. Acoust. Soc. Am* **37**, 348.

[90] THOMPSON, R.B., SMITH, J.F. AND LEE,S.S. (1983) *Rev. Prog. NDE* **2**, ed. D.O. Thompson and D.E. Chimenti, Plenum Press, New York.

[91] THOMPSON, R.B., LEE, S.S. and SMITH, J.F. (1984) *Rev. Prog. NDE* **3**, ed. D.O. Thompson and D.E. Chimenti, Plenum Press, New York.

[92] LEE, S.S., SMITH, J.F. and THOMPSON, R.B., (1985) *Rev. Prog. NDE* **4**, ed. D.O. Thompson and D.E. Chimenti, Plenum Press, New York.

ACOUSTOELASTICITY AND APPLICATIONS

P.P. Delsanto
Politecnico di Torino, Torino, Italy

ABSTRACT

After a short introduction on the physical foundations of the acoustoelastic effect, its applications to the determination of applied and residual stresses and texture are discussed. Available techniques, based on the use of bulk and surface waves, are briefly reviewed. One of the methods, based on a perturbation of the equation of motion, is then applied to the case of Rayleigh waves propagating on the surface of a material plate, made of orthotropic polycrystalline aggregates.

1. INTRODUCTION

The acoustoelastic effect (AE) is defined as the stress dependence of the propagation velocity of ultrasonic waves in deformed elastic media [1-5]. It is important, from a NDE point of view, as an "inverse" tool for the prediction of stress fields. Its optical counterpart, i.e. the photoelastic effect (PE), is usually easier to apply. In fact, although the relative change of propagation velocity is of the same order, due to the much shorter wavelength of light (typically ≥ 3 orders of magnitude), interference effects are easier to evaluate for the PE. Several features, however, contribute to make the AE particularly attractive for NDE purposes [6,7]:

1) Most structural materials are opaque to light, but not to ultrasound.

2) AE techniques allow more flexibility, since both surface and bulk waves may be used.

3) The AE equipment may also be utilized for flaw detection and texture determination.

4) It is also suitable for field measurements and for automatization.

Two major problems have, however, severely hindered the development of practical AE techniques up to recent years:

1) The AE effect is small (of the order of 1°/..). Therefore very sensitive and accurate measurement techniques must be employed.

2) Other effects (e.g. texture in polycrystalline aggregates) are competing with the AE and need to be carefully separated for a meaningful stress determination.

Major advances in the experimental techniques, as discussed by other lecturers, have allowed to overcome these difficulties. As a result, several AE techniques [8-13] are now available for exploitation, both in the laboratory and in the field. Some of them are reviewed in Sec. 3, after a brief discussion (in Sec 2) of the AE, both from a microscopic and macroscopic point of view. Finally, in Sections 4 and 5, one of the techniques, based on a perturbation treatment and the use of Rayleigh waves, is applied to the problem of texture and stress analysis, respectively.

2. THEORY

The AE may be studied both from a microscopic and macroscopic point of view. Microscopically, in the case of polycrystalline aggregates, one can adopt the pseudopotential energy approach [14] and write the cell total energy as a sum of several contributions. E.g.

$$E = E_{es} + E_{fe} + E_{be} + E_r \qquad (1)$$

where E_{es} is the electrostatic Coulomb energy of positive charges, E_{fe} the free electron energy, E_{be} the band structure energy and E_r the ion-core repulsive energy.

Then one defines the (isothermic or isoentropic) second order elastic constants

$$C_{ijkl} = \frac{\partial^2 (E/\Omega)}{\partial \eta_{ij} \partial \eta_{kl}} \qquad (2)$$

where Ω is the cell volume and η_{ij} the lagrangian strain tensor.
The derivatives in Eq. (2) can be easily performed using the formula [15]

$$\frac{\partial}{\partial \eta_{ij}} = \frac{1}{2} \left(X_i \frac{\partial}{\partial x_j} + X_j \frac{\partial}{\partial x_i} \right) \qquad (3)$$

where X_i are the lagrangian coordinates and x_i the eulerian coordinates.

The AE may then be easily understood since, in a deformation, the energy terms in Eq. (1) are affected by a change in the lattice positions, thus affecting the elastic constants and, consequently, the propagation velocity. E.g. in a simple model [16]

$$E_r = 0.5\,\alpha \sum_n \exp(-\beta r_n) \tag{4}$$

where α is the repulsive energy parameter, β the repulsive range parameter and the sum is extended to all the "nearest neighbours" (lattice positions). Since β may be large, even a relatively small deformation may have an appreciable AE effect.

To understand the AE from a macroscopic point of view, we must first recall the definition of "natural" state (i.e. underformed) and "initial" state (i.e. after an "initial" deformation). If the deformed material is probed by an ultrasonic pulse, the consequent additional deformation brings the specimen to the so-called "final" state. The amplitude of the propagating pulse is, however, usually very small, compared with that of the initial deformation and, therefore, its effects may be treated in the framework of a linear theory of elasticity [17].

An exhaustive treatment of the problem of small amplitude waves superposed on finite initial deformations, leading to the acoustoelastic equation of motion, may be found in refs. [18] and [19]. We limit ourselves, here, to show heuristically how the presence of an initial deformation leads to a change in the wave equation. In fact, if the specimen has no initial deformation, the wave equation for the displacements w_i can be written as

$$\partial_l (C_{klmn} \partial_n w_m) = \rho \ddot{w}_k \tag{5}$$

where $\partial_n \equiv \partial/\partial x_n$ and ρ is the density.

The term $C_{klmn}\partial_n w_m$ represents a stress (due to the propagating pulse). If the material is initially deformed, the initial stress σ_{ln} and the direct effect of the strains on the stiffness must be added through a correction

$$C'_{klmn} = \sigma_{ln}\delta_{km} + s_{klmn} \tag{6}$$

where s_{klmn} depends on the initial strains through the third order elastic constants.

The AE can, therefore, be understood as the effect of these correction terms (C'_{klmn}) on the wave equation (Eq.5). Since these terms are usually small, a perturbation treatment may be very convenient [12]. In fact, also other effects, such as due to anisotropy, texture, small temperature changes or external magnetic fields, may be treated as perturbation corrections. Thus all effects are linearized and may be treated independently. Also, explicit formulas for the solution of the inverse problem can be obtained, which is very convenient for NDE purposes [13].

To conclude, we note that a completely general acousto-elastic theory should be in the framework of magneto-termo-acousto-elasto-plasticity and would be extremely complex. Therefore, even if the treatment is not perturbative, several simplifying assumptions are usually made. Among them:

- the initial deformations are considered static (i.e. the body is at equilibrium in its initial state)
- the superposed dynamic motion is very small
- the material is assumed to be homogeneous and hyperelastic (plastic deformations are neglected)
- the process is either isoentropic or isothermic (thermodynamic effect are ignored)

- submicron lattice defects and grain structure are ignored (due to the relatively large wavelength of ultrasonic waves).

3. APPLICATIONS

The propagation velocity of ultrasonic waves depends, in general, on

1)   the direction of particle motion (polarization)

2)   the direction of wave propagation (anisotropy)

3)   the stress field (AE)

4)   the material texture (for polycrystalline aggregates)

5)   any other eventual features, such as defects, irregularities, inhomogeneities, etc.

As mentioned in the Introduction, the competition of different effects (stress, texture, defects, etc.) represents a major difficulty, but also an opportunity since, once the effects are separated, the AE equipment may be used not only for the determination of the stress field, but also of the texture and for the detection of flaws.

Several techniques have been proposed for the application of the AE to the problem of stress and texture determination. They may be based on the transmission of bulk waves or on the propagation of surface waves. The former (bulk waves) are particularly convenient if stress and texture are homogeneous through the thickness of the specimen (e.g. a material plate). In fact, out of a transmission measurement, one can only obtain an average value (e.g. of the stress) through the propagation path. Also, any local variation or inaccuracy in the measurement of the thickness may affect the time-of-flight evaluation by an amount as large (or larger) than the AE itself. To overcome these difficulties, several shear waves techniques [20-2] have been proposed, based on oblique transmission at several angles (to increase the amount of experimental information). As a single example, in the method proposed by Clark and Mignogna [22] in order to obtain specific information on the stress field at a given location P inside the plate, several SH waves are launched from different angles through P , thus enhancing the effect of the local stress in P .

These difficulties may be easily solved if surface waves, e.g. Rayleigh waves (RW), are employed. In fact it is easier, on a plate surface, to control the length of propagation, by keeping fixed the distance between two transducers (e.g. transmitter and receiver). Also different depth ranges may be probed by varying the frequency (and consequently the wavelength) of the surface waves. In fact RW's penetrate into a material up to a depth of approximatively one wavelength.

Surface waves may also be preferable, since they decay as l^{-1} , where l is the length of propagation, while bulk waves, having spherical wavefronts, decay as l^{-2}. Finally, if only one side of the structural element to be inspected is (easily) accessible, surface waves are the obvious choice. In general, however, the choice between bulk and surface waves depends on the particular job to be performed and the equipment available. In some cases, as we will see, it may even be convenient to use a combination of both techniques.

4. TEXTURE ANALYSIS

As an example of application of AE techniques we discuss in this Section a method, based on the use of ultrasonic RW's for the determination of the texture in anisotropic polycrystalline aggregates [13,23]. In the next Section the method will be extended to the determination of applied and residual stresses.

We assume that our specimen consists of an orthotropic distribution of cubic crystallites, which have (in Voigt's notation) only three independent nonzero second order elastic constants:

1) $C_{11} = C_{22} = C_{33}$

2) $C_{12} = C_{13} = C_{23}$ $\hspace{6cm}$ (7)

3) $C_{44} = C_{55} = C_{66}$

We define the anisotropy as

$$c = C_{11} - C_{12} - C_{44} \hspace{6cm} (8)$$

In an isotropic material

$$C_{11} = \lambda + 2\mu$$
$$C_{12} = \lambda \hspace{6cm} (9)$$
$$C_{44} = \mu$$

where λ and μ are the Lamè constants. Therefore

$$c = o \hspace{6cm} (10)$$

The elastic constants are related to the propagation velocity. E.g. in an isotropic material, the longitudinal and transverse velocities are given, respectively, by

$$V_L = \sqrt{(\lambda + 2\mu)/\rho}$$

$$V_T = \sqrt{\mu/\rho} \tag{11}$$

We then recall the definition of texture, as the preferential alignment of crystallographic axes. The texture may be specified in terms of the orientation distribution function (ODF), which represents the normalized probability of finding crystallites with given Euler angles Θ, Ψ and Φ. The ODF can be expanded [24] as a series

$$w\,(\cos\Theta, \Psi, \Phi) = \tag{12}$$

$$= \sum_{l=0}^{\infty} \sum_{m=-l}^{l} \sum_{n=-l}^{l} W_{lmn} Z_{lmn}(\cos\Theta)\exp\,(-im\Psi)\exp\,(-in\Phi)$$

where Z_{lmn} are the generalized Legendre functions [25]. The expansion coefficients W_{lmn} are called orientation distribution coefficients (ODC). Since we are considering elastic waves, due to symmetry considerations, there are only three independent nonzero ODC's that can be measured : W_{400}, W_{420} and W_{440}.

We now assume that our specimen is a material plate with both the principal stress axes and the material symmetry axes parallel to its surface (with an angle χ between them). We then consider a system of axes with x_3 normal to the plate and assume that x_2 is the propagation direction. We call ϕ the angle between x_1 and one of the material symmetry axes.

Following the perturbation treatment of ref. [12], it is possible to express explicitely the change of propagation velocity due to the various perturbative corrections:

$$\Delta v = B_o + B_1 \cos\phi + B_2 \cos 4\phi + B_3 \sin 2\phi \qquad (13)$$

where

$$B_o = b_\sigma \sigma + b_o W_{400}$$

$$B_1 = b_1 W_{420} - b_\Delta \Delta \cos 2\chi$$

$$B_2 = b_2 W_{440}$$

$$B_3 = b_\Delta \Delta \sin 2\chi \qquad (14)$$

Here σ and Δ are the sum and difference, respectively, of the principal stresses. The coefficients b depend on the third order elastic constant and are given explicitely in ref. [13].

If no stress is present in the specimen, $B_3 = o$ and the only unknowns are the ODC's. With three independent time-of- flight measurements (at different propagation angles ϕ), it is possible, in principle, to obtain the B_i coefficients and, hence, the ODC's. Better results can obviously be obtained with a larger number of independent measurements and a least square fitting of the B_i. The procedure has been tested successfully for plates of Al alloys [23,26]; the results have been compared with those of other ultrasonic and neutron diffraction techniques [27]. The experimental set-up is described in detail in ref. [28]. Finally, an application of the method to the determination of the elastic constants in metal alloys is described in refs. [29,30].

5. DETERMINATION OF APPLIED AND RESIDUAL STRESSES

The competition of AE and texture effects, which we have mentioned in the Introduction, can be readily appreciated by inspecting Eqs. (14). In fact both σ and W_{400} contribute to B_o; likewise both Δ and W_{420} contribute to B_1.

It is easy to separate the stress and texture contributions when a reference system with the same texture but no stress is available. E.g. in the measurement of applied stresses, the reference system can be the specimen itself, before any external stress is applied. Then

$$\sigma = (B_o - B_o^{\text{ref}})/b_1$$

$$\Delta = \sqrt{B_3^2 + (B_1 - B_1^{\text{ref}})^2}/b_\Delta \tag{15}$$

$$\chi = 0.5\text{arctan}[B_3/(B_1 - B_1^{\text{ref}})]$$

where B_o^{ref} and B_1^{ref} are the values of B_o and B_1 for the reference specimen.

In the case of residual stresses the situation is much more critical. First, because the presence of residual stresses usually implies some amount of plasticity, which should be properly treated in the framework of a consistent elastoplastic theory [31-3]. Second, because it is not easy to find an adequate reference system.

If, however, plastic deformations are not too large, it is still possible, with some redefinition of the "natural" state and of the concept itself of residual stress [5], to apply the AE technique of Sec.4. The difficulty of obtaining a stress-relieved reference specimen can then be at least partially solved in two different ways [34]. The first method (which, however, is destructive) consists of cutting the plate through its center plane and using the cut plate as a reference specimen. In fact, much of the residual stress is eliminated by the cutting process, with little alteration of the texture (as proved by photomicrographs of the plate edges). The method has been tested for plates of 2024-T351 Al, showing a good agreement with the results of strain gage measurements.

The second method, which has the advantage of being nondestructive, can be implemented if the texture may be assumed to be homogeneous and the stress averages to zero through the thickness of the plate. Then W_{400} and W_{420} can be obtained [8] by measurements with bulk waves, without interference of stress effects, and the corresponding values of B_o and B_1 are used as B_o^{ref} and B_1^{ref}, respectively.

6. SUMMARY

Due to the recent progress in the development of the experimental set-up, ultrasonic techniques based on the exploitation of the AE are now available for both laboratory and field measurements. Pay-offs include texture monitoring and stress measurements, such as around cracks (to determine the stress-intensity factor), in load bearing structural members, in weldings, etc. AE applications may be based on the transmission of bulk waves or on the propagation of surface waves, or even on a combination of both.

For several materials of interest (e.g. Al alloys) a perturbative treatment of the equation of motion can greatly facilitate the analysis, by linearizing the problem and allowing an independent evaluation of different effects, such as stress, anisotropy, texture, thermic effects, etc. A comparison with the results of an exact calculation [35] allows to ascertain the range of validity of the perturbation approach.

Tests performed with several plates of metal alloys confirm, through a comparison with the results of other techniques, the reliability of the AE techniques. Nondestructive measurements can be performed for the determination of the texture and of applied stresses. In the case of residual stresses, however, a destructive procedure is required, unless certain restrictive condictions are verified (homogeneous texture and stress averaging to zero through the thickness of the plate).

REFERENCES

1. Hughes, D.S. and J.L. Kelly, Phys. Rev. 92, 1145 (1953).
2. Crecraft, D.I., J. Sound Vib. 5, 173 (1967).
3. Tokuoka, T. and Y. Iwashimizu, Solids and Structures 4, 383 (1968).
4. Iwashimizu, Y. and Kubomura K. Int J. Solid Struct. 9, 99 (1973).
5. Pao, Y.H., Sachse, W. and H. Fukuoka, in : Physical Acoustics, Vol. 17, 62 (1984).
6. Henneke, E.G. and R.E. Green, J. Acoust. Soc. Am. 45, 1367 (1968).
7. Hsu, N.N., Exp. Mech. 14, 169 (1974).
8. Allen, D.R. and C.M. Sayers, Ultrasonics 22, 179 (1984).
9. Goebbels, K. and S. Hirsekorn, Ultrasonics 22, 338 (1984).
10. Thompson, R.B., Lee, S.S. and J.F. Smith, J. Acoust. Soc. Am. 80, 921 (1986).
11. Husson, D., Bennet, S.D. and G.S. Kino, Mat. Eval. 43, 92 (1985).
12. Delsanto, P.P. and A.V. Clark, J. Acoust. Soc. Am. 81, 952 (1987).
13. Delsanto, P.P., Mignogna, R.B. and A.V. Clark, J. Acoust. Soc. Am. 87, 215 (1990).
14. Wallace, D.C. : Thermodynamics of Crystals, J. Wiley 1972.
15. Delsanto, P.P., Provenzano, V. and H. Überall: Coherency Strain Effects in Metallic Bilayers, to appear in J. Phys.: Condensed Matter.
16. Soma T., J. Phys. F : Met. Phys. 4, 2157 (1974).

17.   Achenbach, J.D. : Wave Propagation in Elastic Solids, North Holland 1980.

18.   Thurston, R.N. in: Physical Acoustics, Vol. 1, 1 (1964).

19.   Eringen, A.C. and E.S. Suhuby: Elastodynamics, Academic Press. 1980.

20.   King, R.B. and C.M. Fortunko, J. Appl. Phys. 54, 3027 (1983).

21.   Thompson, R.B., Smith, J.F. and S.S. Lee, in: NDE of Microstructure for Process Control (Ed. H.N.G. Wadley) ASME 73-80, (1985).

22.   Clark, A,V. and R.B. Mignogna, Ultrasonics 22, 205 (1984).

23.   Delsanto, P.P. Mignogna, R.B. and A.V. Clark, in: Review of Progress in Quantitative NDE (Eds. D.O. Thompson and D.E. Chimenti), Plenum, Vol.5B, 1407-14 (1985).

24.   Sayers, C.M., J. Phys. D 15, 2157-77 (1982).

25.   Roe. R.J., J. Appl. Phys. 37, 2069 (1966).

26.   Delsanto, P.P., Mignogna,R.B. and A.V. Clark, in: Nondestructive Characterization of Materials II (Eds. J.F. Bussière, J. Monchain, C.O. Rudd and R.E. Green), Plenum 535-543, (1987).

27.   Clark, A.V., Reno, R.C., Smith, J.F., Blessing, G.V., Fields, R.J., Delsanto, P.P. and R.B. Mignogna, Ultrasonics 26, 189 (1988).

28.   Mignogna, R.B. Delsanto, P.P. and Clark, A.V. Rath, B.B. and C.L. Vold, ibid. 535-543, (1987).

29.   Delsanto, P.P., Quarati, P., Boschetti, F., Chaskelis, H.H. and R.B. Mignogna, in: Nondestructive Testing (Eds. J. Boogaard and G.M. Van Dijk), Elsevier 1549-51, (1989).

30.   Delsanto, P.P., Quarati, P., Boschetti, F., Chaskelis, H.H. and R.B. Mignogna, in: Monitoring, Surveillance and Predictive Maintenance of Plants and Structures, AIPnD 348-54, (1989).

31.   Johnson, G.C., J. Acoust. Soc. Am. 70, 591 (1981).

32.   Johnson, G.C., J. Appl. Mech. 50, 689 (1983).

33. Pao, Y.H. and U. Gamer, J. Acoust. Soc. Am. 77, 806 (1985).

34. Delsanto, P.P., Mignogna, R.B., Clark, A.V., Mitrakovic, D. and J.C. Moulder, in: Residual Stresses in Science and Technology (Eds. E. Macherauch and V. Hauk), DGM, Vol. 1, 175-81 (1987).

35. Mase, G.T. and P.P. Delsanto, in: Review of Progress in Quantitative NDE (Eds. D.O. Thompson and D.E. Chimenti), Plenum, Vol. 7B, 1349-56 (1988).

INVERSE METHODS AND IMAGING

K.J. Langenberg, P. Fellinger, R. Marklein, P. Zanger, K. Mayer
and T. Kreutter

University of Kassel, Kassel, Germany

Contents

1   Introduction

Ultrasonic nondestructive testing (NDT) of solid materials exploits the scattering of elastic waves by defects such as cracks, voids, inclusions, and other inhomogeneities. The scattered waves carry information about location, size, shape, and orientation of these defects which has to be extracted appropriately from measurements [1, 2]. The ultimate goal is to produce threedimensional images of the interior of the material. In principal, this can be achieved "inverting" the scattering of ultrasound with the aid of inverse scattering theories. Particularly, in three spatial dimensions this turns out to be a complicated and ill-conditioned task even for the much simpler case of scalar acoustic waves. Therefore, approximations and simplifying assumptions are introduced as a trade-off between complexity of algorithms and proper assessment of the integrity of the material [6]. In addition, for practical applications, data recording and processing has to be fast and cheap, which is only available if the algorithms are simple enough; this essentially means, that they *are* capable of producing images — even in three dimensions — but, due to the fact, that they are *not* exact inverse scattering solutions, these images have to be properly assessed and understood. For that reason, apart from making many test experiments, numerical modeling of the radiation, propagation, and scattering of elastic waves for given NDT situations and using the computed data as input for the available imaging schemes seems to be a powerful tool to understand the images better, particularly, if elastic waves are under concern in those imaging algorithms which have been designed for scalar acoustic waves only.

2   Elastic Waves in Solids

If a force is applied to, say, the surface of a solid, "particles" in terms of volume elements are removed from their equilibrium position. The resulting deformation state is characterized by the (second rank) strain tensor $\underline{\underline{S}}(\mathbf{R}, t)$, where $\mathbf{R}$ denotes the vector of position, and t the time. Linear elastodynamics (compare, for instance [3]) relates $\underline{\underline{S}}$ to the vector $\underline{u}(\mathbf{R}, t)$ of the (differential) volume element displacement in terms of

$$\underline{\underline{S}}(\mathbf{R}, t) = \frac{1}{2}[\boldsymbol{\nabla}\underline{u}(\mathbf{R}, t) + \underline{u}(\mathbf{R}, t)\boldsymbol{\nabla}] \ . \tag{1}$$

Here, we have utilized the del-symbol for the gradient operation, $\nabla \underline{u}$ indicating a dyadic product, i.e. it denotes the gradient dyadic of $\underline{u}$; $\underline{u}\nabla$ stands for the transpose of $\nabla \underline{u}$. Obviously, the strain tensor is symmetric. Depending on the elastodynamic properties of the solid, a prescribed strain state results in a stress state, which is described by the (second rank) stress tensor $\underline{\underline{T}}(\underline{R}, t)$; again, linear elastodynamics provides a linear relationship between $\underline{\underline{S}}$ and $\underline{\underline{T}}$ according to Hooke's law

$$\underline{\underline{T}}(\underline{R}, t) = \underline{\underline{c}}(\underline{R}) : \underline{\underline{S}}(\underline{R}, t) \tag{2}$$

where the forth rank stiffness tensor $\underline{\underline{c}}$ reflects the material properties; the double dot indicates double contraction with regard to adjacent indices, i.e. in components, utilizing Einstein's summation convention we have

$$T_{ij}(\underline{R}, t) = c_{ijkl}(\underline{R})S_{kl}(\underline{R}, t) \ . \tag{3}$$

Allowing $\underline{\underline{c}}$ to be dependent on $\underline{R}$ accounts for an inhomogeneous elastic medium, whereas its independence on time is ultimately an issue of loss neglection; otherwise, (2) would have to be replaced by a time domain convolution, and appropriate Kramers-Kronig relations would have to be satisfied. The symmetry of $\underline{\underline{S}}$ by definition, the symmetry of $\underline{\underline{T}}$ due to the conservation of momentum, and the assumption of perfect elasticity require

$$\begin{aligned} c_{ijkl}(\underline{R}) &= c_{ijlk}(\underline{R}) \\ c_{ijkl}(\underline{R}) &= c_{jikl}(\underline{R}) \\ c_{ijkl}(\underline{R}) &= c_{klij}(\underline{R}) \ , \end{aligned} \tag{4}$$

thus leaving 21 elastic constants for the general linear anisotropic medium. The assumption of isotropy reduces $\underline{\underline{c}}$ to

$$\underline{\underline{c}} = \lambda \underline{\underline{II}} + \mu \left(\underline{\underline{II}}^{1342} + \underline{\underline{II}}^{1324} \right) \ , \tag{5}$$

where λ and μ denote Lamé's constants; $\underline{I}$ is the dyadic idemfactor, which has the properties

$$\underline{\underline{A}} \cdot \underline{I} = \underline{I} \cdot \underline{\underline{A}} = \underline{\underline{A}} \tag{6}$$

and

$$\underline{\underline{A}} : \underline{I} = \underline{I} : \underline{\underline{A}} = \text{trace } \underline{\underline{A}} \tag{7}$$

for every second rank tensor $\underline{\underline{A}}$. The upper indices on the forth rank tensor $\underline{\underline{II}}$, i.e. the dyadic product of $\underline{\underline{I}}$ with itself, indicate transposition of elements [4]; accordingly, $\underline{u}\nabla$ can be written as $\nabla\underline{u}^{21}$. Therefore, $\underline{\underline{II}}^{1342}$ is the identity operator with regard to double contraction, i.e.

$$\underline{\underline{A}} : \underline{\underline{II}}^{1342} = \underline{\underline{II}}^{1342} : \underline{\underline{A}} = \underline{\underline{A}} \ , \tag{8}$$

and $\underline{\underline{II}}^{1324}$ double contracts any second rank tensor to its transpose:

$$\underline{\underline{A}} : \underline{\underline{II}}^{1324} = \underline{\underline{II}}^{1324} : \underline{\underline{A}} = \underline{\underline{A}}^{21} \ . \tag{9}$$

Typical materials to be investigated nondestructively can, for instance, be considered as isotropic (ferritic steel, polycristalline materials like aluminum), or as transversely isotropic (noncorrosive steel, fiberreinforced composites, metal matrix composites), the latter ones requiring 5 independent elastic constants. Transverse isotropy means isotropic in all planes perpendicular to a given direction $\hat{\underline{a}}$, and anisotropic in the direction $\hat{\underline{a}}$ only; a coordinate-free representation of the stiffness tensor is then given by [5]:

$$\begin{aligned}
\underline{\underline{c}} \ = \ & \lambda_\perp \underline{\underline{II}} + \mu_\perp \left[\underline{\underline{II}}^{1324} + \underline{\underline{II}}^{1342}\right] + \\
& + \left[\lambda_\perp + 2\mu_\perp + \lambda_\| + 2\mu_\| - 2(\nu + 2\mu_\|)\right] \hat{\underline{a}}\hat{\underline{a}}\hat{\underline{a}}\hat{\underline{a}} + \\
& + (\nu - \lambda_\perp)(\underline{\underline{I}}\hat{\underline{a}}\hat{\underline{a}} + \hat{\underline{a}}\hat{\underline{a}}\underline{\underline{I}}) + \\
& + (\mu_\| - \mu_\perp) \left[\underline{\underline{I}}\hat{\underline{a}}\hat{\underline{a}}^{1324} + \hat{\underline{a}}\hat{\underline{a}}\underline{\underline{I}}^{1324} + \underline{\underline{I}}\hat{\underline{a}}\hat{\underline{a}}^{1342} + \hat{\underline{a}}\hat{\underline{a}}\underline{\underline{I}}^{1342}\right] \ , \tag{10}
\end{aligned}$$

where $\lambda_\|, \lambda_\perp, \mu_\|, \mu_\perp$, and ν are five elastic constants replacing λ and μ of the isotropic case.

With the help of (7), (8), and (9), Hooke's law for isotropic media, i.e. (2) with (5), can be transformed into the more convenient form

$$\underline{\underline{T}}(\underline{R},t) = \lambda\underline{\underline{I}}\nabla \cdot \underline{u}(\underline{R},t) + \mu \left[\nabla\underline{u}(\underline{R},t) + \nabla\underline{u}(\underline{R},t)^{21}\right] \ , \tag{11}$$

if the definition (1) is observed. We write (11) in short-hand notation

$$\underline{\underline{T}}(\underline{R},t) = \underline{\underline{\mathcal{D}}}_{Ho} \cdot \underline{u}(\underline{R},t) \tag{12}$$

introducing the Hooke-operator

$$\underline{\underline{\mathcal{D}}}_{Ho} = \lambda\underline{\underline{I}}\nabla + \mu\nabla\underline{\underline{I}} + \mu\nabla\underline{\underline{I}}^{213} \ . \tag{13}$$

Balancing forces within the solid according to Newton's law leads to Cauchy's equation of motion

$$\varrho \frac{\partial^2 \underline{u}(\underline{R}, t)}{\partial t^2} = \nabla \cdot \underline{\underline{T}}(\underline{R}, t) + \underline{f}(\underline{R}, t) \quad , \tag{14}$$

where ϱ is the mass density; the divergence of $\underline{\underline{T}}$ according to $\nabla \cdot \underline{\underline{T}}$ can be interpreted as a local force density of the solid due to its intrinsic elastic stresses, while $\underline{f}(\underline{R}, t)$ represents a force density applied from "outside" and accounts for the excitation of elastic waves in terms of an inhomogeneity of the differential equation (14).

Inserting Hooke's law for isotropic materials into (14) and observing — Δ denotes the Laplacian —

$$\nabla \cdot \underline{\underline{\mathcal{D}}}_{Ho} = (\lambda + \mu)\nabla\nabla + \mu\Delta\underline{\underline{I}} \tag{15}$$

results in the following second order partial differential equation for the particle displacement vector $\underline{u}$

$$(\lambda + \mu)\nabla\nabla \cdot \underline{u}(\underline{R}, t) + \mu\Delta(\underline{R}, t) - \varrho\frac{\partial^2 \underline{u}(\underline{R}, t)}{\partial t^2} = -\underline{f}(\underline{R}, t) \quad ; \tag{16}$$

utilizing

$$\nabla \times \nabla \times \underline{u} = \nabla\nabla \cdot \underline{u} - \Delta\underline{u} \quad , \tag{17}$$

alternatively results in

$$(\lambda + 2\mu)\nabla\nabla \cdot \underline{u}(\underline{R}, t) - \mu\nabla \times \nabla \times \underline{u}(\underline{R}, t) - \varrho\frac{\partial^2 \underline{u}(\underline{R}, t)}{\partial t^2} = -\underline{f}(\underline{R}, t) \quad . \tag{18}$$

This equation can be reduced to two conventional wave equations for the scalar potential $\Phi(\underline{R}, t)$ and the vector potential $\underline{\Psi}(\underline{R}, t)$ according to — Δ denoting the Laplacian —

$$\Delta\Phi(\underline{R}, t) - \frac{1}{c_P^2}\frac{\partial^2\Phi(\underline{R}, t)}{\partial t^2} = -\frac{1}{\lambda + 2\mu}\Phi_{\underline{f}}(\underline{R}, t) \tag{19}$$

$$\Delta\underline{\Psi}(\underline{R}, t) - \frac{1}{c_S^2}\frac{\partial^2\underline{\Psi}(\underline{R}, t)}{\partial t^2} = -\frac{1}{\mu}\underline{\Psi}_{\underline{f}}(\underline{R}, t) \quad , \tag{20}$$

provided $\underline{u}$ as well as $\underline{f}$ are subject to their respective Helmholtz decompositions

$$\underline{u}(\underline{R}, t) = \nabla \Phi(\underline{R}, t) + \nabla \times \underline{\Psi}(\underline{R}, t) \tag{21}$$

$$\underline{f}(\underline{R}, t) = \nabla \Phi_{\underline{f}}(\underline{R}, t) + \nabla \times \underline{\Psi}_{\underline{f}}(\underline{R}, t) \tag{22}$$

with

$$\nabla \cdot \underline{\Psi}(\underline{R}, t) = 0 \ . \tag{23}$$

In (19) and (20) c_P and c_S denote wave speeds of pressure and shear waves, respectively, where

$$c_P = \sqrt{\frac{\lambda + 2\mu}{\varrho}} \tag{24}$$

$$c_S = \sqrt{\frac{\mu}{\varrho}} \ . \tag{25}$$

Plane wave solutions of the homogeneous equations (19) and (20) are immediately seen to be

$$\Phi(\underline{R}, t) = \Phi_0 \left(t - \frac{\hat{\underline{k}}_P \cdot \underline{R}}{c_P} \right) \tag{26}$$

$$\underline{\Psi}(\underline{R}, t) = \underline{\Psi}_0 \left(t - \frac{\hat{\underline{k}}_S \cdot \underline{R}}{c_S} \right) \tag{27}$$

yielding a longitudinally polarized pressure wave plus a transversely polarized shear wave in terms of the displacement:

$$\underline{u}(\underline{R}, t) = -\frac{1}{c_P} \hat{\underline{k}}_P \Phi_0' \left(t - \frac{\hat{\underline{k}}_P \cdot \underline{R}}{c_P} \right) - \frac{1}{c_S} \hat{\underline{k}}_S \times \underline{\Psi}_0' \left(t - \frac{\hat{\underline{k}}_S \cdot \underline{R}}{c_S} \right) \ . \tag{28}$$

The arbitrary unit-vectors $\hat{\underline{k}}_{P,S}$ determine the direction of propagation, and equally arbitrary are the wave functions $\Phi_0(t)$ and $\underline{\Psi}_0(t)$ with the exception that, due to (23)

$$\hat{\underline{k}}_S \cdot \underline{\Psi}_0(t) = 0 \ , \tag{29}$$

i.e. $\underline{\Psi}_0$ has to be orthogonal to $\hat{\underline{k}}_S$; the primes in (28) indicate derivatives with respect to the argument.

Notice, in solids of infinite extent, *plane*, and only plane, pressure and shear waves are decoupled and, hence, can travel independently. If the solid is made up of two half-spaces with different material properties, boundary conditions — continuity of $\underline{u}$ and $\underline{T} \cdot \underline{n}$, $\underline{n}$ being the unit-normal on the boundary — have to be satisfied. This yields reflection, refraction, and mode conversion phenomena, the latter ones denoting conversion of pressure into shear waves, and vice versa. The mathematical reason for that is the combination of Φ *and* $\underline{\Psi}$ in $\underline{T}$ via Hooke's law and observing (1) and (21). Also, due to the decomposition of (nearly) any applied spatially constrained force $\underline{f}$ into *both* Φ_f and $\underline{\Psi}_f$ according to (22), it always produces pressure *and* shear waves simultaneously; hence, coupled excitation of pressure and shear waves is the general phenomenon one has to cope with, apart from mode conversion by material inhomogeneities.

3   Computational Methods to Model Elastic Wave Propagation and Scattering

Only very few highly idealized canonical NDT problems can be modeled with analytical mathematical methods [7]. Therefore, in order to cope with real life, numerical techniques have to be developed to solve (18) for arbitrary environments, that is to say, for arbitrary forces $\underline{f}(\underline{R}, t)$, i.e. transducers, and arbitrary geometries of specimens and defects, i.e. arbitrary prescription of boundary conditions.

3.1   Direct Numerical Methods

Direct numerical methods operate directly on the fundamental equation of motion. A technique already applied for a long time in geophysics is to replace all derivatives by finite differences, ending up with an FDTD scheme (Finite Difference Time Domain) [8, 9, 10]. For its implementation for NDT purposes the reader is, for instance, refered to [11].

Another popular method borrows the idea of finite elements from solid mechanics and matches it to elastic wave propagation [12]. Here, the functional

$$F(\underline{u}) = \frac{1}{2} \iiint_V \mathcal{L}\{\underline{u}\} \cdot \underline{u}\, dV - \iiint_V \varrho \frac{\partial^2 \underline{u}}{\partial t^2} \cdot \underline{u}\, dV \qquad (30)$$

with the operator

$$\mathcal{L}\{\underline{u}\} = \nabla \cdot \underline{\underline{T}} \qquad (31)$$

is minimized over the volume of interest V; according to the Lax-Milgram lemma this minimum is reached if and only if $\underline{u}$ is a solution of Cauchy's equation of motion. On the other hand, minimizing (30) is equivalent to the requirement

$$\iiint_V \left(\nabla \cdot \underline{\underline{T}} - \varrho \frac{\partial^2 \underline{u}}{\partial t^2} \right) \cdot \underline{w} \, dV = 0 \qquad (32)$$

for arbitrary weighting functions $\underline{w}$. With the help of Gauss' theorem

$$\iiint_V \frac{\partial}{\partial x_i} A(\underline{R}, t) \, dV = \iint_S \underline{e}_i \cdot \underline{n} A(\underline{R}, t) \, dS \quad , \qquad (33)$$

where the x_i, $i = 1, 2, 3$ are cartesian coordinates with the mutually orthogonal unit-vectors $\underline{e}_i$, $i = 1, 2, 3$, and where $A(\underline{R}, t)$ denotes any scalar cartesian component of any tensor of arbitrary rank, the derivatives on $\underline{\underline{T}}$ are transfered to $\underline{w}$:

$$\iiint_V \underline{\underline{T}} : \nabla \underline{w} \, dV + \iiint_V \varrho \underline{w} \cdot \frac{\partial^2 \underline{u}}{\partial t^2} \, dV = \iint_S \underline{n} \cdot \underline{\underline{T}} \cdot \underline{w} \, dS \quad . \qquad (34)$$

Now, $\underline{u}$ and $\underline{w}$ are discretized according to

$$\underline{u}(\underline{R}, t) \simeq \sum_{k=1}^{3} \sum_{i=1}^{N} \phi_i(\underline{R}) u_k^i(t) \underline{e}_k \qquad (35)$$

$$\underline{w}(\underline{R}, t) \simeq \sum_{l=1}^{3} \sum_{j=1}^{N} \phi_j(\underline{R}) w_l^j(t) \underline{e}_l \qquad (36)$$

with the entire domain basis functions $\phi_i(\underline{R})$ and the mesh index $i = 1, \ldots, N$ over V; $\underline{e}_k$ and $\underline{e}_l$ denote unit vectors of an appropriate coordinate system. Hence, (34) is turned into a matrix equation, which can be solved with an explicit marching in time procedure.

Following the ideas developed for electromagnetic waves [13, 14], our own ansatz is to replace the differential formulation (14) and the isotropic version of Hooke's law ((2) together with (5)) by appropriate integral formulations, as it is convenient for Maxwell's equations. These integral formulations are then applied to each cell of a cubic grid superimposed to the test specimen under concern yielding a numerical scheme which we call EFIT for Elastodynamic Finite Integration Technique [15, 16]. Some details and numerical results of the scheme will be discussed in the following.

3.2   EFIT — Elastodynamic Finite Integration Technique

3.2.1   EFIT Scheme

Introducing the particle velocity

$$\underline{v}(\underline{R}, t) = \frac{\partial \underline{u}(\underline{R}, t)}{\partial t} \tag{37}$$

the volume integration of Cauchy's equation of motion and application of Gauss' theorem yields

$$\int\int\int_V \left[\varrho \frac{\partial \underline{v}(\underline{R}, t)}{\partial t} - \underline{f}(\underline{R}, t) \right] dV = \int\int_S \underline{n} \cdot \underline{\underline{T}}(\underline{R}, t) \, dS \ . \tag{38}$$

Doing the same to the time derivative of the isotropic version (11) of Hooke's law we obtain

$$\int\int\int_V \frac{\partial \underline{\underline{T}}(\underline{R}, t)}{\partial t} dV = \int\int_S \left\{ \lambda \underline{\underline{In}} \cdot \underline{v}(\underline{R}, t) + \mu \left[\underline{nv}(\underline{R}, t) + \underline{v}(\underline{R}, t)\underline{n} \right] \right\} dS \ . \tag{39}$$

Equations (38) and (39) are the fundamental EFIT-equations for isotropic media; transducers are either modelled by volume force densities $\underline{f}(\underline{R}, t)$ or prescribing tractions $\underline{n} \cdot \underline{\underline{T}}(\underline{R}, t)$ on specimen surfaces.

We discretize (38) and (39) for $\underline{f} \equiv 0$ via approximation of the volume and surface integrals over a cubic cell of volume Δ^3. Fig. 1 shows such a cell with edges paralell to the 1,2,3-axes of a cartesian coordinate system; the components of the velocity vector are denoted by v_i, the latter ones being dislocated from the local coordinate origin by $\Delta/2$. The computation of the surface integral in Cauchy's equation requires the knowlegde of

$$\underline{n} \cdot \underline{\underline{T}} = \begin{cases} \pm\underline{e}_1 \cdot \underline{\underline{T}} = \left(T_{11}^{r,l}, T_{12}^{r,l}, T_{13}^{r,l} \right) \\ \pm\underline{e}_2 \cdot \underline{\underline{T}} = \left(T_{12}^{f,b}, T_{22}^{f,b}, T_{23}^{f,b} \right) \\ \pm\underline{e}_3 \cdot \underline{\underline{T}} = \left(T_{13}^{u,d}, T_{23}^{u,d}, T_{33}^{u,d} \right) \ , \end{cases} \tag{40}$$

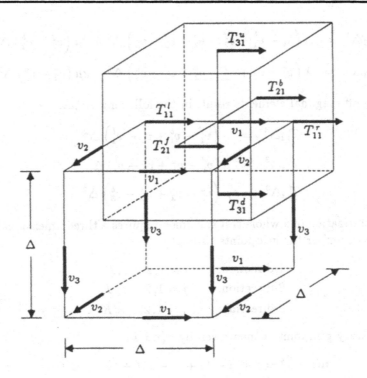

Figure 1: Elementary cubic cell

where we refer to the right and left side, front and back side, and up and down side of the cube via r, l, f, b, u, d. In Fig. 1 we indicated that the off-diagonal elements of the stress tensor are also dislocated to the midpoints of the cell surfaces. This dislocation is equally appropriate for Cauchy's as well as Hooke's law. We obtain as discretized version of (38)

$$\rho \dot{v}_1 \Delta^3 = \left(T_{11}^r - T_{11}^l + T_{12}^f - T_{12}^b + T_{13}^d - T_{13}^u \right) \Delta^2$$

$$\rho \dot{v}_2 \Delta^3 = \left(T_{12}^r - T_{12}^l + T_{22}^f - T_{22}^b + T_{23}^d - T_{23}^u \right) \Delta^2 \qquad (41)$$

$$\rho \dot{v}_3 \Delta^3 = \left(T_{13}^r - T_{13}^l + T_{23}^f - T_{23}^b + T_{33}^d - T_{33}^u \right) \Delta^2 \ ,$$

where the dot indicates the time derivative. For the diagonal elements the discretized Hooke's law reads

$$\dot{T}_{11} \Delta^3 = \lambda \left(v_1^r - v_1^l + v_2^f - v_2^b + v_3^d - v_3^u \right) \Delta^2 + 2\mu \left(v_1^r - v_1^l \right) \Delta^2$$

$$\dot{T}_{22}\Delta^3 \;=\; \lambda \left(v_1^r - v_1^l + v_2^f - v_2^b + v_3^d - v_3^u \right) \Delta^2 + 2\mu \left(v_2^f - v_2^b \right) \Delta^2 \quad (42)$$

$$\dot{T}_{33}\Delta^3 \;=\; \lambda \left(v_1^r - v_1^l + v_2^f - v_2^b + v_3^d - v_3^u \right) \Delta^2 + 2\mu \left(v_3^d - v_3^u \right) \Delta^2 \quad,$$

and the off-diagonal elements result in the following equations

$$\dot{T}_{12}\Delta^3 \;=\; \mu \left(v_1^f - v_1^b + v_2^r - v_2^l \right) \Delta^2$$

$$\dot{T}_{13}\Delta^3 \;=\; \mu \left(v_1^d - v_1^u + v_3^r - v_3^l \right) \Delta^2 \qquad (43)$$

$$\dot{T}_{23}\Delta^3 \;=\; \mu \left(v_2^d - v_2^u + v_3^f - v_3^b \right) \Delta^2 \quad.$$

Discretization of a whole test specimen requires a threedimensional mesh, where we number the gridpoints through

$$
\begin{array}{lll}
\text{1-direction:} & i = 1, 2, \ldots, I \\
\text{2-direction:} & j = 1, 2, \ldots, J & \qquad (44) \\
\text{3-direction:} & k = 1, 2, \ldots, K \quad.
\end{array}
$$

An arbitrary gridpoint is then given by $n(i, j, k)$

$$n(i, j, k) = 1 + (i - 1) + (j - 1)I + (k - 1)IJ \quad, \qquad (45)$$

when we move through the grid in $(1 \rightarrow 2 \rightarrow 3)$-direction. We then define the following vectors as solutions of our discrete elastodynamic equations of motion

$$v_\alpha \;=\; \left[v_\alpha^{(1)}, \ldots, v_\alpha^{(n)}, \ldots, v_\alpha^{(IJK)} \right] \qquad (46)$$

$$T_{\alpha\beta} \;=\; \left[T_{\alpha\beta}^{(1)}, \ldots, T_{\alpha\beta}^{(n)}, \ldots, T_{\alpha\beta}^{(IJK)} \right] \qquad (47)$$

$$\text{with } \alpha, \beta = 1, 2, 3$$

$$\text{and } T_{\alpha\beta} = T_{\beta\alpha} \quad.$$

Discretization of time is performed in full and half time steps Δt according to

$$v_\alpha(t) \;\Longrightarrow\; v_\alpha(l\Delta t) \text{ with } l \in \mathcal{Z} \qquad (48)$$

$$T_{\alpha\beta}(t) \;\Longrightarrow\; T_{\alpha\beta}\left(l\Delta t - \frac{1}{2}\Delta t \right) \quad, \qquad (49)$$

or — in shorthand notation —

$$v_\alpha(l\Delta t) = v_\alpha^l \tag{50}$$

$$T_{\alpha\beta}\left(l\Delta t - \frac{1}{2}\Delta t\right) = T_{\alpha\beta}^{l-\frac{1}{2}} . \tag{51}$$

Causality then requires

$$v_\alpha^l = 0, \quad T_{\alpha\beta}^l = 0 \text{ for } l \le 0 . \tag{52}$$

The time derivative is appropriately approximated by central differences

$$v_\alpha^l = v_\alpha^{l-1} + \dot{v}_\alpha^{l-\frac{1}{2}}\Delta t \tag{53}$$

$$T_{\alpha\beta}^{l+\frac{1}{2}} = T_{\alpha\beta}^{l-\frac{1}{2}} + \dot{T}_{\alpha\beta}^l\Delta t . \tag{54}$$

In (53) v_α^{l-1} is known from the preceding time step, and $\dot{v}_\alpha^{l-\frac{1}{2}}$ can be computed via the discretized Cauchy equation from $T_{\alpha\beta}^{l-\frac{1}{2}}$; for $l = 1$ $T_{\alpha\beta}^{\frac{1}{2}}$ is prescribed as initial condition. In (54) $T_{\alpha\beta}^{l-\frac{1}{2}}$ is also known from the preceding time step — for $l = 1$ the initial condition is inserted —, and $\dot{T}_{\alpha\beta}^l$ enters from Hooke's law via v_α^l.

3.2.2 Numerical Results and Comparison with Experiments

In the following we discuss some numerical results obtained with the two-dimensional version of the EFIT-code; first, various transducer models are investigated, then, scattering of these transducer fields by several canonical defects is computed. One particular example, a bachwall breaking crack illuminated by a 45° angle probe, is carefully modeled by an accompanying experiment; it turns out that the mesurements comply excellently with the numerical results. In addition, output data of the EFIT-code are compared to those of other numerical codes like Finite Elements or Boundary Elements.

Ultrasonic Transducers for Nondestructive Testing
A typical example as computed with the 2D-EFIT-code is shown in Fig. 2 in terms of various wave front time snapshots; an elastic isotropic half-space (steel: $c_P = 5.9$ mm/μs, $c_S = 3.2$ mm/μs) with a stress-free boundary, i.e.

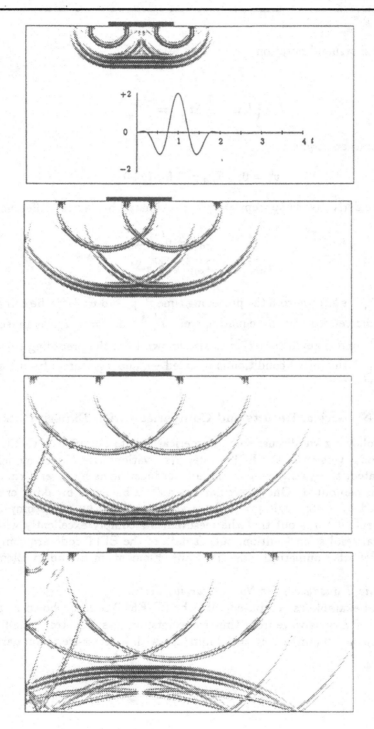

Figure 2: 2D-EFIT-Model of a "Longitudinal" Transducer

with $\underline{\underline{T}} \cdot \underline{n} \equiv 0$ on the surface, is excited by a vertical force homogeneously distributed within a finite aperture of width $2a$ indicated by the bar on the surface. The time dependence of the force is prescribed as a high-frequency pulse with center frequency of 2 MHz; it is also displayed in Fig. 2. The following interpretation of the single wave fronts is available. First, two circular cylindrical pressure wave fronts, which emanate from the two edges of the aperture, have propagated the largest distance into the solid because $c_P > c_S$; they are combined tangentially by a plane pressure wave front parallel to the aperture, which represents the "geometric-optical" wave front of the finite aperture, when diffraction effects are disregarded. The second prominent features of the wave front snap-shot are two circular cylindrical shear wave fronts, equally emanating from the aperture edges — interesting enough, they are missing the plane tangential wave front — and being connected to the surface via plane so-called head wave fronts to fulfill the boundary condition on the surface together with the pressure wave fronts. The stress-free surface of the elastic half-space is responsible for the appearance of another very important wave phenomenon, namely a Rayleigh surface wave; here, four Rayleigh wave pulses are observed, two from each aperture edge, travelling to the right and to the left, respectively, and being only a little slower than the cylindrical shear waves. The surface wave character yields strong amplitudes very close to the surface with an exponential decay into the material. By the way, these Rayleigh surface waves are also found in the elastic wave fields of earthquakes; they carry most of the deformation energy thus being responsible for nearly all the damage. In contrast to that, for non-destructice testing purposes they are most welcome to locate near surface and surface breaking cracks. For instance, the wheels of the high speed ICE trains in Germany are regularly checked with an ultrasonic system based on the properties of Rayleigh waves [18]. Here, to illustrate the performance of algorithmic ultrasonic imaging, we will concentrate on the bulk waves, especially on the plane aperture pressure wave front.

Of course, due to the finite dimensions of the "test specimen", the waves as displayed in Fig. 2, are reflected and mode converted once they reach the stress free boundaries of the specimen.

The plane aperture wave front of Fig. 2 is indeed the most significant wave feature of an ultrasonic transducer radiating vertically into a solid. Mathematically, it is associated with the *scalar* potential $\Phi(\underline{R}, t)$, thus allowing a scalar wave field approximation of such a transducer model, as it is strictly true only for *plane* waves.

The wave vector $\hat{k}_P$ of the plane aperture wave front can be steered applying a linear time-delay of the impulsive force acting onto the material surface, say, increasing from the left to the right edge of the aperture. In practice, this is either achieved cutting the radiating piezoelectric crystal into an array of single elements to be excited independently, or putting the crystal on a solid wedge, thus exploiting Snell's law of refraction for elastic waves [3]. In any case, most interestingly, the plane aperture pressure wave front is now accompanied by a plane aperture shear wave front with a different propagation vector $\hat{k}_S$ versus $\hat{k}_P$; the direction of $\hat{k}_S$ compared to $\hat{k}_P$ is (approximately) given by Snell's law:

$$c_P \sin \vartheta_S = c_S \sin \vartheta_P \quad , \tag{55}$$

ϑ_S and ϑ_P denoting polar angles as indicated in Fig. 3, where the pertinent wave front time snap shots are displayed. For nondestructive testing purposes, such a transducer wave field is not very appropriate as both ultrasonic signals are scattered by defects in the material, and due to the general coupling of pressure and shear waves via mode conversion a real mess of scattered pulses travels back to any appropriately located receiving transducer requiring careful interpretation. Therefore, to produce "clean" ultrasound in a solid with the wave normal making a prescribed tilt angle with respect to the surface, it is convenient to increase the time-delay of the force on the surface beyond *that* value, which enforces the pressure wave to "propagate" beyond its critical angle, i.e. making it an evanescent surface wave with exponential decay into the solid, as it is predicted investigating critical angle reflection phenomena of *plane* elastic waves. As a matter of fact, in Fig. 4 a pertinent choice of parameters has been made, i.e. the time-delay has been chosen *exactly* critical-angle-like. The result is essentially a superposition of the two pressure wave fronts emanating from the edges of the radiating aperture forming the so-called subsurface longitudinal wave [19]; this wave front is accompanied by a 45° shear wave traveling into the bulk material. It was shown that the EFIT-code is a suitable means to identify the details of such a probe [19].

Increasing the time-delay a little bit further than in the SSLW-probe, the two pressure edge waves run out of phase, and, obviously, the only remaining wave front with significant amplitude (apart from Rayleigh wave pulses) is a

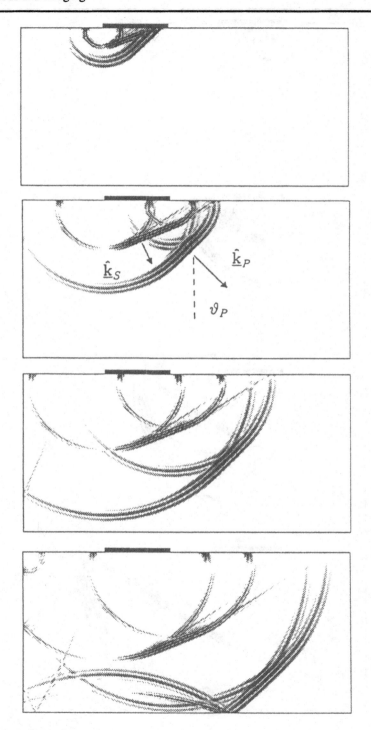

Figure 3: 2D-EFIT-Model of an Angle Transducer Radiating both Pressure and Shear Waves

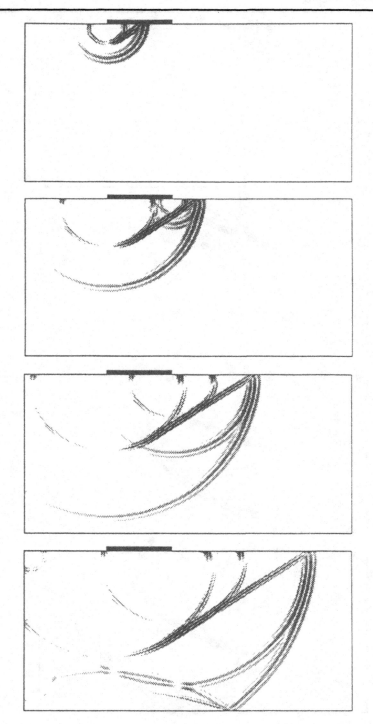

Figure 4: 2D-EFIT-Model of the Subsurface Longitudinal Wave (SSLW)

plane aperture shear wave front with a 45° wave vector $\hat{\underline{k}}_S$ (Fig. 5); a shear wave angle probe is modeled that way.

If steel is considered as the material to be investigated nondestructively, the above principles have lead to commercially available transducers, i.e. so-called longitudinal-vertical- and transverse-angle-probes, the latter ones being separated into 45°-, 60°-, and 70°-degree probes. Hence, polarization properties — longitudinal and transverse — are used synonymously to propagation properties — pressure and shear — to name devices, which is strictly allowed for plane waves only.

Once the critical angle of the shear waves is reached for increasing time-delay within the transducer aperture, only Rayleigh waves remain; Fig. 6 shows the pertinent wave fronts. This is a suitable means to "excite" Rayleigh waves by a numerical code.

Obviously, if special care is applied to the excitation of shear waves (compare Fig. 5), they appear decoupled and independent from pressure waves; they can be — at least approximately — described by a single *scalar* component of the vector potential. This provides the theoretical background for the phenomenological evaluation of ultrasonic algorithmic imaging in terms of a scalar, and, hence, acoustic formulation, even though elastic propagation media are under concern, and, hence, pressure and shear waves have to be distinguished. To a certain extent, even scattering phenomena can be approximated that way.

Notice, the scale of all the above wave front snap shots is logarithmic, and, except for the Rayleigh wave excitation (Fig. 6), each series is normalized to its common absolute maximum, whereas in Fig. 6 each single picture is normalized to its own maximum. The same is true for the figures to follow, Fig. 9 representing once more the exception.

Scattering of Elastic Waves
Next, we consider the scattering of the elastic wave field emanating from the transducer modeled in Fig. 2 by a horizontal plane crack with a stress free boundary. Fig. 7 shows in terms of several time snap shots distinct mode conversion effects as well as Rayleigh surface waves traveling along the crack faces. Of course, the backscattered wave fronts, when "detected" within an appropriate aperture, can serve as synthetic data for an imaging algorithm; this will be discussed in the second part of the paper.

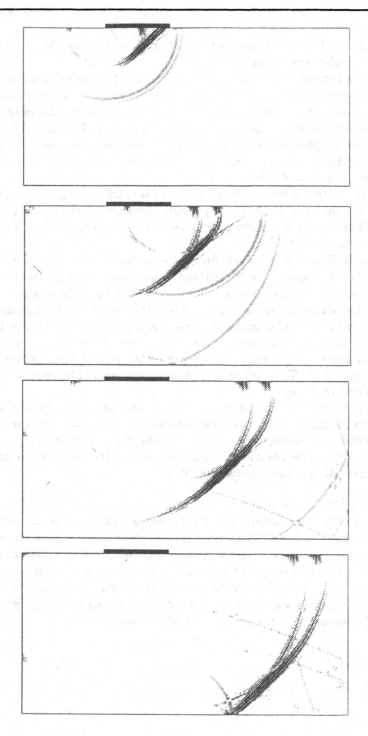

Figure 5: 2D-EFIT-Model of an Angle Shear Wave Transducer

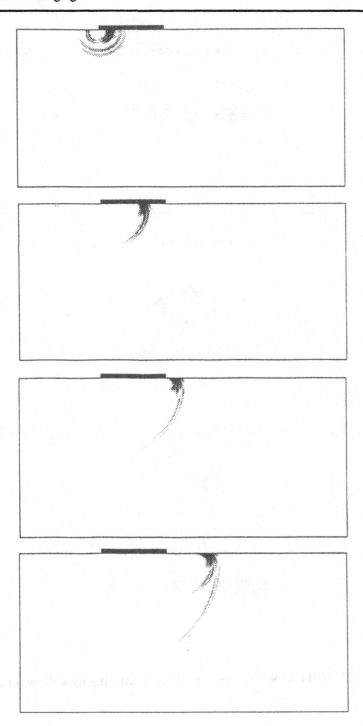

Figure 6: 2D-EFIT-Model of Rayleigh Wave Excitation

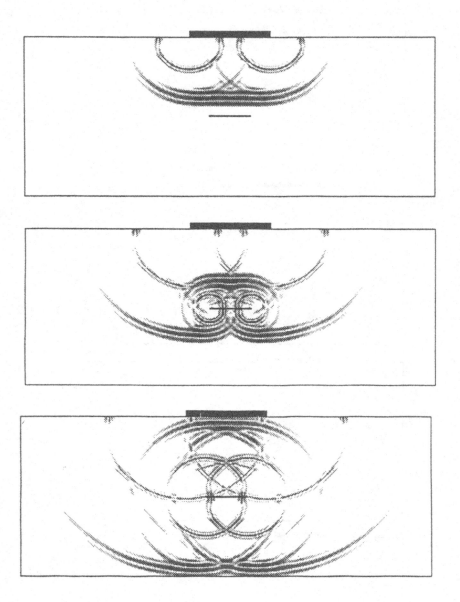

Figure 7: 2D-EFIT-Model of Pressure Wave Scattering by a Horizontal Crack

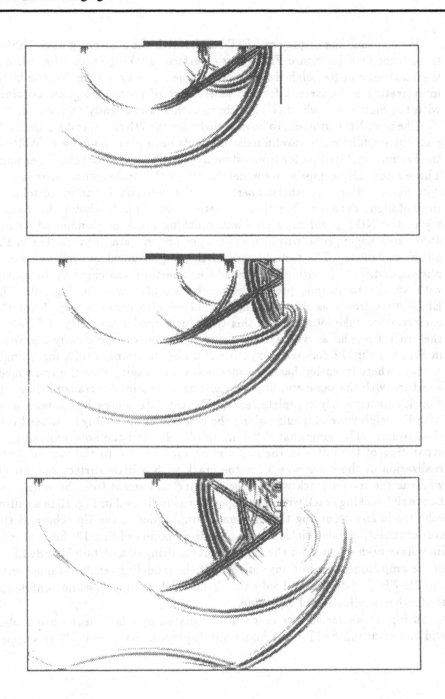

Figure 8: 2D-EFIT-Model of SSLW Wave Field Scattering by a Surface Breaking Crack

We continue the display of EFIT-results with the scattering of the SSLW transducer field (compare Fig. 4) by a surface breaking crack (Fig. 8), and the scattering of Rayleigh waves by the same crack type (Fig. 9); the latter investigation is important for the assessment of reflected signals obtained with the high speed wheel NDT technique that was already mentioned.

The next NDT problem to be modeled with the 2D-EFIT-code is an example, where additionally careful measurements have been made by F. Walte in the Fraunhofer Institute for Nondestructive Testing in Saarbrücken/Germany.. These essentially comprise some details about the probe, single A-scans, rf-data-fields within a synthetic aperture and amplitude dynamics of selected time-of-flight curves within these rf-data-fields. Fig. 10 shows the geometry of the NDT problem; a backwall breaking crack is illuminated with a 45°-degree angle probe within a given aperture, rf-data are recorded in the pulse-echo mode. The time history of the displacement as well as its amplitude distribution within the transducer aperture was carefully measured with an electrodynamic probe, the results are also given in Fig. 10. The EFIT wave fronts as displayed in Fig. 11 — also compare Fig. 5 for the excitation — take into account this transducer model explicitly. Of course, the crack tip echo as well as the corner reflection can be clearly identified in Fig. 11. Fig. 12 compares experimental and modeling results for a single A-scan, where reception has been modeled via averaging over the transducer aperture with the same amplitude weighting as applied for transmission; the coincidence is nearly complete, except for the late pulses being associated with Rayleigh waves traveling along the crack faces and being converted into shear waves. The somewhat different amplitudes and the non-resolved time separation of the latter in the experiment may be due to the fact that the realization of the crack was a fatigue crack with a little surface roughness, whereas the model crack was perfectly plane and stress free. Scanning the backwall breaking crack within the aperture as indicated in Fig. 10 in a pulse-echo mode and recording the maximum amplitudes of the tip echo and the corner reflection pulse yields the amplitude dynamics of Fig. 13. Several scan lines have been selected for the experiment resulting in a certain "bandwidth" of the amplitude distribution; notice, that the modeling result obtained with the 2D-EFIT-code fits well into this bandwidth giving us some confidence for further application of this code.

In Fig. 14 a near surface crack is illuminated by a 60° shear wave probe, and the resulting EFIT wave fronts are displayed. As in Fig. 13, the upper

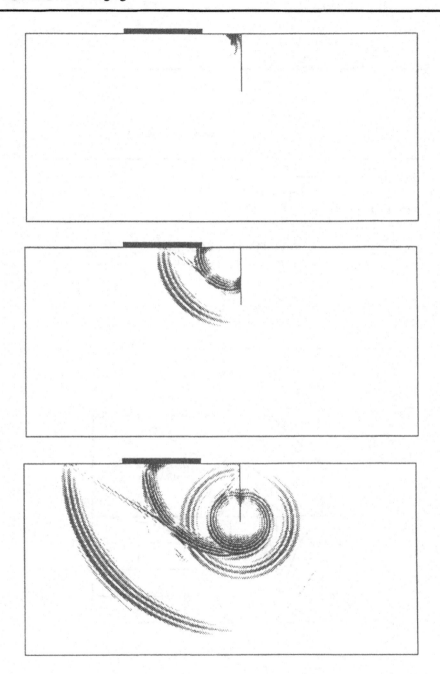

Figure 9: 2D-EFIT-Model of Rayleigh Wave Scattering by a Surface Breaking Crack

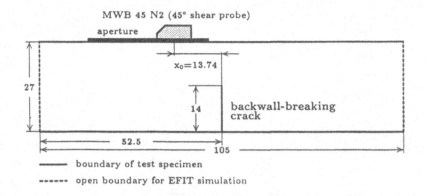

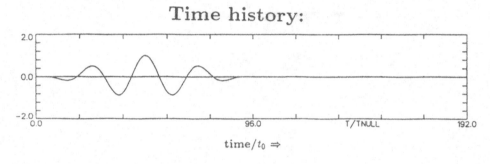

Time history:

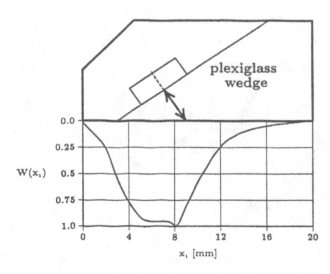

Figure 10: NDT Geometry, Measured Time History of Transducer Response, and Traction Distribution within the Transducer Aperture

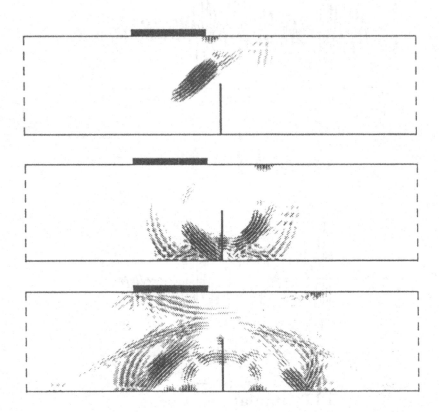

Figure 11: 2D-EFIT-Model of Shear Wave Scattering by a Backwall Breaking Crack; the Measured Transducer Data of the Previous Figure Have Been Accounted for

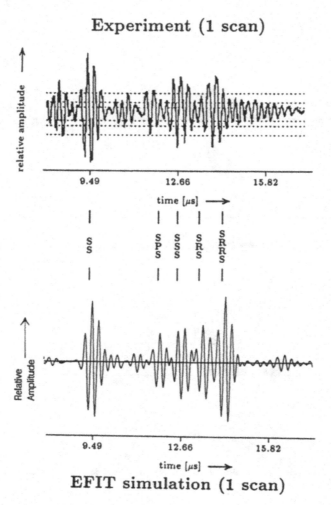

Experiment (1 scan)

EFIT simulation (1 scan)

SS   Shear-Shear-Reflection (crack tip echo)
SPS   Shear-Pressure-Shear-Mode-Conversion
SSS   Shear-Shear-Shear-Reflection (corner reflection)
SRS   Shear-Rayleigh-Shear-Mode-Conversion
SRRS   Shear-Rayleigh-Rayleigh-Shear-Mode-Conversion

Figure 12: Experiment and 2D-EFIT Simulation of a Single A-Scan for a Given Aperture Point, Where the Crack Tip Echo is More Prominent than the Corner Reflection

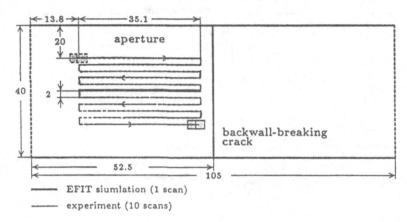

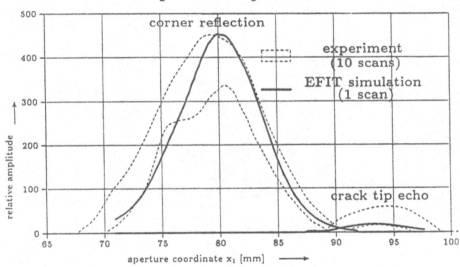

Figure 13: Pulse-Echo Amplitude Scan of Crack Tip Echo and Corner Reflection of Backwall Breaking Crack

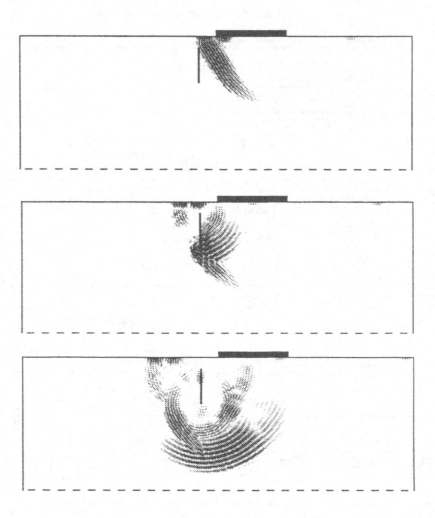

Figure 14: 2D-EFIT-Model of a Near Surface Crack Illuminated by a 60°
Shear Wave Probe

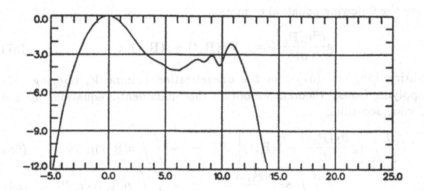

Figure 15: Pulse-Echo Amplitude Scan of Crack Tip Echo for the EFIT-Model of Figure 14

crack tip echo was scanned within a synthetic aperture in a pulse-echo mode, the result is given in Fig. 15. We already point out the good agreement with the result obtained by the boundary element method ignoring the interaction with the stress free specimen surface as it is available from Fig. 19.

3.3   AFIT — Acoustic Finite Integration Technique

As already mentioned in the discussion of Fig. 2 and Fig. 5 it might be reasonable to approximate time separated pressure and/or shear wave fronts by a scalar formulation, particularly because standard imaging algorithms rely on such an approximation. For that reason we developed the scalar acoustic version of the EFIT-code acronymed AFIT for Acoustic Finite Integration Technique.

Indroducing the pressure $p(\underline{R}, t)$ in an acoustic medium with compressibility κ according to Hooke's law

$$\kappa p(\underline{R}, t) = -\boldsymbol{\nabla} \cdot \underline{u}(\underline{R}, t) \tag{56}$$

we have the following equation of motion

$$\varrho \frac{\partial^2 \underline{u}(\underline{R}, t)}{\partial t^2} = -\boldsymbol{\nabla} p(\underline{R}, t) + \underline{f}(\underline{R}, t) \ . \tag{57}$$

Integrating (56) and (57) over the discretization volume V, obeying (37) and applying Gauss' theorem we obtain the fundamental equations for the AFIT-code according to

$$\iiint_V \left[\varrho \frac{\partial \underline{v}(\underline{R}, t)}{\partial t} - \underline{f}(\underline{R}, t) \right] dV = -\iint_S p(\underline{R}, t) \underline{n} \, dS \tag{58}$$

$$\iiint_V \kappa \frac{\partial p(\underline{R}, t)}{\partial t} dV = -\iint_S \underline{v}(\underline{R}, t) \cdot \underline{n} \, dS \ . \tag{59}$$

Discretization, as discussed in Fig. 1 for the EFIT-code, follows similar lines; the discretized pressure is located in the nodes of the cubic mesh, whereas the components of the velocity vector $\underline{v}(\underline{R}, t)$ are dislocated from these nodes as indicated in Fig. 1. Hence, the discrete counterparts of (58) and (59) read for $\underline{f} \equiv 0$

$$\varrho \dot{v}_1 \Delta^3 = -(p^r - p^l) \Delta^2$$
$$\varrho \dot{v}_2 \Delta^3 = -(p^f - p^b) \Delta^2 \tag{60}$$
$$\varrho \dot{v}_3 \Delta^3 = -(p^d - p^u) \Delta^2$$
$$\kappa \dot{p} \Delta^3 = -(v_1^r - v_1^l + v_2^f - v_2^b + v_3^d - v_3^u) \Delta^2 \ . \tag{61}$$

Discretization of time is similar to the EFIT-code.

Fig. 16 repeats the computation of Fig. 7 with the AFIT-code, and, of course, in comparison with Fig. 7, all mode conversion effects have disappeared. Nevertheless, the scattered pressure wave fronts of both figures compare very well, supporting the "argument" for strictly scalar imaging even for elastic waves provided the appropriate wave fronts to be dealt with can be time gated.

Another example is given in Fig. 17; it revisits the NDT problem considered in Fig. 10. This time, the incident shear wave is modeled acoustically

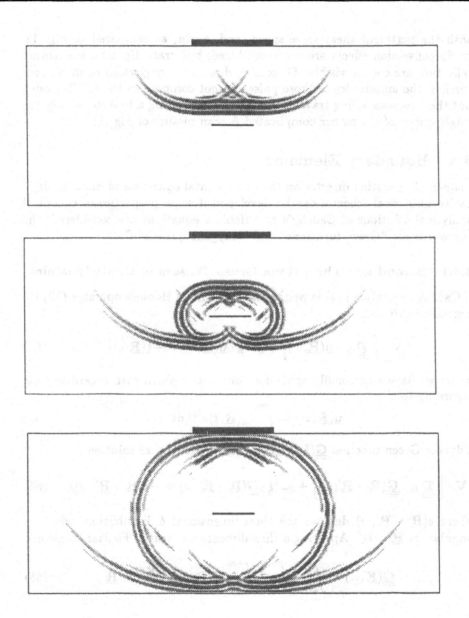

Figure 16: 2D-AFIT-Model of Pressure Wave Scattering by a Horizontal Crack

with the pertinent shear wave speed, and, again, as compared to Fig. 11, mode conversion effects are not reproduced, but crack tip echo and corner reflection are clearly visible. Of course, due to the neglection of mode conversion, the amplitudes of those pulses do not compare to the EFIT-model, but their corresponding travel times do, thus providing a tool to identify the single pulses of the rather complicated A-scan picture of Fig. 12.

3.4 Boundary Elements

Instead of operating directly on the fundamental equations of motion, alternative numerical schemes can be developed if some properties of canonical analytical solutions of Cauchy's and Hooke's equations are considered; the key words are "Green functions" and "Huygens' principle".

3.4.1 Second and Third Rank Green Tensors of Elastodynamics

If Cauchy's equation (14) is written with the aid of Hooke's operator (13) for isotropic media, i.e. as

$$\nabla \cdot \left[\underline{\underline{\mathcal{D}}}_{\text{Ho}} \cdot \underline{u}(\underline{R},\omega) \right] + \varrho\omega^2\underline{\underline{I}} \cdot \underline{u}(\underline{R},\omega) = -\underline{f}(\underline{R},\omega) \quad , \qquad (62)$$

where we have additionally applied a Fourier transform with regard to time according to

$$\underline{u}(\underline{R},\omega) = \int_{-\infty}^{\infty} \underline{u}(\underline{R},t)e^{j\omega t}\,\mathrm{dt} \quad , \qquad (63)$$

a dyadic Green function $\underline{\underline{G}}(\underline{R} - \underline{R}',\omega)$ can be defined as solution of

$$\nabla \cdot \left[\underline{\underline{\mathcal{D}}}_{\text{Ho}} \cdot \underline{\underline{G}}(\underline{R} - \underline{R}',\omega) \right] + \varrho\omega^2\underline{\underline{I}} \cdot \underline{\underline{G}}(\underline{R} - \underline{R}',\omega) = -\delta(\underline{R} - \underline{R}',\omega)\underline{\underline{I}} \quad , \quad (64)$$

where $\delta(\underline{R} - \underline{R}',\omega)$ denotes the threedimensional δ-distribution, which is singular for $\underline{R} = \underline{R}'$. Applying a threedimensional *spatial* Fourier transform

$$\underline{\underline{\tilde{G}}}(\underline{K},\omega) = \int_{-\infty}^{+\infty} \int_{-\infty}^{+\infty} \int_{-\infty}^{+\infty} \underline{\underline{G}}(\underline{R},\omega)e^{-j\underline{K}\cdot\underline{R}}\,\mathrm{d}^3\underline{R} \qquad (65)$$

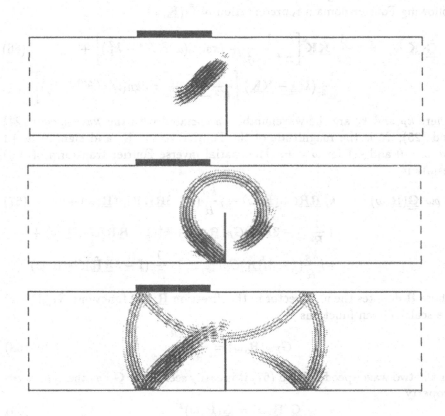

Figure 17: 2D-AFIT-Model of "Shear" Wave Scattering by a Backwall Breaking Crack; the Measured Transducer Data of Figure 10 Have Been Accounted for

to (64) and inverting the resulting algebraic dyadic operator [16] yields the following Fourier domain representation of $\underline{\tilde{\underline{G}}}(\mathbf{K},\omega)$:

$$\underline{\tilde{\underline{G}}}(\mathbf{K},\omega) = \frac{1}{\varrho\omega^2}\mathbf{K}\mathbf{K}\left[\frac{1}{K^2-k_P^2} + j\pi\text{sign}(\omega)\delta(K^2-k_P^2)\right] + \tag{66}$$

$$+\frac{1}{\varrho\omega^2}\left(k_S^2\underline{\underline{I}}-\mathbf{K}\mathbf{K}\right)\left[\frac{1}{K^2-k_S^2} + j\pi\text{sign}(\omega)\delta(K^2-k_S^2)\right] \quad,$$

where k_P and k_S are the wavenumbers associated with the wave speeds (24) and (25); K is the magnitude of the Fourier vector $\mathbf{K}$, and $\text{sign}(\omega)$ is $+1$ for $\omega > 0$ and -1 for $\omega < 0$. The spatial inverse Fourier transform of (66) results in

$$\varrho\omega^2\underline{\underline{G}}(\mathbf{R},\omega) = k_P^2\hat{\mathbf{R}}\hat{\mathbf{R}}G_P(\mathbf{R},\omega) - j\frac{k_P}{R}(\underline{\underline{I}}-3\hat{\mathbf{R}}\hat{\mathbf{R}})G_P(\mathbf{R},\omega) + \tag{67}$$

$$+\frac{1}{R^2}(\underline{\underline{I}}-3\hat{\mathbf{R}}\hat{\mathbf{R}})G_P(\mathbf{R},\omega) + k_S^2(\underline{\underline{I}}-\hat{\mathbf{R}}\hat{\mathbf{R}})G_S(\mathbf{R},\omega) +$$

$$+j\frac{k_S}{R}(\underline{\underline{I}}-3\hat{\mathbf{R}}\hat{\mathbf{R}})G_S(\mathbf{R},\omega) - \frac{1}{R^2}(\underline{\underline{I}}-3\hat{\mathbf{R}}\hat{\mathbf{R}})G_S(\mathbf{R},\omega) \quad,$$

where $\hat{\mathbf{R}}$ denotes the unit-vector in the direction $\mathbf{R}$; the functions $G_{P,S}(\mathbf{R},\omega)$ are scalar Green functions

$$G_{P,S}(\mathbf{R},\omega) = \frac{e^{jk_{P,S}R}}{4\pi R} \tag{68}$$

for the two wave speeds. From (67) it is easily seen, that $\underline{\underline{G}}$ has the symmetry property

$$\underline{\underline{G}}(\mathbf{R},\omega) = \underline{\underline{G}}(\mathbf{R},\omega)^{21} \quad. \tag{69}$$

The dyadic Green function (67) allows for the solution

$$\underline{u}(\mathbf{R},\omega) = \iiint_V \underline{\underline{G}}(\mathbf{R}-\mathbf{R}',\omega)\cdot\underline{f}(\mathbf{R}',\omega)\,\mathrm{d}^3\mathbf{R}' \tag{70}$$

of the inhomogeneous equation (62), where the volume integration extends over the region of non-zero force density $\underline{f}(\mathbf{R},\omega)$.

Applying Hooke's operator to (70) results in a pertinent representation of the stress tensor

$$\underline{\underline{T}}(\mathbf{R},\omega) = \iiint_V \underline{\underline{\Sigma}}(\mathbf{R}-\mathbf{R}',\omega)\cdot\underline{f}(\mathbf{R}',\omega)\,\mathrm{d}^3\mathbf{R}' \tag{71}$$

if a third rank Green tensor $\underline{\underline{\underline{\Sigma}}}$ is defined according to

$$\underline{\underline{\underline{\Sigma}}}(\mathbf{R},\omega) = \underline{\underline{\mathcal{D}}}_{\mathrm{Ho}} \cdot \underline{\underline{G}}(\mathbf{R},\omega) \ . \tag{72}$$

This tensor has the property

$$\underline{\underline{\underline{\Sigma}}}(\mathbf{R},\omega) = \underline{\underline{\underline{\Sigma}}}(\mathbf{R},\omega)^{213} \ . \tag{73}$$

Utilizing (13) and (67) an explicit spatial representation of $\underline{\underline{\underline{\Sigma}}}$ can be derived as well.

3.4.2 Elastodynamic Huygens Principle

With the aid of the second and third rank Green tensors of elastodynamics an appropriate Huygens principle can be evaluated [17]. In the following we give a simpler derivation, which essentially exploits the upper indicial notation.

Let us construct the following second rank tensor

$$\left[\underline{\underline{\mathcal{D}}}_{\mathrm{Ho}} \cdot \underline{u}(\mathbf{R},\omega) \right] \cdot \underline{\underline{G}}(\mathbf{R}-\mathbf{R}',\omega) - \underline{u}(\mathbf{R},\omega) \cdot \left[\underline{\underline{\mathcal{D}}}_{\mathrm{Ho}} \cdot \underline{\underline{G}}(\mathbf{R}-\mathbf{R}',\omega) \right] =$$

$$\underline{\underline{T}}(\mathbf{R},\omega) \cdot \underline{\underline{G}}(\mathbf{R}-\mathbf{R}',\omega) - \underline{u}(\mathbf{R},\omega) \cdot \underline{\underline{\underline{\Sigma}}}(\mathbf{R}-\mathbf{R}',\omega) \ , \tag{74}$$

contract it with the outward normal of a surface S enclosing a volume V, which contains the forces $\underline{f}$, and integrate the result over that surface:

$$\iint_S \underline{n} \cdot \left\{ \left[\underline{\underline{\mathcal{D}}}_{\mathrm{Ho}} \cdot \underline{u}(\mathbf{R},\omega) \right] \cdot \underline{\underline{G}}(\mathbf{R}-\mathbf{R}',\omega) - \right.$$

$$\left. -\underline{u}(\mathbf{R},\omega) \cdot \left[\underline{\underline{\mathcal{D}}}_{\mathrm{Ho}} \cdot \underline{\underline{G}}(\mathbf{R}-\mathbf{R}',\omega) \right] \right\} dS =$$

$$\iiint_V \nabla \cdot \left\{ \left[\underline{\underline{\mathcal{D}}}_{\mathrm{Ho}} \cdot \underline{u}(\mathbf{R},\omega) \right] \cdot \underline{\underline{G}}(\mathbf{R}-\mathbf{R}',\omega) - \right.$$

$$\left. -\underline{u}(\mathbf{R},\omega) \cdot \left[\underline{\underline{\mathcal{D}}}_{\mathrm{Ho}} \cdot \underline{\underline{G}}(\mathbf{R}-\mathbf{R}',\omega) \right] \right\} dV = \tag{75}$$

$$\iiint_V \left\{ \nabla \cdot \left[\underline{\underline{\mathcal{D}}}_{\mathrm{Ho}} \cdot \underline{u}(\mathbf{R},\omega) \right] \cdot \underline{\underline{G}}(\mathbf{R}-\mathbf{R}',\omega) + \right.$$

$$+ \left[\underline{\underline{\mathcal{D}}}_{\mathrm{Ho}} \cdot \underline{u}(\mathbf{R},\omega) \right]^{21} : \nabla \underline{\underline{G}}(\mathbf{R}-\mathbf{R}',\omega) -$$

$$-\nabla \underline{u}(\underline{R}, \omega) : \left[\underline{\underline{\mathcal{D}}}_{\text{Ho}} \cdot \underline{\underline{G}}(\underline{R} - \underline{R}', \omega) \right] -$$

$$-\underline{u}(\underline{R}, \omega) \cdot \left[\nabla \cdot \left[\underline{\underline{\mathcal{D}}}_{\text{Ho}} \cdot \underline{\underline{G}}(\underline{R} - \underline{R}', \omega) \right]^{213} \right] \Bigg\} \, dV =$$

$$- \iiint_V \underline{f}(\underline{R}, \omega) \cdot \underline{\underline{G}}(\underline{R} - \underline{R}', \omega) \, dV + \underline{u}(\underline{R}', \omega) \Gamma_V(\underline{R}')$$

In order to get the last equation we have used Gauss' theorem (33) as well as the identity

$$\left(\underline{\underline{\mathcal{D}}}_{\text{Ho}} \cdot \underline{u} \right)^{21} : \nabla \underline{\underline{G}} = \nabla \underline{u} : \left(\underline{\underline{\mathcal{D}}}_{\text{Ho}} \cdot \underline{\underline{G}} \right) \quad , \tag{76}$$

and the differential equations (62) and (64). In addition, the characteristic function $\Gamma_V(\underline{R})$ of the volume V has been introduced, which is defined as

$$\Gamma_V(\underline{R}) = \begin{cases} 1 & \text{for } \underline{R} \in V \\ 0 & \text{for } \underline{R} \notin V \end{cases} \quad . \tag{77}$$

The first and the last equation of (75) represent Huygens' principle for elastic waves:

$$\underline{u}(\underline{R}, \omega) \Gamma_V(\underline{R}) = \iiint_V \underline{f}(\underline{R}', \omega) \cdot \underline{\underline{G}}(\underline{R} - \underline{R}', \omega) \, dV' + \tag{78}$$

$$+ \iint_S \underline{n}' \cdot \left[\underline{\underline{T}}(\underline{R}', \omega) \cdot \underline{\underline{G}}(\underline{R} - \underline{R}', \omega) - \underline{u}(\underline{R}', \omega) \cdot \underline{\underline{\Sigma}}'(\underline{R} - \underline{R}', \omega) \right] \, dS' \quad ,$$

where, for convenience, we interchanged the role of $\underline{R}$ and $\underline{R}'$; the prime on $\underline{\underline{\Sigma}}'$ indicates derivatives with regard to $\underline{R}'$ now, and notice, we have $\underline{\underline{\Sigma}}(\underline{R} - \underline{R}', \omega) = -\underline{\underline{\Sigma}}'(\underline{R} - \underline{R}', \omega)$.

The surface integral in (78) is extremely useful for the following applications:

- It allows derivation of an integral equation for either the displacement $\underline{u}(\underline{R}', \omega)$ or the traction $\underline{n}' \cdot \underline{\underline{T}}(\underline{R}', \omega)$ on the surface of a scatterer with appropriate boundary conditions once the surface S is identified as the surface S_c of that scatterer and the proper limit for $\underline{R} \longrightarrow S_c$ is calculated [20]. After the (numerical) solution of this integral equation, the particle displacement field is computed via (78). This method is called the boundary element method; we will give an example below.

- For a planar surface S extending to infinity, a proper radiation condition reduces (78) to a twodimensional spatial convolution integral, whose twodimensional spatial Fourier transform with regard to the cartesian coordinates on that plane results in the elastic plane wave spectrum representation of the displacement [16]. Once this representation is further manipulated [16] to yield the displacement in an elastic half space for a given traction boundary condition on the surface of the half space, the NDT transducer problem has found an appropriate model [7].

- The elastodynamic Huygens principle can be interpreted as a forward propagation of elastic waves; in the following it is outlined that imaging — at least within the approximations for practical applications — can be understood in terms of *backpropagation*, which is nothing but Huygens' principle with Green's functions replaced by their complex conjugate.

3.4.3 NDT system model

A given NDT problem usually comprises the transducer radiation, scattering of elastic waves, and their reception. An NDT system model consists of proper models of each of these subsystems. Fig.'s 10 to 13 present results of that kind, where a direct numerical method, i.e. the EFIT-code, has been used; alternatively, Huygens' principle and the boundary element method can be exploited to model the transducer in terms of its plane wave spectrum, scattering in terms of an integral equation for the surface displacement distribution on a stress free scatterer, i.e. a void or a crack, and reception either in terms of appropriate averaging over the receiving aperture or in terms of Auld's reciprocity theorem [21], which is also related to Huygens' principle. A detailed discussion combining these steps into a system model can be found in [7]. Here, we give a numerical example for 70^0 shear wave excitation of a 10 mm plane crack being vertically oriented with regard to the scanning surface with its upper tip 50 mm below that surface. The amplitude variation of the upper crack tip pulse in the pulse-echo mode within a synthetic aperture has been calculated using the above procedure for each frequency within the pulse spectrum and applying an inverse Fourier transform; the result is given in Fig. 18. This figure does not only show our own

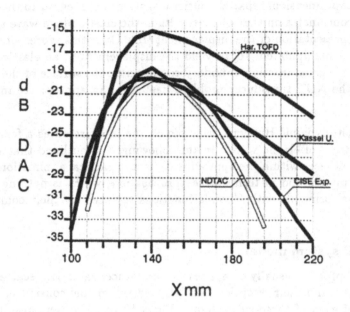

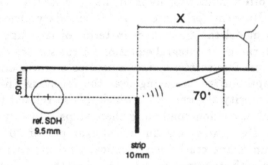

Figure 18: Amplitude Variation of Upper Crack Tip Pulse: Comparison of Three Different System Models and Experiments

results, but it essentially compares three different models with experimental data obtained by CISE in Italy [22]; for that reason, all amplitudes are normalized to the maximum backscattering amplitude of the same sound field by a 9.5 mm side drilled hole at the same location as the crack. The other model results were provided by the Nondestructive Testing Application Centre in Manchester/U.K. (NDTAC Model) and the Harwell Laboratory, also in the U.K. (Har. TOFD). These models essentially apply the Géometric Theory of Diffraction (GTD) [23], and they use beam models for the transducer [22]. It seems that for the case of Fig. 18 all model predictions are fairly good; for a variety of other NDT problems this behavior was more or less confirmed, no particular model was really superior. This is different for defect locations very close to the specimen surface, thus penetrating the near field of the sound beam and causing wave interactions with the stress free surface; an example is given in Fig. 19. Obviously, none of the models really meets the experiment, but it is interesting to notice that our own results exhibit at least a similar amplitude variation even though the absolute values are not correct. We have intentionally selected this particular example, because data obtained with the EFIT-code have already been discussed in Fig. 15; there, reception was modeled via averaging, here via Auld's theorem. The agreement of the amplitude variation is striking[1]; with a careful interpretation of the EFIT wave fronts as displayed in Fig. 14 and using the "help" of the AFIT-code as well, we can even identify the oscillations in the variation obtained with the EFIT-code as the influence of interferences of the upper tip echo with the specimen surface, and, of course, these are not accounted for by the system model based on the boundary element method.

Another remark concerning imaging is in order here: obviously, wave field interactions with a specimen surface are not the main feature in an rf-data field, hence, imaging algorithms, which usually do not account for these interactions, might be successfully applied to that kind of data.

---

[1]The absolute amplitudes cannot be compared, because the EFIT results have not been normalized to the side drilled hole.

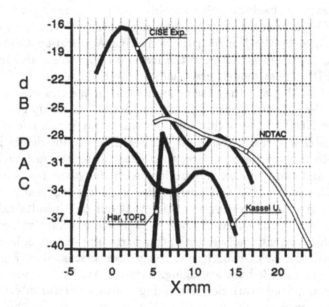

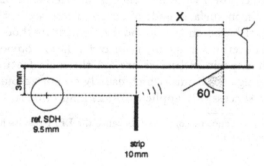

Figure 19: Amplitude Variation of Upper Crack Tip Pulse for a Near Surface Crack

4 Inverse Methods and Imaging

4.1 Scalar Scattering of Ultrasound by Defects

According to the above discussion we consider ultrasound in a solid approximately as a scalar acoustic wave phenomenon, where the propagation medium is characterized by the density ϱ and the wave speed c; if pressure waves are under concern, c stands for c_P, and inserting c_S for c accounts for shear waves.

Let us refer to a material inhomogeneity of finite volume V_c and wave speed $c(\mathbf{R})$ embedded in an otherwise homogeneous medium with wave speed c_0 — densities should be equal. Denoting the scalar potential always as $\Phi(\mathbf{R}, t)$ — even for shear waves — we then have

$$\Delta\Phi(\mathbf{R}, t) - \frac{1}{c_0^2}\frac{\partial^2\Phi(\mathbf{R}, t)}{\partial t^2} = -q(\mathbf{R}, t) \text{ for } \mathbf{R} \notin V_c \qquad (79)$$

$$\Delta\Phi(\mathbf{R}, t) - \frac{1}{c^2(\mathbf{R})}\frac{\partial^2\Phi(\mathbf{R}, t)}{\partial t^2} = 0 \text{ for } \mathbf{R} \in V_c \ , \qquad (80)$$

where the source term $q(\mathbf{R}, t)$ accounts for the excitation of ultrasound; q is considered to be nonzero only within a finite source volume, which does not overlap with V_c. Performing a Fourier transform with respect to time for all time dependent quantities (compare (63))

$$\Phi(\mathbf{R}, \omega) = \int_{-\infty}^{\infty} \Phi(\mathbf{R}, t)e^{j\omega t}\, \mathrm{d}t \ , \qquad (81)$$

and defining subsequently the characteristic function $\Gamma(\mathbf{R})$ of the volume V_c

$$\Gamma(\mathbf{R}) = \begin{cases} 0 & \text{for } \mathbf{R} \notin V_c \\ 1 & \text{for } \mathbf{R} \in V_c \end{cases} \ , \qquad (82)$$

the object function $O(\mathbf{R})$ of the scatterer

$$O(\mathbf{R}) = \left[1 - \frac{c_0^2}{c^2(\mathbf{R})}\right]\Gamma(\mathbf{R}) \ , \qquad (83)$$

and the equivalent volume source $q_c(\mathbf{R}, \omega)$ according to

$$q_c(\mathbf{R}, \omega) = -k_0^2 O(\mathbf{R})\Phi(\mathbf{R}, \omega) \ , \qquad (84)$$

the two equations (79) and (80) can be combined to

$$\Delta\Phi(\underline{R},\omega) + k_0^2\Phi(\underline{R},\omega) = -q(\underline{R},\omega) - q_c(\underline{R},\omega) \text{ for all } \underline{R} \qquad (85)$$

with $k_0 = \omega/c_0$. Therefore, the scatterer can be explicitly accounted for by an equivalent volume source.

Introducing the time harmonic free space Green function $G(\underline{R} - \underline{R}',\omega)$ via

$$G(\underline{R} - \underline{R}',\omega) = \frac{e^{jk_0|\mathbf{R}-\mathbf{R}'|}}{4\pi |\mathbf{R} - \mathbf{R}'|} \qquad (86)$$

we obtain the solution of (85) as superposition of the incident field $\Phi_i(\underline{R},\omega)$ and the scattered field $\Phi_s(\underline{R},\omega)$ according to

$$\Phi(\underline{R},\omega) = \Phi_i(\underline{R},\omega) + \Phi_s(\underline{R},\omega) \quad , \qquad (87)$$

where

$$\Phi_i(\underline{R},\omega) = \int_{-\infty}^{+\infty}\int_{-\infty}^{+\infty}\int_{-\infty}^{+\infty} q(\underline{R}',\omega)G(\underline{R} - \underline{R}',\omega)\,d^3\underline{R}' \qquad (88)$$

$$\Phi_s(\underline{R},\omega) = \int_{-\infty}^{+\infty}\int_{-\infty}^{+\infty}\int_{-\infty}^{+\infty} q_c(\underline{R}',\omega)G(\underline{R} - \underline{R}',\omega)\,d^3\underline{R}' \quad . \qquad (89)$$

If the defect is a void or a crack, it cannot be modelled by a material inhomogeneity with a sound speed variation $c(\underline{R})$; instead, we describe its surface S_c by the so-called singular function $\gamma(\underline{R})$ defined by

$$\gamma(\underline{R}) = -\underline{n} \cdot \nabla\Gamma(\underline{R}) \qquad (90)$$

— $\underline{n}$ being the outward unit-normal on the surface —, which has the sifting property of reducing a volume integral to a surface integral over S_c:

$$\int_{-\infty}^{+\infty}\int_{-\infty}^{+\infty}\int_{-\infty}^{+\infty} \gamma(\underline{R})\Phi(\underline{R})\,d^3\underline{R} = \iint_{S_c} \Phi(\underline{R})\,d^3\underline{R} \quad . \qquad (91)$$

Considering now a representation of the scattered field outside V_c in terms of the scalar Huygens principle

$$\Phi_s(\underline{R},\omega) = \int\int_{S_c} \left[\Phi(\underline{R}',\omega)\frac{\partial G}{\partial n'} - G(\underline{R} - \underline{R}',\omega)\frac{\partial\Phi}{\partial n'}\right] dS' \quad , \qquad (92)$$

where the total field $\Phi(\underline{\mathbf{R}}',\omega)$ on S_c as well as its normal derivative

$$\frac{\partial \Phi}{\partial n'} = \underline{\mathbf{n}}' \cdot \nabla' \Phi(\underline{\mathbf{R}}',\omega) \tag{93}$$

enter as equivalent surface sources to be propagated either as dipole waves $\partial G/\partial n'$ or as spherical waves $G(\underline{\mathbf{R}} - \underline{\mathbf{R}}',\omega)$, we can define equivalent *volume* sources of the scattering *surface* as

$$q_c^s(\underline{\mathbf{R}},\omega) = -\gamma(\underline{\mathbf{R}})\underline{\mathbf{n}} \cdot \nabla \Phi(\underline{\mathbf{R}},\omega) \tag{94}$$

$$q_c^r(\underline{\mathbf{R}},\omega) = -\Phi(\underline{\mathbf{R}},\omega)\nabla \cdot \gamma(\underline{\mathbf{R}})\underline{\mathbf{n}} \ . \tag{95}$$

This is true as (94) or (95) reduce the representation (89) to

$$\Phi_s(\underline{\mathbf{R}},\omega) = - \int\!\!\int_{S_c} \frac{\partial \Phi}{\partial n'} G(\underline{\mathbf{R}} - \underline{\mathbf{R}}',\omega) \, \mathrm{d}S' \tag{96}$$

or

$$\Phi_s(\underline{\mathbf{R}},\omega) = \int\!\!\int_{S_c} \Phi(\underline{\mathbf{R}}',\omega) \frac{\partial G}{\partial n'} \, \mathrm{d}S' \tag{97}$$

on behalf of the sifting property (91). Notice, (96) and (97) are explicit Huygens' type representations of the scattered field provided the scatterer satisfies either the condition of being perfectly soft, i.e. $\Phi(\underline{\mathbf{R}}',\omega) = 0$ for $\underline{\mathbf{R}}' \in S_c$, or being perfectly rigid, i.e. $\partial \Phi/\partial n' = 0$ for $\underline{\mathbf{R}}' \in S_c$, whence the notation q_c^s and q_c^r.

Therefore, the introduction of equivalent volume sources q_c, q_c^s, q_c^r allows us to negotiate cracks, voids or inclusions within the same mathematical formalism exploiting the representation (89); the scattered field $\Phi_s(\underline{\mathbf{R}},\omega)$ can be considered as "data" known for all frequencies ω on some measurement surface S_M surrounding S_c, and, hence, the "inversion" of (89) with regard to the equivalent sources should locate and size a defect quantitatively.

4.2 Ultrasonic Scattering Data for Very Small Scatterers

As it is obvious from Fig. 16, ultrasonic pulsed waves scattered by a defect travel back to the specimen surface carrying information about the scattering geometry to a receiver, which might be used to scan the surface along a line (linear aperture) or within a twodimensional aperture thus recording xt-data — x for the scan coordinate — or xyt-data. Imaging algorithms are bound to invert this process.

The simplest scattering geometry is a point scatterer at some spatial point $\underline{R}_0$ within a solid. Considering the pertinent xt-data field provides us with a phenomenological, but very intuitive imaging algorithm. Therefore, we postulate an equivalent volume source in terms of a delta-function

$$q_c(\underline{R}, t) = \delta(\underline{R} - \underline{R}_0)\delta(t) \quad , \tag{98}$$

which is "switched on" at $t = 0$ as a delta-pulse $\delta(t)$ (in "reality" it would have to be excited by an incident wave). This source yields the following scattered field in the time domain after the inverse Fourier transform of (89) — as a matter of fact, it is just the free space Green function in the time domain —

$$\Phi_s(\underline{R}, t) = \frac{\delta\left(t - \frac{1}{c_0}|\underline{R} - \underline{R}_0|\right)}{4\pi|\underline{R} - \underline{R}_0|} \quad . \tag{99}$$

Suppose, in two spatial dimensions, the observation point $\underline{R}$ varies within a synthetic aperture along a straight line given by the cartesian coordinate x; due to the δ-function behaviour of $\Phi_s(\underline{R}, t)$, nonzero xt-data are then obtained for

$$c_0 t = |\underline{R} - \underline{R}_0| \quad . \tag{100}$$

This equation represents a hyperbola in data space.

Fig.'s 20 to 22 present AFIT-simulations for plane wave scattering by very small scatterers, Fig. 20 showing the geometry under concern, Fig. 21 typical wave fronts, and Fig. 22 xt-data. Three layers of different materials are bonded together, and two prescribed μm-delaminations simulate very small defects. For simplicity, plane wave excitation was assumed in the AFIT simulation, and that way, xt-data have been computed; they typically exhibit the hperbolic structure as predicted above.

4.3 Phenomenological Algorithmic Imaging: Synthetic Aperture Focussing Technique (SAFT)

Fig. 23 illustrates how the hyperbolic time of flight curves can be phenomenologically exploited for imaging purposes. The solid squares simulate two

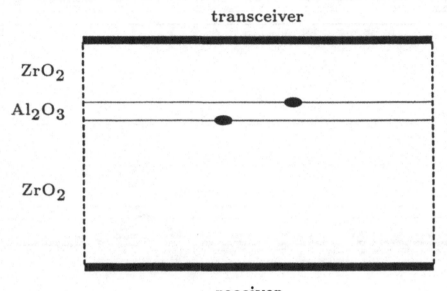

$$ZrO_2 : \quad c_p = 6915\tfrac{m}{s} \quad \lambda_c = 138.3\mu m$$
$$Al_2O_3 : \quad c_p = 10862\tfrac{m}{s} \quad \lambda_c = 217.2\mu m$$

Figure 20: Two Very Small Scatterers — Delaminations — in Bonded Layers of Different Materials

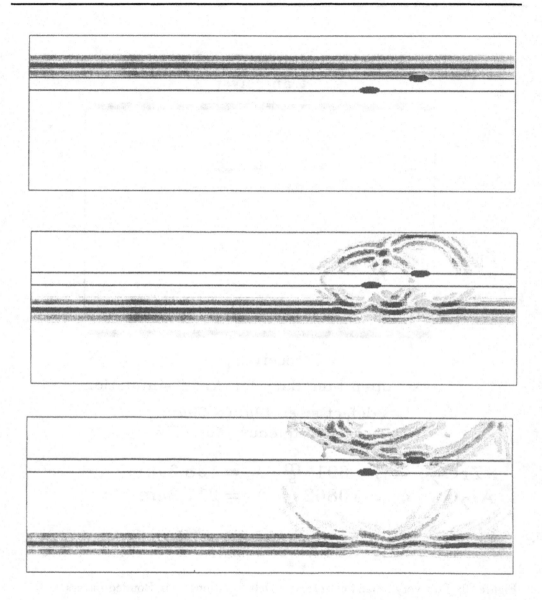

Figure 21: AFIT Wave Fronts for Plane Wave Incidence Parallel to the Layers of the Geometry of Figure 20

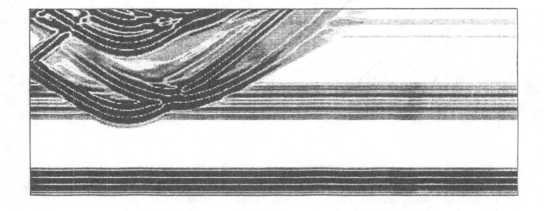

Figure 22: xt-Data "Recorded" from the AFIT Wave Fronts of Figure 21

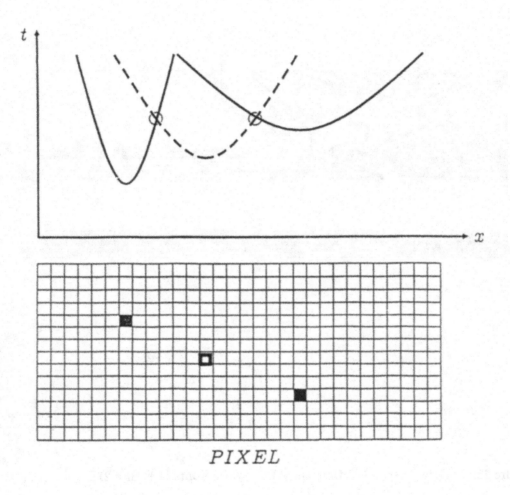

Figure 23: Phenomenological Algorithmic Imaging in Terms of the Synthetic Aperture Focusing Technique

"physically real" point scatterers. Dividing the spatial region to be imaged into pixels, we can define a hypothetic pixel scatterer — given by the empty square — with a pertinent hypothetic dashed "data" hyperbola. Phenomenological algorithmic imaging is based upon the following idea: imagine, the hypothetic pixel is a physically real pixel; we would then have to focus all the data on the pertinent dashed hyperbola into that pixel, but there are no data — the pixel being only a hypothetic scatterer — except at the two points indicated by the circles, where the hypothetic hyperbola intersects the real data hyperbolas. The same is true for *all* possible hypothetic pixels, which can be algorithmically checked one after each other with a computer. Only when the moving "computer pixel" meets physically real scatterers, data summation is over real data hyperbolas, yielding high amplitudes at those image space pixels versus neglectable amplitudes at all hypothetic pixels. Thus, an "image" is obtained via a synthetic aperture focussing technique (SAFT). Fig. 24 demonstrates that this procedure really works; experimental xt-data have been obtained from a crack oriented vertically with respect to the linear synthetic aperture — requiring 45°-degree shear wave excitation — yielding a SAFT-image with a lateral and axial resolution depending mainly upon the center frequency of the pulse and the size of the aperture. Even though numerous experiments have demonstrated the feasability of SAFT for practical applications [25], the fundamental question remains as to what degree and under what assumptions the image of Fig. 24 is "quantitative", i.e. reveals the "real" defect. This question can be answered analytically within the framework of inverse scattering, and, as a byproduct, powerful and fast alternative processing schemes in terms of diffraction tomography — an extension of X-ray computer tomography to wave phenomena — can be applied to the same xt-, and even xyt-data fields to allow for threedimensional algorithmic imaging.

Alternatively, modeling techniques such as the AFIT- or EFIT-code can be applied to a given NDT problem to yield synthetic data, which can then serve as input into the imaging schemes — for instance SAFT — to reveal as to what extent the image agrees with the given defect geometry. We discuss an example in Fig.'s 25 through 27. A "real" crack with a stress free rough surface is illuminated by a 45° shear wave angle probe and scalar acoustic scattering data are computed with the AFIT-code in a pulse echo mode; this means, that for each transducer location within a synthetic aperture a complete AFIT run — several thousand times steps — is computed, but only

Figure 24: SAFT Algorithmic Imaging as Applied to Experimental Data
Obtained by the Fraunhofer Institute for Nondestructive Testing in Germany

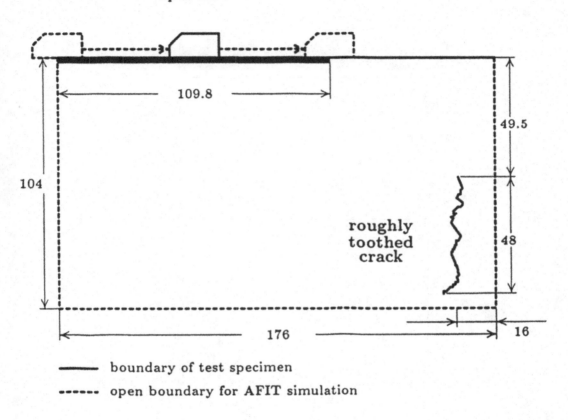

Figure 25: Given Defect Geometry for Modeling and Pulse-Echo SAFT Imaging

Snapshots corresponding to the 150th A-scan   (out of 256)

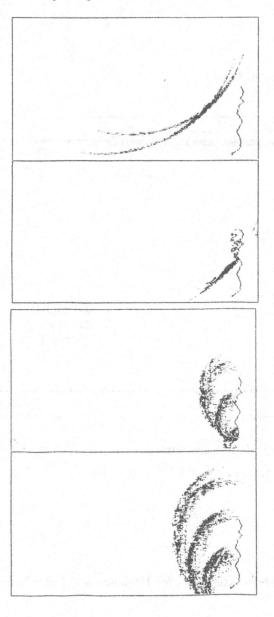

Figure 26: AFIT-Model of Pressure Wave Fronts Scattered by the Defect Geometry of Figure 25; One Particular Transducer Location Has Been Selected within the Synthetic Pulse-Echo Mode Aperture

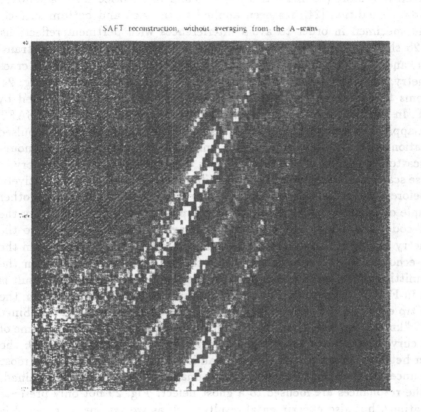

Figure 27: Pulse Echo SAFT Image Obtained with AFIT DATA for the Defect geometry of Figure 25

one particular A-scan at the transducer midpoint location is stored. Alternatively, to simulate the finite size of the transducer, an averaging procedure of several A-scans over the transducer aperture can be performed; then, only the average is store (compare also Fig.'s 13 and 15). Notice, a so-called open boundary condition [24] has been applied to the side and bottom surfaces of the specimen in order to avoid interferences with pertinent reflections. Fig. 26 shows some AFIT wave fronts for a selected position of the transducer, and Fig. 27 displays the SAFT image together with the given crack geometry; obviously, when compared to the experimental result of Fig. 24, it seems that such an experimental set-up has been properly modeled by AFIT. In addition, such a model tells us, that an imaging scheme like SAFT when applied to a single planar measurement surface and relying on a pulsed excitation with a finite bandwidth, is only able to image certain pronounced scattering centers of the rough defect. As mentioned before a theory of inverse scattering is available which identifies these drawbacks quantitatively.

Before we discuss our approach to inverse scattering, we give another example of modeling and measurement comparison; this time we apply the EFIT-code, and again, SAFT is used as imaging scheme. We refer to the geometry of Fig. 10, apply the EFIT code to produce an rf-data field in the pulse-echo mode with proper amplitude tapering of the transducer in the transmitting as well as in the receiving mode (compare Fig. 10); the result is given in Fig. 28. Typically, time of flight curves for the corner reflection, the crack tip echo and the various Rayleigh wave resonances are observed. Since SAFT "knows" nothing about mode conversion and resonances, each time of flight curve is attributed to a distinct defect, whence the image of Fig. 29: the corner between the crack and the backwall is properly imaged from the most pronounced time of flight curve, a weaker image of the crack tip is obtained, and the resonances are focused to a ghost defect. Fig. 29 not only presents simulations, but also experimental results, and, as we assume, our model is well reproduced. When investigated within the framework of inverse scattering, the derivation of SAFT exhibits, that a linearization is involved, and, therefore, in the present version, SAFT is not bound to handle resonances and mode conversions properly; neither is it designed for inhomogeneous or coarse grained, not for anisotropic materials.

Figure 28: Rf-Data Produced with the EFIT-Code for the NDT Problem as Displayed in Figure 10

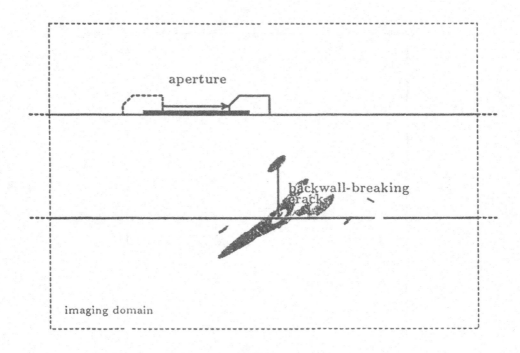

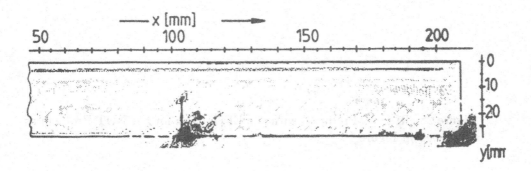

Figure 29: SAFT Image Obtained from the Simulated Data of Figure 28 (Top), and from Experiments Made in the Fraunhofer Institute for Nondestructive Testing (Bottom)

4.4   Linearized Quantitative Algorithmic Imaging

4.4.1   Frequency Diversity of Generalized Holography

Suppose, measurements of $\Phi_s(\underline{R}, t)$ and its normal derivative have been made on a surface S_M enclosing the volume V_M and surrounding the scatterer completely. Then, according to Huygens' principle, for every point $\underline{R}$ *outside* V_M the scattered field is equally known through the integral representation

$$\Phi_s(\underline{R}, \omega) = \int\int_{S_M} \left[\Phi_s(\underline{R}', \omega)\frac{\partial G}{\partial n'} - G(\underline{R} - \underline{R}', \omega)\frac{\partial \Phi_s}{\partial n'} \right] \, dS' \; , \qquad (101)$$

which is valid for every frequency ω, and which can be interpreted as a forward propagation scheme. Unfortunately, for points $\underline{R}$ *inside* V_M, i.e. where the scatterer resides, Huygens' principle yields only

$$\Phi_s(\underline{R}, \omega) = -\Phi_i(\underline{R}, \omega) \; , \qquad (102)$$

which says nothing about the scatterer. A remedy consists in the definition of the so-called generalized holographic field $\Theta_H(\underline{R}, \omega)$ as a backpropagation of the measurements on S_M into the volume V_M through the application of the complex conjugate of the free space Green function G^* in a Huygens' type integral

$$\Theta_H(\underline{R}, \omega) = -\int\int_{S_M} \left[\Phi_s(\underline{R}', \omega)\frac{\partial G^*}{\partial n'} - G^*(\underline{R} - \underline{R}', \omega)\frac{\partial \Phi_s}{\partial n'} \right] \, dS' \; . \qquad (103)$$

In contrast to the scattered field, the generalized holographic field $\Theta_H(\underline{R}, \omega)$ can be essentially computed *inside* V_M; the question is now, how it relates to the equivalent volume sources representing the scatterer. The answer is given in terms of an integral equation associated with the names of Porter and Bojarski [6], which can be derived applying Green's theorem to (103) and recognizing (89). It is of convolutional type and involves the imaginary part G_I of the free space Green function as kernel

$$\Theta_H(\underline{R}, \omega) = 2j \int_{-\infty}^{+\infty}\int_{-\infty}^{+\infty}\int_{-\infty}^{+\infty} q_c(\underline{R}', \omega)G_I(\underline{R} - \underline{R}', \omega) \, d^3\underline{R}' \; . \qquad (104)$$

The advantage of (104) over (89) is the accessability of Θ_H throughout *all* space, the disadvantage is, that it cannot simply be solved by deconvolution, because of the distributional properties of the multidimensional spatial

Fourier transform of $G_I(\underline{R}, \omega)$, denoted by $\tilde{G}_I(\underline{K}, \omega)$, the vector $\underline{K}$ being the Fourier vector variable corresponding to $\underline{R}$ according to

$$\tilde{G}_I(\underline{K}, \omega) = \int_{-\infty}^{+\infty} \int_{-\infty}^{+\infty} \int_{-\infty}^{+\infty} G_I(\underline{R}, \omega) e^{-j\underline{K} \cdot \underline{R}} \, d^3\underline{R} \ . \tag{105}$$

As a matter of fact, we obtain

$$\tilde{\Theta}_H(\underline{K}, \omega) = \frac{j\pi}{k_0} \tilde{q}_c(\underline{K}, \omega) \delta(K - k_0) \tag{106}$$

where $K = |\underline{K}|$; (106) reveals, that, for a single frequency ω, $\tilde{\Theta}_H(\underline{K}, \omega)$ is nonzero only on a sphere of radius k_0 in spatial Fourier space, the so-called Ewald-sphere. Fortunately, data outside the Ewald-sphere *are* available on behalf of our time domain experiment, which ensures ω to vary within the bandwidth of the exciting ultrasound. Hence, sweeping frequency in (106) should enable us to invert $\tilde{\Theta}_H(\underline{K}, \omega)$ from a finite support in $\underline{K}$-space to yield the equivalent volume sources explicitly. Unfortunately, a closer look at $q_c(\underline{R}, \omega)$ in terms (83), (94) or (95) reveals, that all the q_c's are dependent upon the *total* field $\Phi(\underline{R}, \omega)$ localized in or on the scatterer. Hence, their variation with frequency is unknown; the complete spatial structure of $\Phi(\underline{R}, \omega)$ — as a function of $\underline{R}$ and, therefore, as a function of $\underline{K}$ for its Fourier transform — is changed if ω is changed. This tells us that frequency sweeping yields an uncontrollable variation of $q_c(\underline{R}, \omega)$, or $\tilde{q}_c(\underline{K}, \omega)$; the inverse problem is nonlinear with respect to the equivalent sources [26]. Physically, this is due to the fact, that local scattering centers on the defect are not independent of each other but interact via radiation.

Until today, to end up with working algorithms, the only remedy to the nonlinear inversion problem is its *linearization*. Here, essentially the first order Born approximation for the inclusion, or the Kirchhoff or physical optics approximation for the void (crack) come in. For simplicity, let us discuss the case of weak scattering by an inclusion being associated with the name of Born; the Kirchhoff approximation is discussed in detail in [27] . Weak scattering applies intuitively if the scatterer is present in the real world to produce a scattered field *outside*, but simultaneously, it is absent as not to disturb the incident field while propagating through it, i.e. the scattered field is zero *inside*. Mathematically, the Born approximated scattered field is the first term in its Neumann series expansion. Therefore, instead of (83) we put

$$q_c(\underline{\mathbf{R}}, \omega) \simeq -k_0^2 O(\underline{\mathbf{R}}) \Phi_i(\underline{\mathbf{R}}, \omega) \ , \tag{107}$$

which obviously requires some specification concerning the incident field now in order to be inserted into (106). The previous discussion of ultrasonic transducer models has revealed that plane wave representations of incident fields are quite appropriate, i.e. we assume

$$\Phi_i(\underline{\mathbf{R}}, \omega) = \Phi_0(\omega) e^{jk_0 \hat{\mathbf{k}} \cdot \underline{\mathbf{R}}} \ , \tag{108}$$

where $\Phi_0(\omega)$ is the frequency spectrum of the excitation (compare Fig. 2). After some mathematical manipulation [6], insertion of (108) into (106) finally yields

$$O(\underline{\mathbf{R}}) = -\frac{1}{\pi} \Re \int_0^\infty \frac{1}{k_0^2 \Phi_0(\omega)} \hat{\mathbf{k}} \cdot \nabla \left[\Theta_H(\underline{\mathbf{R}}, \omega) e^{-jk_0 \hat{\mathbf{k}} \cdot \underline{\mathbf{R}}} \right] dk_0 \ , \tag{109}$$

where the k_0-integration accounts for the frequency sweeping in our experiment. The above equation is called the frequency diversity version of generalized holography [6]; it is an algorithmic recipe of data processing to obtain an image of the scatterer in terms of its object function. This recipe is quantitative in the sense, that any approximations and assumptions, which have been made to formulate it, are well defined; as such, it is part of a unified theory of linearized inverse scattering [28]. The assumptions comprise:

- "Scalarization" of the elastic wave field in a solid handling it as a truly acoustic wave field

- Linearization of the inverse scattering problem treating the scatterer as only weak

- Plane wave illumination

- Deconvolution with the frequency spectrum $\Phi_0(\omega)$ as it appears in the denominator of (109)

- Closed measurement surface.

By the way, the corresponding linearization for the perfectly soft (rigid) scatterer, simulating a void or a crack, is in terms of the physical optics

(Kirchhoff) approximation, which says, that the equivalent surface sources
are proportional to the incident field on the illuminated side being zero on
the dark side, as it is *strictly* true for plane surfaces only [27].

Of course, any violation of the above-mentioned items will cause degra-
dations and blurring of the image, but with equation (109), in contrast to
phenomenological algorithmic imaging, one is able to study these influences
quantitatively, particularly in combination with appropriate modeling codes.

4.4.2 Time Domain Backpropagation

Surprisingly enough, the backpropagation ansatz of (103) comprises the exact
mathematical formulation of the phenomenological SAFT algorithm. This
can be seen considering the k_0-integral in (109) as an inverse Fourier integral
with regard to frequency for $t = 0$; the result is an *exact* time domain
backpropagation algorithm [6], where the introduction of certain additional
approximations yields the algorithmic procedure as described in terms of
Fig. 23. Because we have approached SAFT in the present paper from pulse-
echo experiments, we will give the pulse-echo, or, in radar terminology, the
monostatic version of (109), before we interpret it in the time domain.

We return to (84) and (89) and combine both equations to the so-called
Lippmann-Schwinger equation

$$\Phi(\mathbf{R}, \omega) = \Phi_i(\mathbf{R}, \omega) - k^2 \iiint_{V_c} O(\mathbf{R}')\Phi(\mathbf{R}', \omega)G(\mathbf{R} - \mathbf{R}', \omega)\, \mathrm{d}^3\mathbf{R}' \ , \quad (110)$$

which is an integral equation for the total field $\Phi(\mathbf{R}, \omega)$. It transforms into
an integral *representation*, if we introduce the Born approximation for the
equivalent source of the penetrable scatterer inserting

$$q_c^{\mathrm{B}}(\mathbf{R}, \omega) = -k_0^2 \Phi_0(\omega) O(\mathbf{R}) \frac{e^{jk_0|\mathbf{R} - \mathbf{R}_0|}}{4\pi|\mathbf{R} - \mathbf{R}_0|} \ , \quad\quad (111)$$

where we have assumed point-source illumination from $\mathbf{R}_0$. For the monostatic case we select $\mathbf{R} = \mathbf{R}_0$ to observe the scattered field and hence

$$\Phi_s(\mathbf{R}, \omega) = -k_0^2 \Phi_0(\omega) \iiint_{V_c} O(\mathbf{R}') \frac{e^{2jk_0|\mathbf{R} - \mathbf{R}'|}}{(4\pi|\mathbf{R} - \mathbf{R}'|)^2}\, \mathrm{d}^3\mathbf{R}' \ , \quad\quad (112)$$

which gives rise to a modified scattered field

$$\Phi_s^{mo}(\underline{R}, \omega) = 2\pi j \frac{\partial}{\partial k_0} \left[\frac{\Phi_s(\underline{R}, \omega)}{k_0^2 \Phi_0(\omega)} \right] \tag{113}$$

resulting in the differential equation

$$\Delta \Phi_s^{mo}(\underline{R}, \omega) + 4k_0^2 \Phi_s^{mo}(\underline{R}, \omega) = -O(\underline{R}) \quad . \tag{114}$$

This equation defines the monostatic equivalent sources in terms of the object function itself; their field dependence has disappeared, because we already had to introduce the linearizing Born approximation in order to verify the *concept* of the equivalent sources. Equation (114) defines a Green function

$$G^{mo}(\underline{R}, \omega) = \frac{e^{2jk_0|\underline{R}-\underline{R}'|}}{4\pi|\underline{R} - \underline{R}'|} \quad , \tag{115}$$

a pertinent Huygens principle, and the modified Porter-Bojarski integral equation for $k_0 \geq 0$ according to

$$\tilde{\Phi}_H^{mo}(\underline{K}, \omega) = \frac{j\pi}{2k_0} \tilde{O}(\underline{K}) \delta(K - 2k_0) \quad . \tag{116}$$

Frequency diversity yields as inversion scheme

$$O(\underline{R}) = \frac{4}{j\pi} \int_0^\infty k_0 \Theta_H^{mo}(\underline{R}, \omega) \, dk_0 \quad . \tag{117}$$

The right-hand side of (117) can be interpreted as an inverse Fourier integral with regard to frequency for $t = 0$; by definition, it involves inversion of positive frequency data only, hence, the resulting time function is complex with an imaginary part being the Hilbert transform of the real part. In order to see this explicitly we rewrite (117) introducing a unit-step function $u(\omega)$

$$O(\underline{R}) = \frac{4}{\pi c^2} \int_{-\infty}^\infty (-j\omega) \Theta_H^{mo}(\underline{R}, \omega) u(\omega) \, d\omega \tag{118}$$

and realizing that its inverse Fourier transform is given by

$$\mathcal{F}^{-1}\{u(\omega)\} = \frac{1}{2}\delta(t) + \frac{1}{2\pi jt} \quad , \tag{119}$$

which results in

$$O(\mathbf{R}) = \frac{4}{c^2} \left[\frac{\partial}{\partial t} \Theta_H^{mo}(\mathbf{R},t) + j\mathcal{H}_t \left\{ \frac{\partial}{\partial t} \Theta_H^{mo}(\mathbf{R},t) \right\} \right]_{t=0} . \qquad (120)$$

Here, $\mathcal{H}_t$ indicates the Hilbert transform with regard to t according to

$$\mathcal{H}_t\{f(t)\} = -\frac{1}{\pi} \int_{-\infty}^{\infty} \frac{f(\tau)}{t-\tau} \, d\tau . \qquad (121)$$

Per definition the object function is real valued, and, therefore,

$$\mathcal{H}_t \left\{ \frac{\partial}{\partial t} \Theta_H^{mo}(\mathbf{R},t) \right\}_{t=0} = 0 \qquad (122)$$

has to hold. Of course, this is only so for exact Born data, which leads to the conclusion that the degree of nonzeroness of the left-hand side of (122) indicates the deviation from the assumptions and approximations which are involved in the inversion scheme.

From (122), the monostatic time domain inverse scattering algorithm reads

$$O(\mathbf{R}) = \frac{4}{c^2} \frac{\partial}{\partial t} \Theta_H^{mo}(\mathbf{R},t) \Big|_{t=0} . \qquad (123)$$

With the help of (103) we can compute $\Theta_H^{mo}(\mathbf{R},t)$ explicitly in terms of the monostatic scattered field resulting in an exact — within the Born approximation — time domain backpropagation scheme of data to recover the object function [6].

For practical purposes it seems more appropriate to utilize some assumptions, which are in general satisfied in ultrasonic imaging applications. First, let us evaluate (113) asymptotically for high frequencies according to

$$\Phi_s^{mo}(\mathbf{R},\omega) \simeq \frac{2\pi j}{k_0^2} \frac{\partial}{\partial k_0} \Phi_s^I(\mathbf{R},\omega) , \qquad (124)$$

where the upper index "I" refers to "impulse response", i.e. to the $\Phi_0(\omega)$-filtered field; second, we ignore the normal derivative of the data in (103) — it is hardly available except for planar and circular cylindrical measurement surfaces — and, third, we ignore terms of the order $|\mathbf{R} - \mathbf{R}'|^{-2}$ resulting from the normal derivative of Green's function in (103). Then we obtain from (117)

$$O(\underline{R}) = \frac{4}{j\pi} \int_0^\infty dk_0 \int\int_{S_M} \frac{\partial}{\partial k_0} \left[\Phi_s^I(\underline{R}', \omega) \right] \frac{e^{-2jk_0|\underline{R}-\underline{R}'|}}{|\underline{R}-\underline{R}'|} \frac{\underline{n}' \cdot (\underline{R}-\underline{R}')}{|\underline{R}-\underline{R}'|} dS' \ ,$$

$$(125)$$

or in the time domain

$$O(\underline{R}) = 4 \int\int_{S_M} \left(t + \frac{2|\underline{R}-\underline{R}'|}{c} \right) \frac{\Phi_s^I \left(\underline{R}', t + \frac{2|\underline{R}-\underline{R}'|}{c} \right)}{|\underline{R}-\underline{R}'|} \frac{\underline{n}' \cdot (\underline{R}-\underline{R}')}{|\underline{R}-\underline{R}'|} \Bigg|_{t=0} dS'.$$

$$(126)$$

For closely paraxial data and image points we can further ignore the factor

$$\frac{\underline{n}' \cdot (\underline{R}-\underline{R}')}{|\underline{R}-\underline{R}'|}$$

to obtain

$$o(\underline{R}) = \int\int_{S_M} \Phi_s^I \left(\underline{R}', t = \frac{2|\underline{R}-\underline{R}'|}{c} \right) dS' \ . \qquad (127)$$

Notice that we switched intentionally from the object function to an "image function" — it also absorbs all prefactors —, because we can no longer expect that an imaging scheme relying on all the above approximations still yields the object function, or the singular function for the perfect scatterer case.

As a matter of fact, (127) is precisely the heuristic SAFT time domain backpropagation scheme, which was already discussed in Section 4.3; a fixed time sample for a fixed observation point $\underline{R}'$ in data space has to be backpropagated to all image space points $\underline{R}$ satisfying

$$t = \frac{2|\underline{R}-\underline{R}'|}{c_0} \ , \qquad (128)$$

which is obviously a circle. Therefore, linearized inverse scattering theory is the quantitative framework for SAFT imaging.

In Fig. 30 we once more processed the experimental data already used to produce Fig. 24, but this time we applied time domain backpropagation without paraxial approximation; obviously, the result compares very well with that of Fig. 24.

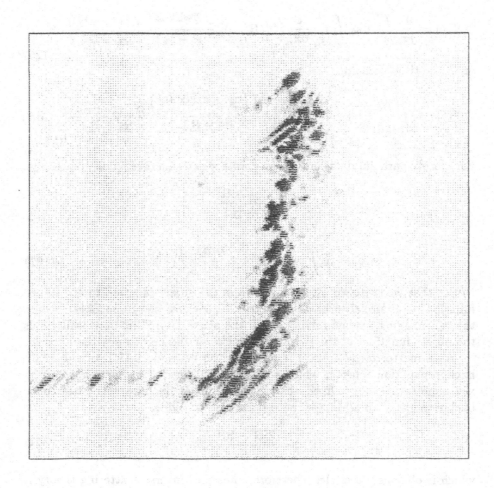

Figure 30: Time Domain Backpropagation Without Paraxial Approximation Imaging as Applied to Experimental Data Obtained by the Fraunhofer Institute for Nondestructive Testing in Germany

4.4.3 Frequency Diversity Diffraction Tomography or FT-SAFT

Quantitative algorithmic imaging based on inverse scattering theory has by now two advantages: it provides quantitative algorithms, and it contains phenomenological imaging as a well defined special case. Furthermore, and this might be the most important feature, (109) can be very effectively computed, at least when planar or circular cylindrical measurement surfaces are involved, which is often the case. This is readily seen, if a particular representation of the generalized holographic field in terms of the far-field scattering amplitude $H(\hat{\mathbf{R}}, \omega)$ is introduced. The latter one is defined by (89) for $R \gg R'$ and $k_0 R \gg 1$ through

$$\Phi_s^{far}(\mathbf{R}, \omega) = H(\hat{\mathbf{R}}, \omega)\frac{e^{jk_0 R}}{R} \tag{129}$$

with

$$H(\hat{\mathbf{R}}, \omega) = \frac{1}{4\pi} \int_{-\infty}^{+\infty} \int_{-\infty}^{+\infty} \int_{-\infty}^{+\infty} q_c(\underline{\mathbf{R}}', \omega)e^{-jk_0\hat{\mathbf{R}}\cdot\underline{\mathbf{R}}'}\, d^3\underline{\mathbf{R}}' \;, \tag{130}$$

where $\hat{\mathbf{R}} = \mathbf{R}/R$. Notice, (130) says that the scattering amplitude is related to the spatial Fourier transform of the equivalent sources on the Ewald-sphere, i.e. (130) can be rewritten as

$$H(\hat{\mathbf{R}}, \omega) = \frac{1}{4\pi}\tilde{q}_c(\mathbf{K} = k_0\hat{\mathbf{R}}, \omega) \;, \tag{131}$$

because $\mathbf{K}$ is restricted to a sphere with radius k_0 via $\mathbf{K} = k_0\hat{\mathbf{R}}$.

From (103) we can compute the following representation of the generalized holographic field

$$\Theta_H(\mathbf{R}, \omega) = \frac{jk_0}{2\pi} \int\int_{S^2} H(\hat{\mathbf{R}}', \omega)e^{jk_0\hat{\mathbf{R}}'\cdot\mathbf{R}}\, d^2\hat{\mathbf{R}}' \tag{132}$$

for $\mathbf{R}$ taken in the far-field [28, 26]; here, S^2 denotes the unit-sphere. Insertion into the imaging recipe (109) yields the triple integral

$$\int\int_{S^2} \int_0^\infty H(\hat{\mathbf{R}}', \omega)\hat{\underline{\mathbf{k}}}\cdot(\hat{\mathbf{R}}' - \hat{\mathbf{k}})e^{jk_0(\hat{\mathbf{R}}'-\hat{\mathbf{k}})\cdot\mathbf{R}}\, dk_0 d^2\hat{\mathbf{R}}' \;, \tag{133}$$

which, due the exponentials, looks very much like a spatial inverse Fourier integral, provided the Fourier vector $\underline{\mathbf{K}}$ is defined by

$$\underline{\mathbf{K}} = k_0(\hat{\underline{\mathbf{R}}}' - \hat{\underline{\mathbf{k}}}) \ . \tag{134}$$

Computation of the required Jacobian results in

$$\mathrm{d}^3\underline{\mathbf{K}} = -k_0^2\hat{\underline{\mathbf{k}}} \cdot (\hat{\underline{\mathbf{R}}}' - \hat{\underline{\mathbf{k}}})\, \mathrm{d}k_0\mathrm{d}^2\hat{\underline{\mathbf{R}}}' \ , \tag{135}$$

making it possible to process (133) with threedimensional FFT-techniques. Hence, $O(\underline{\mathbf{R}})$ can be effectively computed as soon as $H(\hat{\underline{\mathbf{R}}}',\omega)$, i.e. far-field data, is available. This might a priorily not be the case; then, from (101) we find

$$H(\hat{\underline{\mathbf{R}}},\omega) = -\frac{1}{4\pi} \int\!\!\int_{S_M} \left[\frac{\partial \Phi_s}{\partial n'} + jk_0\underline{\mathbf{n}}' \cdot \hat{\underline{\mathbf{R}}}\Phi_s(\underline{\mathbf{R}}',\omega) \right] \mathrm{e}^{-jk_0\hat{\underline{\mathbf{R}}}\cdot\hat{\underline{\mathbf{R}}}'}\, \mathrm{d}S' \ , \tag{136}$$

which is a computational scheme to obtain the scattering amplitude from measurements on an arbitrary surface S_M. As a matter of fact, for planar and circular cylindrical surfaces, FFT-techniques can be used as well to evaluate (136) numerically [27, 29, 26]. First of all, this results in extremely fast algorithmic schemes making threedimensional imaging possible, and, as a byproduct, the normal derivative of the scattered field on the measurement surface is not required in terms of explicit measurements as it can be computed from the field itself. From a conceptual viewpoint, the mapping of the "data" $H(\hat{\underline{\mathbf{R}}}',\omega)$ into $\underline{\mathbf{K}}$-space via (134) is a procedure, which closely resembles algorithms applied for X-ray computer tomography, except that diffraction effects of the acoustic wave field are properly accounted for. Therefore, quantitative algorithmic imaging exploiting Fourier transform relationships explicitly has been named diffraction tomography; here, for obvious reasons, we call it F(ourier)T(ransform)-SAFT.

In Fig. 31 we revisit the data utilized to obtain the SAFT image of Fig. 24 and apply the FT-SAFT technique; obviously, there are only marginal differences in the two images confirming that SAFT and FT-SAFT are in fact imaging alternatives yielding the same result. Notice, FT-SAFT does not rely on any further than the linearizing approximation, therefore, there might be cases where FT-SAFT is more "exact", even though we have not yet encountered one in our daily NDT business.

Figure 31: FT-SAFT Image with Experimental Data Obtained by the Fraunhofer Institute for Nondestructive Testing in Germany; Compare Also Figures 24 and 30

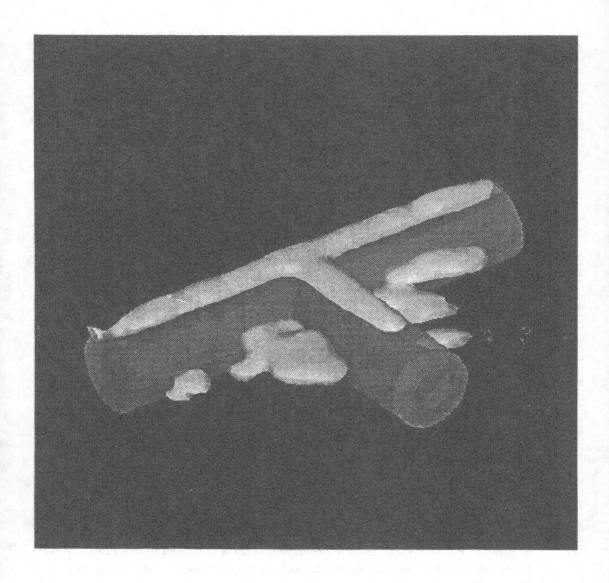

Figure 32: 3D FT-SAFT Image from Experimental Data Obtained by Scanning a T-Shaped Tube in a Solid with Nearby Crack Fields

Fig. 32 shows an example for threedimensional FT-SAFT imaging. A T-shaped tube in a solid is displayed via geometry rendering, and, simultaneously, an image obtained from an xyt-data field taken on the planar top surface of the pertinent specimen is added as an isocontour surface. A clear indication of the top surface of the tube as it is "visible" from the transducer is obvious; additional crack fields close to the tube have been imaged, making it possible to evaluate their size, shape and orientation nondestructively in a quantitative manner.

Finally, we discuss three figures confirming that it might be necessary to use in fact a threedimensional imaging scheme instead of a twodimensional one. Fig. 33 shows the "real" 3D FT-SAFT image; of course, the flat bottoms of several flatbottom holes drilled into the backwall of a specimen are appropriately imaged and resolved according to the bandwidth of the ultrasonic pulse and according to the extent of the scan aperture. In contrast, Fig.'s 34 and 35 show pseudo 3D images obtained from the same xyt-data set; in Fig. 34 the y scan lines were used as a discrete number of xt-data sets, from which a pertinent number of twodimensional FT-SAFT images have been produced, and in Fig. 35 the x scan lines were used as a discrete number of yt-data sets, again resulting in pertinent twodimensional images. All these twodimensional images were then displayed using the same isocontour software as for Fig. 33 revealing that interferences between the single flat bottom holes can remarkably degrade the image assessment if only a twodimensional imaging scheme is used.

4.5   Algorithmic Imaging Yesterday, Today, and Tomorrow: Elastodynamic Diffraction Tomography

With the advent of storage and processing devices to record and handle large time domain ultrasonic data fields from one- or twodimensional scans within a synthetic aperture, algorithmic imaging for nondestructive testing purposes experienced a big jump ahead as appropriate intuitive ideas of synthetic aperture data focussing could be exploited. Today, these ideas have been found to be a rigorous mathematical consequence of inverse scattering theories, providing not only a quantitative basis for algorithmic imaging, but allowing especially for the development of fast threedimensional imaging systems. The availability of the theoretical background is now going to sti-

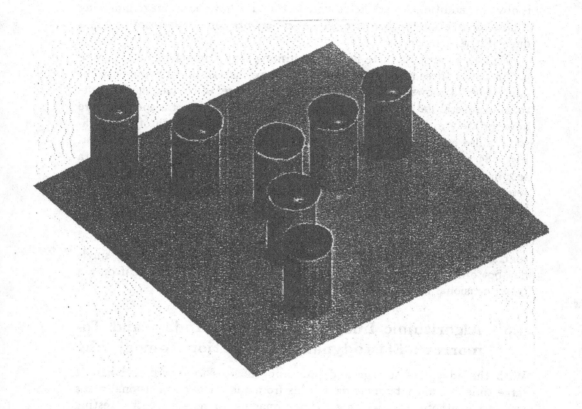

Figure 33: 3D FT-SAFT Image from Experimental Data Obtained by Scanning Several Flatbottom Holes Drilled into the Specimen from the Backwall

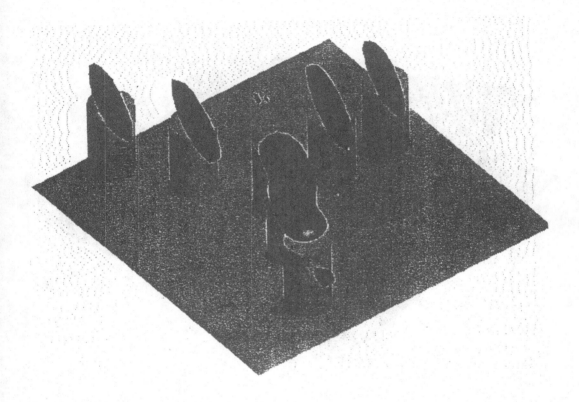

Figure 34: Pseudo 3D FT-SAFT Image Obtained from the Same Data that Were Used for Figure 33

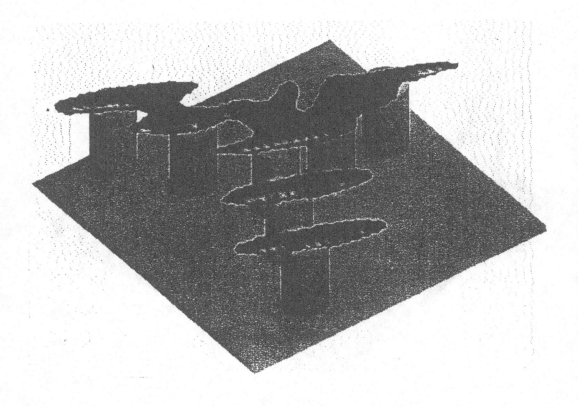

Figure 35: Pseudo 3D FT-SAFT Image Obtained from the Same Data that Were Used for Figure 33

mulate the evaluation of algorithms accounting for the full elastodaynamic features of ultrasound in solids — even in anisotropic solids [5] — as well as for the linearization to be overcome [25].

4.5.1 Elastodynamic Far-Field Inversion for Scatterers with Stress Free Boundaries

A Huygens-type solution of (18) for the displacement $\underline{u}_s$ scattered by a defect with a stress-free surface S_c is given by (compare (78))

$$\underline{u}_s(\mathbf{R}, \omega) = \int\!\!\int_{S_c} \underline{n}'\underline{u}(\mathbf{R}', \omega) : \underline{\underline{\underline{\Sigma}}}(\mathbf{R} - \mathbf{R}', \omega) \, dS' \ , \tag{137}$$

where $\underline{n}'$ denotes the outer normal on S_c; we do not give the third rank Green tensor $\underline{\underline{\underline{\Sigma}}}$ explicitly, because we only need its far-field approximation

$$\underline{\underline{\underline{\Sigma}}}^{far}(\mathbf{R} - \mathbf{R}', \omega) = -\frac{j\mu k_S^3}{\varrho\omega^2}\left[\hat{\mathbf{R}}\underline{\underline{I}} + (\hat{\mathbf{R}}\underline{\underline{I}})^{213} - 2\hat{\mathbf{R}}\hat{\mathbf{R}}\hat{\mathbf{R}}\right] G_S^{far}(\mathbf{R} - \mathbf{R}', \omega)$$

$$-\frac{jk_P^3}{\varrho\omega^2}(\lambda\underline{\underline{I}}\hat{\mathbf{R}} + 2\mu\hat{\mathbf{R}}\hat{\mathbf{R}}\hat{\mathbf{R}})G_P^{far}(\mathbf{R} - \mathbf{R}', \omega)$$

$$= \frac{e^{jk_S R}}{4\pi R}e^{-jk_S\hat{\mathbf{R}}\cdot\mathbf{R}'}\underline{\underline{\Sigma}}_S(\hat{\mathbf{R}}, \omega) + \frac{e^{jk_P R}}{4\pi R}e^{-jk_P\hat{\mathbf{R}}\cdot\mathbf{R}'}\underline{\underline{\Sigma}}_P(\hat{\mathbf{R}}, \omega) \ . \tag{138}$$

We obtain the far-field approximation of (137) through

$$\underline{u}_s^{far}(\mathbf{R}, \omega) = \frac{e^{jk_S R}}{4\pi R}\int\!\!\int\!\!\int_{V_c}\underline{\underline{U}}_c(\mathbf{R}', \omega) : \underline{\underline{\Sigma}}_S(\hat{\mathbf{R}}, \omega)e^{-jk_S\hat{\mathbf{R}}\cdot\mathbf{R}'} \, d^3\mathbf{R}' +$$

$$\frac{e^{jk_P R}}{4\pi R}\int\!\!\int\!\!\int_{V_c}\underline{\underline{U}}_c(\mathbf{R}', \omega) : \underline{\underline{\Sigma}}_P(\hat{\mathbf{R}}, \omega)e^{-jk_P\hat{\mathbf{R}}\cdot\mathbf{R}'} \, d^3\mathbf{R}'$$

$$= \frac{e^{jk_S R}}{R}\underline{C}_S(\hat{\mathbf{R}}, \omega) + \frac{e^{jk_P R}}{R}\underline{C}_P(\hat{\mathbf{R}}, \omega)$$

$$= \underline{u}_{sS}^{far}(\mathbf{R}, \omega) + \underline{u}_{sP}^{far}(\mathbf{R}, \omega) \ , \tag{139}$$

where we have introduced dyadic equivalent surface sources $\underline{\underline{U}}_c(\mathbf{R}', \omega)$ through

$$\underline{\underline{U}}_c(\mathbf{R}', \omega) = \gamma(\mathbf{R}')\underline{n}'\underline{u}(\mathbf{R}', \omega) \tag{140}$$

with the singular function $\gamma(\underline{R})$ of S_c; the introduction of the latter allows the extension of the surface integral in (137) to a volume integral over the volume V_c of the scatterer. The quantities $\underline{C}_\beta(\hat{\underline{R}}, \omega)$ with $\beta = P, S$ play the role of vectorial scattering amplitudes in the β-mode, and it is easily verified that the following properties hold

$$\underline{C}_P(\hat{\underline{R}}, \omega) \times \hat{\underline{R}} = 0 \tag{141}$$

$$\underline{C}_S(\hat{\underline{R}}, \omega) \cdot \hat{\underline{R}} = 0 . \tag{142}$$

This permits computational selection of a special mode in the far-field

$$\underline{u}_{s\beta}^{far}(\underline{R}, \omega) = \frac{e^{jk_\beta R}}{R} \underline{C}_\beta(\hat{\underline{R}}, \omega) \tag{143}$$

to be considered separately for inversion. We linearize in terms of physical elastodynamics (PE) [3]

$$\underline{U}_c(\underline{R}', \omega) \Rightarrow \underline{U}_{c\alpha}^{PE}(\underline{R}', \omega)\gamma_u(\underline{R}')$$

$$= \underline{U}_{0\alpha}(\underline{n}', \hat{\underline{k}}_i, \omega)\gamma_u(\underline{R}')e^{jk_\alpha \hat{\underline{k}}_i \cdot \underline{R}'} \tag{144}$$

with

$$\underline{U}_{0\alpha}(\underline{n}', \hat{\underline{k}}_i, \omega) = \underline{u}_{0\alpha}(\underline{n}', \hat{\underline{k}}_i, \omega)\underline{n}' , \tag{145}$$

where k_α is the wave number of the incident plane wave in the mode $\alpha = P, S$; $\underline{u}_{0\alpha}(\underline{n}', \hat{\underline{k}}_i, \omega)$ denotes the surface displacement "induced" by the incident wave as computed by PE, whence the occurrence of only *that* part of the singular function, which is illuminated by that wave, i.e. $\gamma_u(\underline{R}')$, the index u indicating multiplication of $\gamma(\underline{R}')$ with a step-function. We obtain

$$\underline{u}_{s\alpha\beta}^{far}(\underline{R}, \omega) = \frac{e^{jk_\beta R}}{4\pi R} \int\!\!\int\!\!\int_{V_c} \gamma_u(\underline{R}')\underline{U}_{0\alpha}(\underline{n}', \hat{\underline{k}}_i, \omega) : \underline{\underline{\Sigma}}_\beta(\hat{\underline{R}}, \omega)e^{-j(k_\beta \hat{\underline{R}} - k_\alpha \hat{\underline{k}}_i)\cdot \underline{R}'}d^3\underline{R}' . \tag{146}$$

Computing a "stationary phase normal" $\underline{n}' = \underline{n}_{\alpha\beta}$ according to

$$\underline{n}_{\alpha\beta} = \frac{k_\beta \hat{\underline{R}} - k_\alpha \hat{\underline{k}}_i}{|k_\beta \hat{\underline{R}} - k_\alpha \hat{\underline{k}}_i|} , \tag{147}$$

we get rid of the $\hat{\underline{R}}$-dependence of the source term in the integral (146) replacing $\underline{n}'$ by $\underline{n}_{\alpha\beta} = \underline{n}_{\alpha\beta}(\hat{\underline{R}}, \hat{\underline{k}}_i)$ to end up with

$$\underline{u}_{s\alpha\beta}^{far}(\underline{R}, \omega) = \frac{e^{jk_\beta R}}{4\pi R}\underline{U}_{0\alpha}(\hat{\underline{R}}, \hat{\underline{k}}_i, \omega) : \underline{\underline{\Sigma}}_\beta(\hat{\underline{R}}, \omega) \int\!\!\int\!\!\int_{V_c} \gamma_u(\underline{R}')e^{-j\underline{K}\cdot\underline{R}'} d^3\underline{R}' \tag{148}$$

with the Fourier vector

$$\mathbf{K} = k_\beta \hat{\mathbf{R}} - k_\alpha \hat{\mathbf{k}}_i \quad , \tag{149}$$

and

$$\underline{\mathbf{U}}_{0\alpha\beta}(\hat{\mathbf{R}}, \hat{\mathbf{k}}_i, \omega) = \underline{\mathbf{u}}_{0\alpha}(\underline{\mathbf{n}}_{\alpha\beta}, \hat{\mathbf{k}}_i, \omega) \underline{\mathbf{n}}_{\alpha\beta} \quad . \tag{150}$$

The double contraction of the dyad $\underline{\mathbf{U}}_0$ with the triad $\underline{\underline{\boldsymbol{\Sigma}}}_\beta$ yields a vector

$$\underline{\mathbf{U}}_{0\alpha\beta} : \underline{\underline{\boldsymbol{\Sigma}}}_\beta = \underline{\mathbf{C}}_{0\alpha\beta}(\hat{\mathbf{R}}, \hat{\mathbf{k}}_i, \omega) \quad . \tag{151}$$

Hence, in shorthand notation (148) reads

$$\underline{\mathbf{u}}_{s\alpha\beta}^{far}(\mathbf{R}, \omega) = \frac{e^{jk_\beta R}}{4\pi R} \underline{\mathbf{C}}_{0\alpha\beta}(\hat{\mathbf{R}}, \hat{\mathbf{k}}_i, \omega) \, \tilde{\gamma}_u(\underline{\mathbf{K}}) \quad , \tag{152}$$

where $\tilde{\gamma}_u(\underline{\mathbf{K}})$ denotes the threedimensional Fourier transform of $\gamma_u(\underline{\mathbf{R}})$; (152) is our desired far-field inversion scheme, because, due to the knowlegde of $\underline{\mathbf{C}}_{0\alpha\beta}(\hat{\mathbf{R}}, \hat{\mathbf{k}}_i, \omega)$ from experiments, we can easily compute the Fourier transform of the "visible" singular function from

$$4\pi Re^{-jk_\beta R} \frac{\underline{\mathbf{C}}_{0\alpha\beta} \cdot \underline{\mathbf{u}}_{s\alpha\beta}^{far}}{|\underline{\mathbf{C}}_{0\alpha\beta}|^2} = \tilde{\gamma}_u(\underline{\mathbf{K}}) \quad , \tag{153}$$

and, therefore, a threedimensional Fourier inversion will get us back to the spatial domain. Obviously, the result of the far-field inversion scheme (153) is neither dependent on the incoming mode nor dependent on the observed wave mode, hence, it might be called a mode-matched inversion scheme. From the scalar theory, we know that the time domain counterpart of (153) is a backprojection algorithm [6], as it is the case for far-field SAFT, so the name mode-matched FT-SAFT for (153) seems appropriate, where "FT" indicates that spatial Fourier transforms are involved.

In case we do not account for the compensation of the elastodynmaic nature of the experiment by taking explicitly care of $\underline{\mathbf{C}}_{0\alpha\beta}(\hat{\mathbf{R}}, \hat{\mathbf{k}}_i, \omega)$ we could instead compute

$$\hat{\mathbf{R}} \cdot \underline{\mathbf{u}}_{s\alpha P}^{far} = \frac{e^{jk_P R}}{4\pi R} \hat{\mathbf{R}} \cdot \underline{\mathbf{C}}_{0\alpha P}(\hat{\mathbf{R}}, \hat{\mathbf{k}}_i, \omega) \, \tilde{\gamma}_u(\underline{\mathbf{K}}) \tag{154}$$

$$\hat{\mathbf{R}} \times \underline{\mathbf{u}}_{s\alpha S}^{far} = \frac{e^{jk_S R}}{4\pi R} \hat{\mathbf{R}} \times \underline{\mathbf{C}}_{0\alpha S}(\hat{\mathbf{R}}, \hat{\mathbf{k}}_i, \omega) \, \tilde{\gamma}_u(\underline{\mathbf{K}}) \quad , \tag{155}$$

which would, via Fourier inversion, result in a weighted singular function depending on the selection of the outgoing mode, which, in a real experiment, could be realized for instance by time gating. At least the inversion according to (154) is strictly scalar, and, in fact, it is nothing but a "mode-dependent" P-wave FT-SAFT, i.e. a "scalarized" FT-SAFT, if, for example, α is chosen as P.

4.5.2 Elastodynamic Near-Field Far-Field Transformation

For measurement surfaces S_M in the near-field of the scatterer we can apply the complete version of the elastodynamic Huygens principle, which, in contrast to (137), also involves the stress tensor $\underline{\underline{T}}$ and a dyadic Green function $\underline{\underline{G}}$, because no particular boundary conditions are imposed:

$$\underline{u}_s(\underline{R},\omega) = \iint_{S_c} \{\underline{u}(\underline{R}',\omega)\cdot[\underline{n}'\cdot\underline{\underline{\Sigma}}(\underline{R}-\underline{R}',\omega)] - [\underline{\underline{T}}(\underline{R}',\omega)\cdot\underline{n}']\cdot\underline{\underline{G}}(\underline{R}-\underline{R}',\omega)\}dS'.$$
(156)

For observation points $\underline{R}$ in the far-feld of S_M we compute the β-mode scattering amplitude from (156) as

$$\underline{C}_\beta(\hat{\underline{R}},\omega) = \frac{1}{4\pi}\iint_{S_M}\left[\underline{u}\underline{n}':\underline{\underline{\Sigma}}_\beta - (\underline{\underline{T}}\cdot\underline{n}')\cdot\underline{\underline{G}}_\beta\right]e^{-jk_\beta\hat{\underline{R}}\cdot\underline{R}'}\,dS' \qquad (157)$$

with

$$\underline{\underline{\Sigma}}_P(\hat{\underline{R}},\omega) = -\frac{jk_P^3}{\varrho\omega^2}(\lambda\underline{\underline{I}}\hat{\underline{R}} + 2\mu\hat{\underline{R}}\hat{\underline{R}}\hat{\underline{R}}) \qquad (158)$$

$$\underline{\underline{\Sigma}}_S(\hat{\underline{R}},\omega) = -\frac{j\mu k_S^3}{\varrho\omega^2}\left[\hat{\underline{R}}\underline{\underline{I}} + (\hat{\underline{R}}\underline{\underline{I}})^{213} - 2\hat{\underline{R}}\hat{\underline{R}}\hat{\underline{R}}\right] \qquad (159)$$

$$\underline{\underline{G}}_P(\hat{\underline{R}},\omega) = \frac{k_P^2}{\varrho\omega^2}\hat{\underline{R}}\hat{\underline{R}} \qquad (160)$$

$$\underline{\underline{G}}_S(\hat{\underline{R}},\omega) = \frac{k_S^2}{\varrho\omega^2}(\underline{\underline{I}} - \hat{\underline{R}}\hat{\underline{R}}) \ . \qquad (161)$$

If the data have been produced with an α-mode, we can explicitly account for that adding that index, which yields

$$\underline{u}_{s\alpha\beta}^{far} = \frac{e^{jk_\beta R}}{R}\underline{C}_{\alpha\beta}(\hat{\underline{R}},\omega) \ , \qquad (162)$$

where

$$\underline{\underline{C}}_{\alpha\beta}(\hat{\mathbf{R}},\omega) = \frac{1}{4\pi} \int\int_{S_M} \left[\underline{u}_\alpha \underline{n}' : \underline{\underline{\Sigma}}_\beta - (\underline{\underline{T}}_\alpha \cdot \underline{n}') \cdot \underline{\underline{G}}_\beta \right] e^{-jk_\beta \hat{\mathbf{R}} \cdot \mathbf{R}'} \, dS' \ . \quad (163)$$

Equ. (162) can then be used as input into the inversion scheme (153).

4.5.3 Elastodynamic Backpropagation

Similar to the scalar case, we can define a displacement vector holographic field in terms of the elastodynamic Huygens principle introducing complex conjugate Green's tensors, thus accounting for elastodynamic backpropagation. The resulting Porter-Bojarski equation can be integrated with regard to frequency, provided the approximation of physical elastodynamics is introduced, but the subsequent dyadic inversion turns out to be very cumbersome; hence, we prefer the above inversion schemes.

Today's drawback of this elastodynamic inversion scheme is obvious: it requires the measurement of all three components of the displacement vector on the measurement surface, and there is no device available to perform this task. On the other hand, the development of an appropriate new transducer seems promising because time gating of measurements into pressure and shear waves is no longer necessary: the above scheme backpropagates elastodynamically!

References

[1] M. Lorenz, L.F. van der Wal, A.J. Berkhout: Ultrasonic Imaging with Multi-SAFT; Nondestructive Characterization of Defects in Steel Components, Nondestr. Test. Eval. 6 (1991) 149

[2] L. Caineri, H.G. Tattersall, J.A.G. Temple, M.G. Silk: Time-of-Flight Diffraction Tomography for NDT Applications, Ultrasonics 30 (1992) 275

[3] J.D. Achenbach: *Wave Propagation in Elastic Solids*, North-Holland, Amsterdam 1973

[4] A. Ben-Menahem, S.J. Singh: *Seismic Waves and Sources*, Springer-Verlag, New York 1981

[5] M. Spies, P. Fellinger, K.J. Langenberg: Elastic Waves in Homogeneous and Layered Transversely Isotropic Media: Gaussian Wave Packets and Green Functions, in: *Review of Progress in Quantitative Nondestructive Evaluation*, Eds.: D.O. Thompson, D.E. Chimenti, Plenum Press, New York 1992

[6] G.T. Herman, H.K. Tuy, K.J. Langenberg. P. Sabatier: *Basic Methods of Tomography and Inverse Problems.*, Adam Hilger, Bristol 1987

[7] K.J. Langenberg, U. Aulenbacher, G. Bollig, P. Fellinger, H. Morbitzer, G. Weinfurter, P. Zanger, V. Schmitz: Numerical Modeling of Ultrasonic Scattering, in: *Mathematical Modelling in Nondestructive Testing*, Eds.: M. Blakemore, G.A. Georgiou, Clarendon Press, Oxford 1988

[8] Z.S. Alterman: Finite Difference Solutions to Geophysical Problems, Journal of Physics of the Earth 16 (1968)

[9] A. Bayliss, K.E. Jordan, B.J. Le Mesurier, E. Turkel: A Forth-Order Accurate Finite-Difference Scheme for the Computation of Elastic Waves, Bulletin Seism. Soc. Am. 75 (1986)

[10] J. Virieux: P-SV Wave Propagation in Heterogeneous Media: Velocity-Stress Finite-Difference Method, Geophysics 51 (1986) 889

[11] L.J. Bond, M. Punjani, N. Saffari: Ultrasonic Wave Propagation and Scattering Using Explicit Finite Difference Methods, in: *Mathematical Modelling in Nondestructive Testing*, Eds.: M. Blakemore, G.A. Georgiou, Clarendon Press, Oxford 1988

[12] R. Ludwig, W. Lord: A Finite-Element Formulation for the Study of Ultrasonic NDT Systems, IEEE Trans. Ultrasonics, Ferroel., and Frequ. Contr. 35 (1988) 809

[13] T. Weiland: On the Numerical Solution of Maxwell's Equations and Applications in the Field of Accelerator Physics, Particle Accelerators 15 (1984)

[14] K.S. Yee: Numerical Solution of Initial Boundary Value Problems Involving Maxwell's Equations in Isotropic Media, IEEE Trans. Ant. Prop. AP-14 (1966) 302

[15] P. Fellinger, K.J. Langenberg: Numerical Techniques for Elastic Wave Propagagtion and Scattering, in: *Elastic Waves and Ultrasonic Nondestructive Evaluation*, Eds.: S.K. Datta, J.D. Achenbach, Y.S. Rajapakse, North-Holland, Amsterdam 1990

[16] P. Fellinger: *Ein Verfahren zur numerischen Behandlung elastischer Wellenausbreitungsprobleme im Zeitbereich durch direkte Diskretisierung der elastodynamischen Grundgleichungen*, Ph.D. Thesis, University of Kassel, Kassel/Germany 1991

[17] Y.H. Pao, V. Varatharajulu: Huygens' Principle, Radiation Conditions, and Integral Formulas for the Scattering of Elastic Waves, J. Acoust. Soc. Am. 59 (1976)

[18] H.J. Salzburger, W. Schmidt: Automatische wiederkehrende Ultraschall-Prüfung der Laufflächen von Hochgeschwindigkeitsschienenfahrzeugen, in: *Mit vernetzten, intelligenten Komponenten zu leistungsfähigeren Meß- und Automatisierungssystemen*, Eds.: G. Schmidt, H. Steufloff, Oldenburg, München 1989

[19] K.J. Langenberg, P. Fellinger, R. Marklein: On the Nature of the So-Called Subsurface Longitudinal Wave and/or the Surface Longitudinal Creeping Wave, Res. Nondestr. Eval. 2 (1990) 59

[20] B.A. Auld: General Electromechanical Reciprocity Relations Applied to the Calculation of Elastic Wave Scattering Coefficients, Wave Motion 1 (1979) 3

[21] F. Lakestani: *Validation of Mathematical Models of the Ultrasonic Inspection of Steel Components*, Report PISC DOC (90)12, NDE Lab., JRC-Ispra/Italy 1990

[22] J.D. Achenbach, A.K. Gautesen, H. McMaken: *Ray Methods for Waves in Elastic Solids*, Pitman, Boston 1982

[23] R. Marklein: *Die Akustische Finite Integrationstechnik (AFIT) — Ein numerisches Verfahren zur Lösung von Problemen der Abstrahlung, Ausbreitung und Streuung von Akustischen Wellen im Zeitbereich*, Master Thesis, University of Kassel/Germany, Kassel 1992

[24] V. Schmitz, W. Müller, G. Schäfer: Practical Experiences with L-SAFT, in: *Review of Progress in Quantitative Nondestructive Evaluation*, Eds.: D.O. Thompson, D.E. Chimenti, Plenum Press, New York 1986

[25] K.J. Langenberg, M.Brandfaß, K. Mayer, T. Kreutter, A. Brüll, P. Fellinger, D. Huo: Principles of Microwave Imaging and Inverse Scattering, Advances in Remote Sensing (1992) (to be published)

[26] K. Mayer, R. Marklein, K.J. Langenberg, T. Kreutter: Threedimensional Imaging System based on Fourier Transform Synthetic Aperture Focussing Technique, Ultrasonics 28 (1990) 241-255

[27] K.J. Langenberg: Introduction to the Special Issue on Inverse Problems, Wave Motion 11 (1989) 99-112

[28] T. Kreutter, S. Klaholz, A. Brüll, J. Sahm, A.Hecht: Optimierung und Anwendung eines schnellen Abbildungsalgorithmus für die Schmiedewellenprüfung, Seminar "Modelle und Theorien für die Ultraschallprüfung" der Deutschen Gesellschaft für zerstörungsfreie Prüfung , Berlin 1990

Printed in the United States
By Bookmasters